EXECUTIVE STRATEGY

EXECUTIVE STRATEGY
Strategic Management and Information Technology

FREDERICK BETZ

JOHN WILEY & SONS, INC.
New York / Chichester / Weinheim / Brisbane / Singapore / Toronto

This publication is designed to provide accurate and authoritative information in regard to the subject matter covered. It is sold with the understanding that the publisher is not engaged in rendering professional services. If professional advice or other expert assistance is required, the services of a competent professional person should be sought.

Library of Congress Cataloging-in-Publication Data

Betz, Frederick.
    Executive strategy: strategic management and information technology / Frederick Betz.
        p. cm.
    ISBN 0-471-38402-X (cloth : alk. paper)
    1. Strategic planning.   2. Industrial management—Data processing.   3. Knowledge management.   4. Management information systems.   I. Title.

HD30.28 .B473 2001
658.4'012—dc21   2001017912

Printed in the United States of America

10 9 8 7 6 5 4 3 2 1

CONTENTS

PREFACE

This is a concepts and processes book on strategy, adapted to the challenge of strategic management of companies in an age of information. After gaining understanding of the foundation ideas of strategy covered in this book, you will be able to better and more efficiently guide the processes of

- Strategic planning
- Implementing strategic plans
- Formulating strategic business models
- Developing comprehensive planning scenarios
- Formulating innovation strategy
- Formulating competitive strategy
- Formulating business-driven information strategy
- Strategically managing business diversification
- Formulating knowledge-assets strategy.

Traditionally, the challenges of strategic management were in its centrality, complexity, and difficulty of practice, but now we must add the new challenge of information.

CENTRALITY OF STRATEGIC THINKING

Strategy is a central intellectual activity of management as it provides for the long-term leadership for any group, organization, or business. Thus strategy is essential to effective leadership. Without a leader providing long-term direction of an organization, the organization will be static and buffeted by change. Without a manager providing long-term direction for a business, that business in vulnerable to the surprises of changing competition.

COMPLEXITY OF STRATEGIC THINKING

Strategy happens to be a complex idea because it is about the vagaries and obscurities of any long-term future. In contrast to the idea of "strategy," the idea of "planning" seems more understandable because it is about immediate certainties and the direct needs of the short-term future. Accordingly and frequently in business practice, real strategic thinking may not occur at all, crowded out by the immediate needs of operational planning. Sometimes strategy even gets confused with planning (e.g., in the U.S. government, all agencies and Congress are run only by an annual budget, completely without strategy, and even after 1982, the annual budget bills of Congress ceased to be all passed in time for the beginning of the federal fiscal year). In all organizations, business as well as government, budgeting is the most direct and familiar process. Next comes planning, and least familiar is strategy.

DIFFICULTY OF STRATEGIC THINKING

Since business is a practical activity, any intellectual leadership in business (such as strategy) is useful only as it contributes to the success of the practical activity. This makes strategy more difficult in practice than planning, because of strategy's mediated, indirect, and unclear connection of ideas to the outcome of action. In a plan, one has an immediate goal, a direct target, and a clear operation to seek the goal and gain the target. And soon it becomes clear (quarterly or yearly) whether one has succeeded or failed. In strategy, the time between implementing a strategy and accomplishing its objective is so long and the goals and targets so changed along the way—often even the leadership has changed—that it may not be at all clear how strategy contributed to a success or failure. The evaluation of any strategy is always a *historical re-creation* of many past events.

All this long-term thinking, indirectness, and lack of clarity about the long-term future makes strategy a difficult thing to practice. For successful strategy, a good strategic leader must have good strategic vision and have this successfully translated into good operational plans. How is all this possible? These are the practical issues about strategy and planning that we will address in this book.

Is strategy even necessary? The answer is a clear, immediate, and direct affirmation of "yes, certainly." All the actual cases and practical instances of planning and strategy that we will review in this book will show that strategy has played a major and essential role in the long-term success of any business, firm, or government (or in failure, the lack of proper strategy). Strategic leadership may be complex and difficult, but it remains central and essential to long-term business success and survival.

INTEGRATING INFORMATION STRATEGY INTO BUSINESS STRATEGY

When the twentieth century ended (and in addition to these traditional challenges about strategy), there had occurred the beginning of electronic commerce (and with it what many then were calling a "new economy" and "virtual commerce" and "virtual corporation.)" Certainly this was something strategically very new in the world of business yet it was also something strategically very old—another of the many impacts of basic innovation upon commerce during the long history of economic development.

What was new in strategy was the need to *explicitly* integrate information strategy into business strategy. What was old in strategy was the continuing and universal pattern of how change in technology has always created major change in business strategy. The Internet was just one more in a series of major innovations that rocked the business worlds of the nineteenth and twentieth centuries and drove economic developments for 200 years.

We will reconstruct traditional strategy theory to include information and knowledge strategy. We will find that information technology has altered traditional strategic management specifically as to focus, techniques, tools, and implementation:

1. In strategic focus, information technology has added electronic commerce to the purview of all business strategy.
2. In strategic techniques, information technology has moved strategic business models to the center of strategic thinking.
3. In strategy tools, progress in information technology has made the formulation of information strategy a major instrument of modern business strategy.
4. In strategy implementation, information technologies are making business integration an important feature of strategic change.

To include these kinds of changes in strategic management, we will use both a survey of the older business literature on strategy and also many new cases of electronic commerce—in order to construct a modern theory of executive strategy.

CHAPTER 1

STRATEGY PROCESS

PRINCIPLE

An effective modern strategy process uses both planning scenarios and strategic business models.

STRATEGIC TECHNIQUE

1. Form top-down and bottom-up strategic-planning teams.
2. Schedule interactions between teams.
3. Construct a planning scenario and a strategic business model.
4. Formulate an intuitive, synthetic strategic vision.
5. Construct an analytical long-term strategic plan.
6. Construct short-term operational plans in the direction of the strategic plan.

CASE STUDIES

Merger of AOL and Time Warner

Rakuten

Barnes and Noble Faces Amazon.com

Hewlett-Packard's Strategy Challenges

INTRODUCTION

The world keeps changing. It always has and always will. This is the fundamental importance of strategic management, for the use of strategic planning is to make *decisions now* to guide an organization's *future directions.*

In terms of future directions, the basic problem of any company (or, for that matter, of any living thing) is *survival.* And to survive over the long term, as Lowell Steele of General Electric succinctly summarized, a company must have two strategic capabilities: the ability to *prosper* and the ability to *change* (Steele 1989).

Prosperity

The failure to prosper imperils survival because when expenses exceed income over a long enough period a company fails in bankruptcy. Moveover, prosperity now requires not only profitability but long-term growth. Modern stock markets often value long-term asset growth over short-term dividends. In these days of corporate takeovers, continual corporate growth in earnings and sales is necessary for management to retain control. Together, this combination of continuing profitability and continual growth presents a tough strategic problem because all markets eventually mature and growth in a company's business is limited by the growth of its markets. In the second half of the twentieth century, this need for continual corporate growth not only created both the driving force for corporate diversification but was also a major cause of the dissolution of large companies. Successful management of a portfolio of different businesses in the same company became a major top corporate leadership challenge.

An illustration of this was the successful growth of General Electric in the last two decades of the twentieth century. The CEO who provided strategic management for GE during that time, Jack Welch, became well known in the business world growing GE in two decades from a market value of $12 billion to $500 billion—then a rare corporate feat.

In his intended last year at GE, Welch bought more growth for GE by acquiring Honeywell:

> "It was vintage Jack Welch. At the Oct. 23 press conference announcing General Eletric Co.'s $45 billion acquisition of the aeospace and industrial conglomerate Honeywell International Inc, the GE chairman and CEO strutted around the stage, boasting of the promise of the deal . . . Welch spoke bullishly of the acquisition—'It's exciting. . . .' "
> —(Barrett et al., 2000, p. 41)

In the fall of the year before Welch's intended retirement, Welch had learned that the Honeywell was agreeing to be acquired by United Technologies, and

Welsh rushed in with a higher offer ($8 billion more) to buy Honeywell. Honeywell would add 7% growth to GE's earnings. GE had revenues of $131 billion with operating profits of $19 billion. Honeywell had $25 billion revenue and $4 billion profits. Growth was important to GE to maintain a high stock value—through continuing growth.

Then Welch's business fame was so extensive that he had earlier received a $ 7.1 million advance for his projected memoirs. It was one of the highest book advances in publishing history; and the publisher, Time Warner's Doubleday, would have to sell at least 1.6 million copies in North America to make a profit:

> Executives of Doubleday had prepared a complete book jacket and marketing plan to pitch to Mr. Welch. Sitting in his shirt sleeves at a conference table, Mr. Welch preferred discussing his ideas about management over marketing details . . . At one point, the conversation turned to Lee Iacocca, the legendary chairman of Chrysler Motors whose autobiography sold 2.6 million copies in hardcover and 3.5 million in paper back in North America. Mr. Welch told them that he would consider a book like Mr. Iacocca's a failure, because it was about a personality rather than ideas. Mr. Welch said he preferred the 1964 book by Alfred P. Sloan, *My Years with General Motors* . . .
>
> —Kirkpatrick (2000)

Strategic management is about ideas. *Worth* magazine asked several successful CEOs about what they thought was important in the job of the CEO. Two of them, Koichi Nishimura of Solectron and Eric Schmidt of Novell, responded:

> *Nishimuara:* Four things, I think, are important. First, communicate a vision of where the company is and what you are doing. The second is that when you communicate, you want to be able to motivate people. Third, you want feedback. And fourth, you want to take action.
>
> *Schmidt:* I think the job breaks down into three parts. First, setting a strategic vision that is implementable. That's number one. Second is recruiting and leading great human beings. The third is worrying about shareholder value. If you follow those three rules, then everything else sort of works. . . . I think the question from an investor should be, Does the CEO have a strategy that you believe can win?
>
> —*Worth* (2000, p. 183)

Strategic vision, communicating vision, recruiting and motivating good people, obtaining feedback, and taking action that creates shareholder prosperity—these are essential elements in corporate leadership. Strategy is ideas about the long-term future.

Change

Attaining the kind of growth and prosperity Welch had achieved at GE required him to make great changes within the company. On taking office back in 1981, Welch first sold off a hundred of GE's businesses and then consolidated the rest into 14 business groups. This massive restructuring gave Welch a tough reputation:

> Neutron Jack, as he is sometimes called, is widely regarded as one of the world's most ruthless managers. The truth is more complex. Some of his actions are indeed harsh, and he antagonized people inside the company and out by fixing something they didn't think was broke. What is becoming clear only now is how those moves fit into a larger plan to strengthen the enterprise and to make its remaining employees more secure.
>
> —Sherman (1989, p.39)

Welch's success at GE was to strategically change GE from primarily manufacturing businesses into primarily financial service businesses. In 1980, manufacturing produced 70% of GE's revenues, with services contributing 30%. By 1999, manufacturing produced only 26% of revenues, with services having grown to 74%. Welch's strategy was to retain only a selective group of manufacturing businesses and grow financial services: "Chairman Welch unloaded the consumer electronics division and built financial services into a powerhouse, while keeping GE dominant in turbines and jet engines." (Teitleman, 2001, p. 31)

Periodic change in a large organization is necessary to help the firm adapt to new times, for new *times* keep on occurring. One of Jack Welch's most widely quoted strategic precepts was:

"Control your own destiny, or someone else will."

The failure to make appropriate changes at the right time, imperials survival because the company may become competitively obsolete in its products, services, and value to customers. Change requires an ability to anticipate the external dynamics of the environments in which a company operates—markets, competition, innovation, government regulation, economic conditions, globalization, and so on. Change also requires an ability to alter a company's directions (e.g., in products, production, marketing, organization, personnel, businesses, etc). Lowell Steele nicely summarized the emphasis of change as the focus of strategic thinking:

> Strategy is concerned overwhelmingly with questions of change. How much must the enterprise change in order to survive and to continue to prosper? How much change can it finance and manage? How fast can it change? These are profoundly difficult questions.
>
> —Steele, (1989, p. 178)

Strategic management is about the difficult questions of future business—whether it should go and how should it change—a particularly risky set of questions for a large business that is already successful.

In many of the older books on strategy, it was presumed that the logic of strategy should begin with a "mission and vision" statement. A mission statement is a statement of what kind of business is the organization; and a vision statement is what kind of business the organization would like to become. However, this older kind of "strategic logic" is not really useful for an ongoing organization, unless the mission changes. What is useful in vision for an organization is foresight on change to the mission. The "vision thing" is: how should the mission change to take advantage of future market opportunities and meet future competitive threats? Strategic thinking is not about the mission of the business but *changes to the mission.*

Strategic Thinking

CEOs like Jack Welch become successful and famous because of their ability to think and act strategically. And because of the fundamental importance of strategy to long-term corporate survival, this ability to think strategically became recognized as an important leadership skill for executives. For example, one can often see specifications for the ability to strategize in common recruitment advertisements for executive positions, such as the following ad, which appeared in *The New York Times* in May 2000:

VICE PRESIDENT—GLOBAL SOURCING
. . . (X) Corporation, a publicly traded manufacturer of products . . . has an excellent opportunity in its Engineered Products headquarters office. . . . The successful candidate will have hands-on experience in sourcing components. The position will report to Group VP and be responsible for purchasing in (5) divisions. . . . The individual must thrive on multitasking, have outstanding negotiating skills, be a good manager of people and projects, and be a strategic thinker. Highly competitive compensation package. For confidential consideration, forward resume and salary requirements to. . . .

This is the fundamental management skill which we address in this book. What is a strategic thinker? How can hands-on experience improve a manager's ability to think strategically? Which practical techniques facilitate effective strategic planning in a large organization? What important strategic concepts used by successful leaders such as Sloan and Welch?

Change in a company environment always forces strategic redirection. Changes in automobile technology provided the strategic ground for Sloan's successful management of General Motors, and changes in services and medical technologies provided some of the impetus for Welch's successful strategic management of General Electric.

Information Technology

By the end of the twentieth century, progress in information technology (IT) had become the strongest and most pervasive force for strategic change in businesses throughout the world. One example of IT's impact was Thomas Middelhoff's strategic exhortation to his company in *2000* (he was then chief executive of Bertelsmann with corporate headquarters in a small German city and 78,000 employees around the world):

> We have to reach every brain to explain that we have nothing less than an industrial revolution. That makes it necessary to change how we see and run our business. That means speed is king. That means we have to be decentralized on the one hand and also more corporate. We have without any question a generational change at Bertelsmann.
>
> —Carvajal, 2000, p. 1

The growth of the Internet was a rapid phenomena. For example, in the United States, 14% of the population used the Internet in 1996, jumping to 22% in 1997, 31% in 1998, 38% in 1999, and 44% in 2000 (Elliott and Rutenberg, 2000). In 2000, the average monthly hours a user spent on line was 19 hours. U.S. consumer spending online had grown from a few million in 1996 to $3 billion dollars in 1997, $7 billion in 1998, $19 billion in 1999, and $36 billion in 2000. Of the $36 billion spent in 2000, $11.0 billion was for travel, $7.7 billion for PCs, $2.4 billion for clothes, $13.4 billion for books, and $13.4 billion for other merchandise. In October 2000, advertising revenues of the Internet in the U.S. totaled $600 million (with portals receiving $150 million of this, search engines $34 million, travel $7 million and local maps $25 million, business and finance $25 million, computing and technology $23 million, Incentive $22 million, shopping and auction $21 million, and news $19 million).

In business history, the decade of the 1990s will likely be called the decade of the Internet. Its innovation and rapid impact on business made it an interesting and challenging time—that brought to everyone's immediate attention the great importance of progress in information technology upon all business strategy. The dramatic experience of that decade was nicely summarized by Joseph Nocera and Time Carvell:

> The Internet decade has seen the unscrupulous rewarded, the dimwitted suckered, the ill-qualified enriched at a pace greater than at any other time in history. The Internet has been a gift to charlatans, hypemeisters, and merchants of vapor . . . and despite all that, it still changes everything.
>
> —(Nocera and Carvell, 2000, p. 137)

The Internet was an example of a larger class of phenomena in business history called *pervasive innovations.* William Abernathy and Kim Clark even introduced a new term to strategic management—*transilience of innovation*—to emphasize the importance of a pervasiveness of an innovation upon the operations of a firm (Abernathy and Clark, 1985). *Transilience* means the ability to pass through a system, and it emphasizes the range of business impacts that a transilient innovation may have upon the value-adding capabilities of a firm, passing through its activities to make changes in

- The kinds of products and way the firm produces products (and/or services)
- The kinds of customers and markets the business serves

Abernathy and Clark classified the types of transilient innovation impacts upon a firm by the innovation's potential to alter either product/production or market/customer competencies:

1. In product/production competency, innovations may alter
   a. Product design
   b. Production systems
   c. Technical skills and knowledge base
   d. Materials and capital equipment
2. Under market/customer competency, innovations may alter
   a. Customer bases
   b. Customer applications
   c. Channels of distribution and service
   d. Customer knowledge and modes of communication

For any of these factors the impact of innovation may range from strengthening existing competencies to making existing competencies obsolete. Accordingly, Abernathy and Clark also classified innovations:

1. A technological innovation that conserved both existing production and market competencies was called a *regular innovation.*
2. A technological innovation that conserved existing production competency but altered market competency was called a *niche-creation innovation.*
3. A technological innovation that made an existing production competency obsolete but preserved existing market competency was called a *revolutionary innovation.*
4. A technological innovation that obsoleted both existing production and market competencies was called an *architectural innovation.*

The innovation of the Internet was an architectural innovation.
Historically, many firms have usually successfully exploited regular or niche-

creation innovations, for they sustain current operations. But many large firms have perished during revolutionary or architectural innovations. For example, Clayton Christensen (2000) examined reasons why large U.S. firms historically have often failed to profit from revolutionary or architectural innovations:

1. Resource dependence in large firms, influenced by investors and current customers.
2. The emergent markets of radical innovations are early-on perceived as too small for big firm growth needs.
3. The ultimate use of radical innovations are often not known early.
4. The performance and features of radical new innovations are often not attractive to current markets.

The architectural impact information technology made on business strategy was summarized by Bill Miller (2000), who spent a long career (at Intel) and argued that IT was:

1. Altering the competitive dynamics of both products and services, leading to the new importance of dominant designs and platforms in product/service strategy
2. Flattening organizational hierarchy or even dissolving boundaries into networked forms, such as "virtual enterprises"
3. Impacting management styles through introducing (at the same time) both a "transparency" of the business model to all levels of employees and making their jobs more complex, through the increased need for teaming and direct attention to the bottom-lines of business goals.

Information technologies could strategically impact businesses in different ways:

- A business can be *in the information technology business*, providing information technology goods and services (e.g., Hewlett-Packard)
- A business can *use information technology as a core technology* in its production of goods and delivery of services (e.g., Amazon)
- A business can *use information technology as a supporting technology* in its design of products/services (e.g., Ford Motor Company);
- A business can *use information technology as a marketing tool* to attract customers to its product/services (e.g., the CNN News Web page).

In this book, we will address this new challenge of IT to strategic management, using a strategy process in which information technology integrates with business strategy.

Strategic Management

Strategy, planning, budgeting, and knowledge are the four *forward-looking activities* of management, and it is important that they be clearly distinguished.

Budgeting is the allocation of resources for the future operations of an organization. Budgeting is neither planning nor strategy. All organizations budget, but they do not necessarily plan nor strategize. Organizations annually budget in order to allocate the resources for continuing operations. Thus the managers of all organizations do formulate budgets; but not all managers plan or formulate strategy.

Planning is thinking out of tactics for the continuing operations of an organization. Planning is not strategy but the implementation of strategy. Managers plan when the tactics of operations change from year to year. Organizations that need to annually plan are those with significant changes in tactics and operations from one year to the next.

Strategy is neither planning nor budgeting. Strategy is the perspective for long term change. Many organizations do not even begin to formulate strategy until an immediate emergency requires change; but by then, it may be too late to formulate effective strategy. Effective strategy requires looking out ahead, anticipating the need for change and preparing for it. A strategy is a *change in the direction* of the objectives of the operations over a course of years (such as the acquisition of new businesses or innovation of new product lines). Few organizations do strategy when external conditions, markets, and competition all are stable. Strategy is needed when external conditions change—change in technology, change in markets, change in competitors.

Knowledge is the basis for improving and controlling the future value-adding operations of a business enterprise. Progress in information technology provides new tools for managing the development of the knowledge assets of the business. Knowledge has been and continues to be the major force in strategic changes in business.

Thus in a modern theory of strategy:

- strategy is change in direction,
- planning is future tactics,
- budgeting is allocation of resources,
- strategic knowledge is future innovation.

CASE STUDY: Merger of AOL and Time Warner

The first historical case we will examine happened just as the twenty-first century began. It is important in illustrating the rapid advance of new forms of business practice due to progress in information technology. One of the new companies in what then was called the "new economy" took over an older and larger business in the "old economy". The case illustrates how innovation that creates rapid market growth can be exceedingly highly valued by a stock market. In this case, the market valued the new markets being created by AOL over the older markets then being served by Time Warner.

The historical setting was when the then-new electronic commerce, or e-commerce, had spawned a whole new raft of companies and media industry. The booming U.S. stock market of that decade had priced most of these new companies exceedingly high. America Online (AOL) was one of these, providing service access to the Internet to subscribers. In January, it used its very high market value to merge with an older media company, Time Warner. The business community took this merger as the first sign that e-commerce companies were beginning to mature. For example, Richard Siklow and Catherine Yang of *Business Week* wrote:

> On the surface, it looked like just another awesome megadeal . . . America Online is the acquirer. The trading symbol for the new company, tellingly, is AOL. Given the realities of the New Economy, it could hardly be otherwise. By now, the pattern is clear: the digital will prevail over the analog, new media will grow faster than old, and the leaders of the Net economy will become the 21st century Establishment.
>
> —(Siklow and Yang 2000, p. 37)

On December 10, 1999, the market capitalization of America Online was about $250 billion dollars, whereas the market capitalization of Time Warner was about $85 billion (Loomis, 2000). The difference was in the stock markets multiplication of their relative price-to-earnings (P/E) ratios. In the last 12 months, AOL had earnings of about $1 billion, so that its P/E ratio was 250/1. Time Warner's earnings were about $1.3 billion, so that its P/E ratio was about 65 to 1. Thus AOL's P/E was being valued over Time Warner's P/E at a multiple of 250/65 (so that AOL stock was 3.8 times more valuable than Time Warner's stock, based on earnings). This was the heart of the deal. AOL's vast P/E ratio gave it the leverage to take over Time Warner, and Time Warner was willing to be acquired, hoping the resulting company would have a PE ratio more like AOL's than Time Warner's.

Even comparing the evaluation on the basis of EBITDA (per-share earnings before interest, taxes, depreciation, and amortization) Time Warner was trading at a multiple of 14; whereas AOL's EBITDA was trading at 55. Time Warner had a major debt load, which AOL did not have. Moreover, AOL was in a rapidly growing new market, e-commerce, into which Time Warner had tried to enter but failed.

Yet in terms of assets—valuable products and steady, proved earnings—Time Warner had a much larger asset base. For example, Time Warner had 73 million consumer subscriptions compared to AOL's 24 million. Time Warner product brands included:

1. *Time, People, Sports Illustrated, Fortune, Money* magazines
2. The cable companies of HBO, Cinemax, CNN, TNT,
3. The movie and music production companies of Warner Bros.

In contrast, America Online had AOL, Netscape Navigator, and stakes in several companies.

Time Warner brought to the merger a powerhouse of media content-producing companies, whereas America Online principally brought success in the new electronic businesses of the time. Barry Schuler, then president of AOL Interactive Services, characterized AOL's strengths: "We (AOL) are good at aggregating eyeballs and delivering services (on the Internet)" (Nocera, 2000, p. 68).

AOL purchased Time Warner for $183 billion, but with AOL having just one-fifth Time Warner's revenue and only 15% of its employees. Time Warner, an upstart in the 1920s, had become a major media establishment company by the 1990s.

The deal was to have AOL shareholders receive one share in the new company for current AOL shares, and for Time Warner shareholders to receive 1.5 shares for each of theirs. AOL shareholders ended up with 55% of the new company, and Time Warner shareholders with 45% of the new company. At the time, the Time Warner shareholders expected a market premium of 70% for their shares.

What were the strategies of the two CEOs of AOL and Time Warner in creating the merger?

Steven Case, founder and then CEO of AOL, had two major strategies. The first was to transmute AOL's high trading multiple of the booming market stock market of 1999 into assets and revenue stream, which would survive any drop in the high-tech companies valuation of that time:

> Time Warner stood out as the only company with the content, distribution, global reach and customers. Case wants it all: The branded content from Warner Music, Turner cable networks, and Time Inc. Magazines that can be digitized and sold online. The cable pipes to speed delivery of AOL. A global promotional platform that will save AOL a fortune in ad spending. Relationships with about 73 million subscribers to Time Warner cable systems, HBO, and Time Inc. Magazines. . . . Time Warner's old-fashioned media properties deliver a stable stream of revenues, about $27.1 billion in 1999, and cash flow, about $6 billion . . . that are shielded from the vagaries of the Internet world.
>
> —(Gunther, 2000, p. 74)

From a bottom-up kind of strategic perspective (business-up to the larger world), Case's strategic perspective on the cash-flows of Time-Warner's major publication and television empire would provide AOL a steady and major source of income over the long term.

Also from a bottom-up strategic perspective, the acquisition of Time Warner's businesses would provide a step in the direction solving AOL's bandwidth problem. AOL had been providing Internet service of connection through existing copper telephone lines of customers—slow and technically limited to 54 Kbit modem connections. The market demand for Internet connections was broadband. Time Warner owned a major cable company that could provide a much faster broadband connection to its cable customers. Through the merger Time Warner gave AOL access to a market of 20 million cable customers.

The CEO of Time Warner before the merger was Gerald Levin. His strategic perspective for Time Warner also involved kinds of top-down and bottom-up perspectives of strategy.

From the top-down—looking out on the growing importance of the Internet and electronic commerce—Levin saw the need to continue moving Time Warner into the digital world:

> Levin can empathize (with Case's vision of the Internet world), as a cable and tech guy stuck atop a content giant. . . . Before Ted Turner dreamed up CNN, Levin made his reputation by putting HBO onto a satellite in 1975. He's also been burned by technology, notably when Time Warner spent upwards of $100 million on a prototype interactive TV network in Orlando. But his biggest tech bet, on the potential of two-way cable lines, paid off handsomely . . . Time Warner's stock a so-so performer for much of the 1990s, surged . . . during the period since Levin took over in 1993.
>
> —Gunther (2000, p. 74)

From the bottom-up perspective of Time Warner's recent business capabilities, Levin saw a strategic advantage for immediately merging Time Warner into one of the biggest successful players in electronic commerce. Levin's ventures for Time Warner into the Internet world had not been strategically successful:

> But Levin's hard-won reputation as a tech-savvy executive has faded since then. He passed up the opportunity to buy a portal like Lycos or Excite, and Time Warner's own Internet hub, called Pathfinder, flopped. . . . So when Case called to offer him the chance to be CEO of AOL Time Warner—the biggest game in cyberspace and media!—why, how could Levin resist?
>
> —Gunther (2000, p. 74)

In July 2000, shareholders of both companies approved the merger, but its success was still not certain. As Gretchen Morgenson, of *The New York Times* commented:

For months, if not years, the virtual has trounced the real in the stock market valuations of Internet concerns vastly exceed the values that investors assigned to companies unlucky enough to own tangible assets . . . Last week . . . the tables turned . . . and investors are about to experience the Great Internet Shakeout . . . A big indication that the tectonic plates of the virtual world were shifting was the bid of the high flying America Online to acquire the landlocked Time Warner . . . 'If calendar 1999 was one of discovery of the internet, 2000 is going to be characterized by much more rigorous scrutiny of the business models . . . ' "

—Morgenson (2000, p. 1)

Before the stock market bubble of dot.com burst, Case had transformed equity of AOL into more equity by acquiring Time Warner.

Case Analysis

In this case, we see two important theoretical ideas about strategy. The first is the importance of information strategy to business strategy. Both AOL and Timer Warner were in the businesses of information. AOL was in the business of being an information channel provider as an Internet service provider. Time Warner was both in the business of providing information channels (television, movies, and magazines) and creating information content in these channels. Progress in information technology was bringing both firms into similar business strategies—channels and content.

The second idea about strategy is how strategy was formulated by both ÇEOs, using two kinds of perspectives on their company's future—a top-down perspective from the big picture of the Internet innovation and from a bottom-up perspective of the little picture of the companies' businesses future operations. Strategic thinking by both CEO's required two kinds of views on the future: (1) a perspective on changes in the *larger* environment of the business and (2) a perspective on future business operations about their *current* strengths and weaknesses to changes in operations for *future* strengths.

TOP-DOWN AND BOTTOM-UP PERSPECTIVE

Let us first examine the idea that there are two basic perspectives in strategic thinking—strategic views from the top of the organization and strategic views from the bottom. As illustrated in Figure 1.1, these different perspectives create different views and even different kinds of logics in stratregic thinking:

- A big-picture view with a logic of proceeding from the general to the specific changes of the future

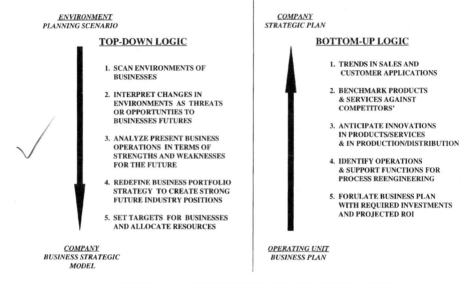

FIGURE 1.1   PERSPECTIVES ON STRATEGIC LOGIC

- An operational-reality view with a logic of proceeding from the specific to the general changes of the future

In the logic of strategic thinking, one can always look to the future by describing the big picture of everything and then deducing how changes there can impact upon the particular situation of one's own future action. For example, in this way one can see oneself as a member of a general economic class, cultural class, or generation and ask how one's particular life is impacted by trends and changes happening to these general categories of people and life. This is the deductive approach to strategy—going from the general trends to the particular descriptions of future life.

Also and conversely in the logic of strategic thinking, one can always look at changes in the particular situation of one's future and then generalize that similar changes are happening to others like oneself. For example, in this way one can generalize changes in one's own life as an exemplar of the kinds of general categories of other people and lives. This is the inductive approach to strategy—going from the particular examples to the general descriptions of future life.

At the top of an organization, information to see the big picture is more readily available there than at the bottom of an organization. Conversely, at the bottom of an organization, information to see the reality of operations is more readily available there than at the top of the organization.

A famous and bitter example of these differences in information between the big-picture-of-the-world and the reality-of-operations was the difference in perspective between the generals and the soldiers in the First World War in Europe in the early twentieth century. From 1914 to 1918, the war stagnated into trench warfare, with the generals on both sides planning one more great battle to win the war. Each battle resulted in thousands of deaths with no substantial gain in territory or weakening of the ability of the other army to fight. The view from the general's perspective was the massing of artillery and soldiers for an attack. But the view from the soldier's perspective was that the new machine gun and artillery made every attack impossible to win, resulting only in the slaughter of the attackers. From the big picture, the view of the war was simply the massing of the attack forces. But from the reality of the trenches, the view of the war was simply devastation and destruction under the sustained and withering fire of machine guns, which would finally halt all attacks. After four years of trench war, both the German/Austrian armies and the British/French armies were too exhausted to win, and a new fresh army of Americans was brought into battle to finish the war. But throughout that war, the perspective of the generals in all armies was that the failure of their massed attacks was due to lack of spirit in their soldiers.

Not only is the experiential base of the two perspectives of top-down and bottom-up different, so too are the logics appropriate to top-down and bottom-up strategy. Figure 1.1 also summarizes the differences as deductive and inductive kinds of logic of the two perspectives.

The top-down perspective of strategy uses a deductive logic that begins with the great and goes down toward the small. In formulating strategy, top leadership should look around at the environments of the firm and its businesses to:

1. Scan the environments of a firm to identify major future trends and changes in government, the economy, territorial markets and competitors, and in the scientific and technological culture.
2. Interpret the changes as threats and opportunities to the businesses of the firm.
3. Analyze the present firm's activities in terms of strengths and weaknesses to face such threats or seize such opportunities.
4. Redefine the missions of the firm's businesses to match the future operations to future threats and opportunities.
5. Set goals and targets for businesses to meet in a proper time horizon.

In contrast, at the operating levels of businesses in a firm, managers should look to the strategic immediacy of the business's markets, competitors, operations, and knowledge:

1. Examine the trends in sales in the markets of the businesses of the firm and identify innovations that can alter these markets.

2. Benchmark a firm's products and processes against competitors' products and processes and identify changes needed to maintain or surpass any competitor's current advantages.
3. Investigate progress in information technology and in the knowledge bases of the business's product and production processes.
4. Reexamine current operations and control, and identify innovations in operations and control of operations needed to adapt to changes in market, competition, and new information and knowledge capabilities.
5. Formulate a business plan, with targets for market-share and profits along with required investments and resources needed to achieve the plan.

These different perspectives of the big-picture and trench-reality are both vital to a good strategy process. Therefore, what is critical to good strategy formulation is the *interaction of these perspectives*. Now, although the top-down and bottom-up perspectives in strategic thinking are important to formulating good strategy, coordinating them is extremely difficult to pull off in a large organization because of the hierarchy of authority.

In large organizations, these perspectives become quite different because of the hierarchical nature of authority. For example in a diversified firm, there are usually at least four levels of management hierarchy:

1. Firm level: board, CEO and firm executive team
2. Business level: president and business executive team
3. Department level: department head and staff
4. Office level: office manager and assistant

The hierarchical levels of authority in a firm usually begin at the top of the firm level with a board of directors and a chief executive officer (CEO). The CEO and his or her executive team are responsible for the strategy of the whole firm. This strategy includes what businesses are and should be within the firm and the overall financial performance of the firm. The planning scenario needed at this level should include anticipation of change in all the industries of the firm's businesses. The bottom-up input to the firm's strategy should be provided by the participation of the business's presidents to the CEO in the strategy process.

At the next organizational level, below that of the firm, is the business unit, and its president is responsible for strategy for the business as a whole. Part of the business's planning scenario is the industrial context of the firm (economy and government) as well as the territories and cultures in the markets to which the business sells. Another part of its planning scenario are goals and targets specified for it by the firm-level strategy, the strategic firm model. The outcome of strategic planning for the business will be a strategic business model specifying changes for its future policies.

The final two levels within a company, department and office, are levels in which managers should provide bottom-up information to the business level in formulating the strategic business model. Policies of this model then provide the guidance for planning and improving operations and activities of the departments and offices.

Since it is organizationally natural for the bottom of the organization to listen more closely to the top than for the top to listen to the bottom, two kinds of misunderstandings are common in strategic thinking in large organizations:

- *The managers of operating units frequently do not think that the top executives understand the strategic problems and challenges in operations.*
- *The top executives frequently do not believe the operating units are trying hard enough to implement the strategic goals they formulate.*

This is the first challenge of strategic thinking in large organizations, to encourage real and accurate communication of strategic perspectives between the top and the bottom.

Therefore, in a good strategy process, one needs to formalize these two perspectives as two views of a firm's totalities, those of the environments of the firm and those of the operations of the firm. The top-to-down perspective looks at the big picture and formulates strategic policies for long-term direction. The bottom-to-top perspective looks at the specific nuts and bolts of the company's operations to try to carry out the desirable long-term direction.

The critical problem in the strategy process of any organization is to facilitate a positive, constructive, and creative interaction between the two perspectives in strategic thinking.

Moreover, this problem is exacerbated by the periodic and noncontinuous requirements of strategic thinking. The actual process of formulating strategy is infrequent but recurrent, exploratory and interactive with the different experiential bases of the top management and of lower management levels. While the results of strategic planning may look as if created by a linear process (either linear in a top-down deductive logic or linear in a bottom-up inductive logic), the strategic process is nonlinear and recursive and interactive with different experiential bases of the company top and bottom.

For example, Arthur A. Thompson and A. J. Strickland nicely summarized the recurrent nature of the strategy process:

The march of external and internal events guarantees that a company's vision, objectives, strategy, and implementation approaches will have to be revisited, reconsidered, and eventually revised. This is why the task of evaluating performance and initiating corrective adjustments is both the end and the beginning of the strategic management *cycle.*

—Thompson and Strickland (1998, p. 16)

Since the strategic management process is interactive and cyclic, the information flow must be recurrently both bottom-up and top-down between the views of the environment and the business. The cyclic nature of strategic planning is also coupled into the budget cyclic of any business, going from yearly planning to yearly planning.

Furthermore, in a multibusiness firm, there are two kinds of the top-down and bottom-up perspectives:

- firm-to-business perspectives
- businesses-to-business divisions perspectives

This makes the top-down and bottom-up communications in a multibusiness firm even more challenging than in a single business firm.

Using an interactive approach to the strategy process is important in the practice of strategic management because the principle cause of failure in the strategy of large organizations has often been due to a lack of proper internal interaction between the two strategic perspectives:

- Inadequate top-down perspective of innovative changes in environments
- Inadequate bottom-up perspectives of the need for new business models for innovative change
- Inadequate communication between executive levels and operational levels about needed strategic change

How can leadership in a large organization avoid these common kinds of mistakes in strategy processes? Good interactions between the strategic perspectives from the top and from the bottom are necessary to create a potentially profitable vision of the future—a strategic vision of the challenges, opportunities, and direction of the future business and how operations need to change to succeed in that future.

The planning process in a large organization can use two kinds of strategic techniques to assist this:

1. A strategic technique for effectively summarizing the changes in the environment's future (i.e., the "big picture") is called a "planning scenario."
2. A strategic technique for effectively summarizing the desirable changes in operations for such a future environment is called a "strategic business model."

CASE STUDY: Rakuten

We next look at a case that illustrates the importance of both an environmental scenario and a business model in strategic planning. Change in business environments has always created opportunities and threats. As noted in the case of AOL, the rise of electronic commerce in the 1990s affected many businesses, not just in the United States but all over the world. A striking example in Japan was a new retail e-commerce web site for Rakuten. This site began impacting retail business in Japan through new electronic marketing via the Internet. Oda-San Mikitani founded Rakuten in 1997 as an electronic marketplace:

> . . . www.rakuten.co.jp draws about 95 million hits a month as a gateway to about 2,800 merchants around the country selling everything from eggs to kimonos. "I want Rakuten to be the best place to sell anything," said Mr. Kikitani. . . . "The marketplace is what we focus on, the exchange between buyers and sellers, while most others focus on being a shop or a collection of shops."
> —(Strom, 2000, p. 32)

Mikitani graduated from Hitotsubashi University in Tokyo and was then employed as a banker at the Industrial Bank of Japan. In 1991, he took a leave to matriculate in the Harvard Business School for an MBA.

> "Before I went to busiess school, I thought my choices were to either work my way up Japanese style to become a director of the bank, or to go into investment banking at a Western firm and make a lot of money," Mr. Mikitani said. But he opted for a third approach, striking out on his own to become a consultant, which led to his foray into e-commerce.
> —(Strom, 2000, p. 32)

At first, Mikitani thought of selling educational services or financial advisory services but saw no economies of scale in these ventures He then hit upon the idea of creating an electronic marketplace for merchants. His inspiration (and the source of the name for the site) came from a famous historical event in Japan from the 1600s. A warlord named Nobunaga Oda changed marketplaces in Japan by taking away control of the market from feudal trade associations and opening trading to all merchants for a small fee. This open market idea was so important in medieval Japan that Oda's city became a dominant commercial center, and this policy was called "rakuichi torakuza" (free markets and free guilds).

Hiroshi Mikitani used this idea in his business model for his web site. He charged merchants about $469 dollars a month (50,000 yen) to link their web sites to Rakuten, so that visitors to Rakuten can find their way to merchant web sites. Some vendors sell their stale inventory on the Rakuten site to avoid undercutting regular sales, and others are avoiding the several layers in a distribution system. Mikitani's strategy considered that Japan might even be a better market for e-commerce than in the United States—since the traditional Japanese retail situation had many layers in the distribution system contributing to high prices. Rakuten provides a way for merchants to avoid layers of middlemen in the traditional distribution system in Japan:

> For example, if you buy an obi, the sash worn with a kimono, at the retailer Kyoto Kimono Ichiba, it will cost 138,000 yen, or about $1,300. Buy it from Kyoto Kimono's shop on Rakuten, and it costs 13,800 yen, or an affordable $130.
>
> —(Strom, 2000, p. 32)

Rakuten also provided a venue for community dicussion, whose customers can chat online about vendors and also participate in online auctions. Mikitani had foreseen the Internet has having an extraordinary impact upon business in Japan and thought it would restructure the Japanese economic system. With Japan's relatively homogeneous population, relatively flat distribution of wealth and strong sense of community, Mikitani saw important strengths in the Internet: "Japanese people like to communicate with each other, and it's much easier to do that on the Net." (Strom 2000, p. 32)

This perception of the cultural importance of communication in Japan's use of the Internet was also shared by Jiro Kokuryo, a professor in the study of information technology and systems at Keio University's business school in Yokohama who commented that e-commerce had a great aspect of social communication in Japan (Strom, 2000, p. 32).

Mikitani had designed Rakuten to be profitable from the beginning, and in 1999, it earned $1 million (dollars) on sales of $5.5 million. This provided a 17% percent profit margin which then was extraordinary in e-commerce at the time.

Mikitani wished to avoid the investments and costs of holding product inventory and of distribution and devised his business model to let retailers sell directly, using his service. Rakutan sold the use of its site and servers to retailers, with 80% of its revenues from monthly membership fees of merchants. An additional 10% came from their advertising on Rakuten, and an additional 10% from auctions. In adding service value to Rakutan's customers, he monitors the shops using the site; and if he sees any shady business practices, he asks those retailers to leave.

It was in the year 2000 that Rakuten successfully went public and gave Mikitani fame, becoming a celebrity, who then was being consulted by government officials and politicians about e-commerce.

Case Analysis

We see in this case how a good business model is important for profitably seizing and exploiting new business opportunities when major innovations change the environments of business. Planning scenarios are a formal technique for anticipating major changes in the world; and strategic business models are ways to profitably exploit such changes.

PLANNING SCENARIOS AND STRATEGIC BUSINESS MODELS

It is always important to strategically think about the big picture of changes in the environments of business (e.g., the Internet). It is equally important to strategically think about the smaller picture of how a particular business (e.g., Rakutan) can exploit the business opportunities in that change.

For strategic thinking in a large organization, we need to consider what kinds of techniques can formally assist groups of managers reach to consensus about what is important: (1) about changes of the future and (2) their implications for future operations. To help a group to strategically think about the big picture, the technique of the planning scenario is effective. To help a group strategically think about operational realities of the future, the technique of a strategic business model is effective. These two techniques can help a large organization describe the two key totalities of strategic thinking—future environments of the company and the future company itself.

Planning Scenario

Strategic thinking needs to grasp the big picture of changes in the environments of a company. For example two of the CEOs *Worth* interviewed in 2000, Raymond Gilmartin CEO of Merck and Koichi Nishimura CEO of Solectron, commented:

> *Gilmartin:* Part of leadership is saying in touch with what's going on outside your company. . . . You need to gather information to see the patterns, to tell if you're on the wrong track, to take risks and make decisions that go against the grain.
>
> *Nishimura:* Getting it right comes from pattern recognition. You integrate information and you go "humm."
>
> —*Worth* (2000, p. 183)

Gathering information and constructing patterns of trends and changes in the environments of business is the purpose of scenario planning in the strategy process. To systematically gather information and create insightful "humm patterns," the *strategy technique of scenario narrative* is very useful.

All strategy is based upon assumptions about the future and its business opportunities and challenges. A modern technique for exploring and expressing these pictures of the future is called a *scenario*, and when used for planning, a *planning scenario*. Scenario planning uses scenario narratives and societal models. Scenario narratives provide a method for describing and thinking about the possible impacts of the future upon a current business.

Strategic stories envision adventures of the business in the future. The future will be an adventure that will challenge business. Experience is always of the present, with memories and stories of the past. The future consists of anticipations and/or surprises and plans for the future conceived in the present. All existence is always in the present.

In the mind only exists intelligent perception of the past and imagination of the future.

As we saw in the case of Rakuten, the major changes in the environments of a business are changes in structures of the society in which the business operates. To build a planning scenario that captures this kind of complexity and completeness of possible future change, one needs to use a general classification of all the societal environments of a business. In all human societies, traditional or modern, there have been general classes of social patterns that create societal structures. These include categories of territory, culture, economy, government. Planning scenarios should address issues of change in the large patterns of society, such as:

- Will there be changes in how the control of territory is decided in the future (e.g., the break-up of the country of Yugoslavia in the 1990s)?
- Will there be changes in the culture of the nation (e.g., changes in science, demographics, etc.)?
- Will there be changes in the economy of the nation, world (e.g., business cycles, innovation of new technologies, etc.)?
- Will there be changes in government of a nation (e.g., changes in taxes, government regulations, etc.)?

Strategic Business Model

Strategic thinking also needs to think about what kind of business model can meet the challenges and exploit the future opportunities in the environments of the company. So the second strategic totality to be considered in the planning process is the future of the business (or businesses) of the corporation. The strategic technique effective for this is a model of how one's company now operates but should change in the future, a strategic business model. Strategic models of the business of a corporate summarize the future policies of the company which will prepare it to perform in the future.

The strategic importance of the concept of a 'business model' was nicely expressed by Geoffrey Colvin, commenting on the troubles Xerox was having in 2000:

> "The quote of the year for 2000 comes from Xerox CEO Paul Allaire . . . He gets the Distinguished Service Cross for extraordinary executive heroism because he told analysts in a conference call, 'We have an unsustainable business model.' In the past CEOs of big, established companies didn't say things like that. They didn't tell the people who rate their stock that the way they make money doesn't work anymore . . . The largest fact of life in business today is that virtually every company . . . has to change its business model to make it sustainable in the Internet worked, infotech-based world."
>
> —(Colvin, 2001, p 54).

A business model is an abstraction of a business identifying how that business makes money. Business models are abstracted about how inputs to an organization are transformed to value-adding outputs. As we will later review in the third chapter, all models of organizations are models of kinds of open systems (Betz, 1968). One important version of this was the now famous value-added model of Michael Porter (Porter, 1981). As a value-adding open-system model, an organization is described as taking resources from its environment and transforming them to value-added outputs sold back into its market environment. The transformation of input resources into output products/services is performed by the processes and operations of the business. The Porter model is only one of several kinds of business models, one can use in strategic planning (and which we will cover in the third chapter).

A strategic business model abstracts the basic value-adding transformation that describes how a business makes its money.

Strategic thinking about how a business now makes money and how it must change to continue making money is the 'bottom-line' for strategic management. For example, in thinking about the future capabilities of an organization, Clayton Christensen and Michael Overdorf emphasized the need to consider resources, processes, and values in an existing organization compared to the challenge of needed change (Christensen and Overdorf, 2000). The resources of a business consist of tangible resources (e.g., personnel, equipment, facilities, cash flows, location, etc.) and intangible resources (e.g., design capability, brand names, relationships to customers and suppliers, etc.). The processes of a business consist of the activities and procedures with which a business procures resources, adds value, and produces and sells products and services. Thus dealing with change also requires determining what changes in processes and procedures are necessary to produce new kinds of value and/or address the needs of new kinds of customers. The value dimension of an organization (sometimes called "corporate values") are the standards by which management and other employees set priorities and judge the importance of activities and results. Values are standards about how resources are used and how processes are run, as Christensen and Overdorf comment:

> An organization's values [are] the standards by which employees set priorities. . . . A company's values reflect its cost structure or its business model because they define the rules its employees must follow for the company to prosper.
> —Christensen and Overdorf (2000, p. 69).

A strategic business model is a systematic list of the policies that will guide the future specification of inputs, outputs, processes and values of the complete operations of the business of the corporation.

The importance of conceiving of a good business model was emphasized by the experience of the many new companies (dot.coms) begun in the Internet growth years of 1996–2000. Then hundreds of these dot.coms were begun with extensive venture capital funding, and many without having a viable business model. Next in the year 2000, over 125 of these companies folded as they ran out of capital and had not yet become profitable (and found new financing difficult to achieve). The often repeated moral then was that a good business model was necessary for profitability and survival.

Strategic business models need to be constructed around strategic issues of change in markets and innovation, competition and structure, operations and control, information and knowledge, asking strategic questions such as:

- What are likely to be the changes in the *markets* that a business will serve and *innovations* that will impact these markets?
- What are likely to be changes in *competition* against which a business will compete and in the structure of the industrial sectors in which competition occurs?
- What is likely to be the progress in *information* technologies that a business can strategically exploit and how can this improve the *knowledge* assets of a business?
- What must be the changes in *operations* and *control* of business processes that the business needs to implement to be efficient and effective in its future value-adding processes?

CASE STUDY: Barnes and Noble Faces Amazon.Com

In a strategic planning process, after a planning scenario and a strategic business model have been constructed in the interactions of the top-down and bottom-up perspective, the next requirement of strategic thinking in a large organization is to create a strategic vision for the future. A strategic vision provides the *direction* for the organization to pursue *prosperity* under the conditions of *change*. A good example of this was Steve Bezos's vision of Amazon in the mid-1990s.

As we saw in the case of the rise of AOL in the time of the innovation of e-commerce in the 1990s, many entrepreneurs created new business visions from the opportunities in the big picture of the Internet and from changing the operational realities of traditional businesses. Bezos' vision was to replace traditional operations in book retail with new kinds of operations through the Internet.

Now, for strategic thinking, the importance of imaginative, creative, and correct vision cannot be overstated. Yet vision remains the most perplexing principle in strategic management. Successful new visions can blindside and totally frustrate competitors. And this was well illustrated by the impact of Amazon.com upon a then-dominant competitor, Barnes and Noble in the late 1990s.

Leonard Riggio was CEO of Barnes and Noble and had grown the nationwide book retailer as a traditional bricks-and-mortar retailer:

> December 16, 1998, was not a good day for Leonard Riggio. . . . Sitting in his cramped windowless conference room at Barnes and Noble's headquarters in lower Manhattan, Riggio just picked at his lunch . . . and shook his head in disbelief. Amazon, an upstart with sales of $600 million and losses that grow bigger every year was now worth seven times more than Barnes and Noble Inc, a chain of 1,000 bookstores with sales of $3 billion."
>
> —(Munk, 1999, p. 50)

Riggio was reacting to a stock announcement that Amazon.com stock had risen from $150 a share to $400 a share. By the end of the day 17 million shares of Amazon changed hands. When the market closed the value of Amazon.com had increased by 20 percent to $15 billion. The value of the stock held by founder of Amazon's, Jeff Bezos, was worth $5.7 billion, $914 million more than 24 hours earlier.

For thirty-five years, Reggio had been selling books (compared to the roughly five years Bezos had begun selling books through the Internet), and Reggio was disturbed: "I am sitting here, hammering away day after day, to come up with new ideas for my stores, and then, in an instant with just a single press release, Jeff Bezos is worth another $1 billion." (Munk, 1999, p. 50)

Riggio had begun selling books as a college student, while attending night school at New York University. During the days, he worked as a clerk at the NYU bookstore. Deciding the could do a better job than the university bookstore, he dropped out of college in 1965 and started the Student Book Exchange) (SBX), near the NYU bookstore. In six years, he had expanded to five campus bookstores in New York City. Next he bought Barnes and Noble, (an unprofitable seller of textbooks on Fifth Avenue at 18th Street). Riggio was 30 years old and ready to innovate. He loaded tables in Barnes and Noble with remaindered books. He installed wood benches for people to sit on and peruse books. He gave away free copies of *The New York Times Book Review*. He adopted techniques to book selling from other mass merchants, using aggressive advertising: "If you paid full price, you didn't get it at Barnes & Noble."

Next, in 1986, using junk bonds for financing, he bought a chain of 37 bookstores, 142 college stores, and B. Dalton, a chain of 800 bookstores. Suddenly. Barnes and Noble was the biggest bookseller in the country. His next strategy was to put bookstores in shopping malls. He continued to expand, buying small bookstore chains, one after another (e.g., Scribner's, Bookstop, and Doubleday Book Stores).

In the early 1990s, he changed strategy again, abandoning his mall-based strategy to build book "superstores." Barnes and Noble's super book stores were conceived as places to gather and spend time. They featured comfortable chairs, served Starbucks coffee, and stayed open until 11 P.M. In addition, he began building a big brand name, using celebrity authors and selling designer shopping bags, bookmarks, and advertisements with illustrations of Ernest Hemingway and Virginia Woolf. Although the idea of the superstore was not original (Borders was the first to build gigantic stores) Riggio moved faster and more nimbly.

So just a year earlier—in 1998—Lenny Riggio had been dominant. Riggio was 58, and until then he had been the most important player in the book retailing industry. In the United States, Barnes and Noble had the most bookstores and a bigger market share than any competitor, and it was profitable. In July 1998 Barnes and Nobles stock price hit $48 dollars, a 220 percent increase over the prior 18 months.

Suddenly, Riggio found he had a new competitor to battle—Jeff Bezos—just as times were again changing and new business strategies emerging. For example, Suzanne Zak, then head of a money management group called Zak Capital (and a large Barnes and Noble shareholder) attended a meeting for analysts and money managers on July 24, 1988, hosted by Amazon. "Initially, like a lot of people, we were skeptical of Amazon," she explained. "But at that meeting, listening to Bezos a light bulb went off. I said 'We're going to have a problem here.' " (Munk, 1999, p. 51)

Zak sold all 400,000 of her Barnes and Noble shares. Others also reduced their holdings, and Barnes and Noble's stock tumbled from $48 to the mid-20s.

The Internet had provided a strategic competitive advantage in retailing. Riggio needed to join the e-commerce strategy to try to catch up. He launched barnesandnoble.com, which in 1998 brought in just 320,000 new customers while Amazon.com added millions. In 1999, Barnes and Noble's share of the U.S. book retail market was 15 percent, while Amazon's was just 2%. Amazon.com had 8.4 million registered customers and sold 75 percent of all books ordered online, while barnesandnoble.com had only 1.7 million, selling 15 percent online. The only problem was that Amazon was not yet profitable.

Case Analysis

This case illustrates the dramatic change that the Internet began to make upon businesses in the middle of the 1990s and also illustrates the importance of strategic vision. The business model of a whole retail sector needed to be rethought in terms of the Internet. Operations had to be changed and improved to take advantage of new opportunities and to meet the challenges of competitors who leap to the challenge.

In this case, the innovation of the Internet created the challenges of change to Barnes and Noble and provided the opportunities to Amazon. Strategy is about change over the long term. When Reggio entered the book retailing business in college textbooks, he saw the opportunities, in the short term, of providing better service and lower prices than the college bookstores. He expanded by perceiving opportunities in long term change in retailing books through expansion into shopping malls and super bookstores. However, Riggio did not at first see the long-term opportunity in the Internet. Accordingly, competition in book retailing was dramatically altered by Bezos's business start-up.

Change is always possible. *Even in a well-established industry, strategic repositioning can and often does occur—over time.* Riggio saw the opportunity to provide better textbook service than the existing NYU bookstore and eventually established a chain of textbook sellers on many campuses. He next saw an opportunity in trade book retailing to discount retail prices and entered that market. There Riggio saw opportunities to build large book retailing chains and position them in shopping malls with high customer traffic. He also saw the opportunity to use the junk bond financing of the 1980s to build a national chain. Next he saw the opportunity of refashioning book retailing into superstores as places to gather and spend time.

No business strategy is forever. This case shows that even as Riggio was attaining a major success in restructuring in the book retail industry, a new business opportunity occurred in the Internet, and it was aggressively exploited by a his competitor Jeff Bezos.

New competitive advantage can occur from different sources. Riggio saw strategic opportunities in the traditional practices of book retailing, such as small inventories, large price margins, central city locations. Bezos saw strategic opportunities in new information technologies, the Internet, and retailing without the bricks-and-mortar store. Visions of change and opportunity are fundamental to strategy.

STRATEGIC VISION

In a large organization, strategic vision needs to arise from (1) strategic thinking exercises of constructing planning scenarios of the future environments of the company and (2) a strategic business model of the company's business(es). The reason this is so important in a large organization is that the experiences of being at the top and at the bottom are so different that both top and bottom perspectives are severely limited. From the top, it is really hard to see the real problems in the trenches; and from the experiences at the bottom, it is equally hard to see the pressures of control on the organization.

A good strategic vision (created from a top-down perspective and properly informed by a bottom-up perspective) is essential in strategic planning because strategic change cannot occur without top leadership's having a vision of and a commitment to change. The relationship of leadership to strategic vision and change is critical in the strategy process:

1. Strategic vision is the fundamental responsibility of leadership since only top management has the authority to make major changes in operating organizations.
2. Strategic change is only periodically necessary; but to be effective such change must be envisioned, anticipated, and planned.
3. Sources for strategic vision are either external in the environments of the organization or internal as opportunities developed within the organization.

For example, *Worth* magazine interviews with some successful CEOs in 2000 also showed their concern with the importance of providing visionary leadership, such as in comments by Koici Nishimura of Solectron and Raymond Gilmartin of Merck and by Eric Schmidt of Novell:

Nishimura: When you are leading a company, you have to figure out, conceptually, what you are trying to do. Once you have decide that, and you think it's okay, the second thing you have to figure out is: What tactics are you going to use . . . ? You continually have to ask: Are the assumptions I made still good? My job is to continually reassess the assumptions or the foundation that the company is built on.

Gimartin: You need to have a vision that is the anchor point for what you're doing. . . . There needs to be some form of overarching statement that makes sense and on which the ECO stakes his or her job.

Schmidt: Leadership is defined about perception, not just reality. So there's always this tension in leadership to overhype. And to make promises that you can't keep and articulate things that can't happen. . . . You want to do some level of overselling, but the problem is that the people you're communicating to are smart. If they think you're a snake-oil salesman, then your whole credibility goes to zero. So leadership is also defined by credibility.

—Worth (2000, pp. 184–186)

A strategic vision summarizes the need and purposes for strategic change

Why leadership in large organizations often fails to envision and prepare for change arises from the nature of leadership in large organizations. In large organizations, leaders are usually selected as those who are committed to doing more of the same. Managers often rise to leadership because they embody a vision of the organization's past.

The vision of the past represented a tested story of success. Past leadership built organizational structures and culture that evolved into a successful company. Later when the business environment changes, the earlier structures and cultures and leadership became ineffectual in the new conditions.

Yet despite the tendency for management not to make changes, still the need for long-term change is indigenous in organizations because organizations have little control over change in their environments—and thus they may be forced by competition to change or die.

Many students of strategy and organization have argued that instability is a periodic experience for all organizations. For example, Michael Tushman and Elaine Romanelli argued that organizations experience periods of relative stability interrupted by sharp strategic reorientations (Tushman and Romanelli, 1985). They and others, (e.g., Norman, 1977; Miller and Friesen, 1984), have seen organizational change as a kind of evolution, stimulated by responding to change in business and economic structures.

Michael Tushman and Elaine Romanelli with Beverly Virany nicely summarized the connection between strategic vision, environmental (structural) change and organizational decline:

At least part of the reason for substantial organization decline in the face of environmental change lies with the executive team. A set of executives who have been historically successful may become complacent with existing systems and/or be less vigilant to environmental changes. Or, even if an executive team registers external threat, they may not have the energy and/or competence to effectively deal with fundamentally different competitive conditions. The importance of an effective executive team is accentuated in industries where the rate of change in underlying technologies is substantial.

—(Tushman et al., 1985, p. 298)

Also students of innovation have documented that a major source of changes in the business environment and within the economic structure consists of applied knowledge discontinuities, which Tushman called "technology discontinuities" (Tushman et al., 1985) and later Clayton Christensen called "disruptive change" (Christensen, 2000). Progress in information technology has created many disruptive changes.

As an example, Tushman and his colleagues looked at the mini computer industry, focusing on fifty-nine firms started between 1967 and 1971. They compared firm records of success and failure over a subsequent 14-year period. In their analysis of the reasons some firms survived and many failed, they argued that one must understand how the *conduct* of the firm in the *context* of the changing economic structure affected the *performance* of the firm.

The conduct of firms consists of the strategic, tactical and organizational activities guided by the executive team (chief operating officer (COO) and other principal executive officers). Conduct must alter as the context of the firm changes when alterations in the economic structure affect competition. Changes can arise from:

1. Technological changes
2. Market changes
3. Resource changes
4. Regulation changes
5. Competitive changes

Successful leadership performance depends upon the executive team's ability to envision, anticipate a correct future and formulate a correct strategy and organization for the future operations of the business. Such vision, strategy and organization correctly anticipates technological opportunities for new products, market changes for new needs and applications, resource changes that affect the availability and cost of materials and energy, and changes in government regulations that affect safety, monopolies, taxes, and so on.

When industries faced changes, Tushman et al. found that those firms whose executive teams lacked vision and made no changes in strategy and organization and product failed after the change occurred. Even those firms with correct vision but whose executives constantly made changes in strategy, organization, and product failed.

The firms that survived and prospered through a competitive discontinuity, a disruptive change, were those whose executive teams:

1. Envision and correctly anticipate the discontinuity and prepare for it with appropriate product strategy and reorganization
2. After making the appropriate strategic change, hold a steady course to produce proper products/services with quality and low costs

It is particularly difficult for a company to formulate a new product strategy when it hits a competitive discontinuity generated by new applied knowledge. The principal reasons in the difficulty of formulating a new product strategy in a competitive discontinuity are:

1. The technical uncertainties of a new applied knowledge vision
2. The differing perspectives among the different product-group managers and the technical staff about that vision

For a company to develop a new next-generation product-line strategy, the whole company must fight out different visions about the product plans in a new applied knowledge situation. To formulate a next-generation product plan for an applied-knowledge competitive-discontinuity, it is necessary for a high-level executive to envision and force the strategic issue and to organize the effort necessary to formulate and implement a new strategy for the whole company.

Competitive discontinuities are a common problem for a firm initially successful in a radically new industry because applied knowledge in the industry continues to progress for a time. The reason for the crisis is that competitive discontinuities due to rapid progress in applied knowledge force not only changes in product strategy but changes in business strategy to exploit the changing market. This is why competitive discontinuities are strategically challenging. And this is why *strategic vision* that foresees discontinuities and *strategic planning* that prepares for discontinuities are the key challenges of strategic thinking.

CASE STUDY: Hewlett-Packard's Strategy Challenges

A good example of the challenge of strategic planning occurred in Hewlett-Packard when the twenty-first century began. The Internet and growing importance of e-commerce had created competitive discontinuities to most existing businesses, including HP's businesses. The strategic challenge to business leadership of existing large firms (as also earlier seen in the case of Time Warner's merger with AOL) was reformulating strategy to survive and prosper upon the discontinuities of information technology innovations. To get more feeling of the atmosphere of the times, even in large firms in the information technology industry, we will look at the case of how Hewlett-Packard's leadership was rethinking strategy.

Hewlett-Packard was founded in 1938 by Bill Hewlett and David Packard, then graduate students in engineering at Stanford University in Palo Alto, California. They developed an electronic measurement business that continued through the years to be a core part of HP's business. In the 1970s, HP entered the minicomputer market and grew significantly, not as a technology leader but gaining a significant market share. However, the minicomputer product become obsolete, and was replaced by personal computers, workstations, and servers. HP, continuing as a technology follower, managed to gain small percentages of the market in each of these product lines. Although HP's market share was small, the markets grew so fast and big that computers did contribute significant growth for HP through the 1980s. John Young, CEO from 1978 until 1992, presided over this growth: "Young oversaw HP's rise into a major computer company. . . . But as the 1990s began, Young's efforts to corral HP's independent units led to bureaucracy that got HP badly bogged down" (Burrows and Elstrom, 1999, p. 84).

The next CEO to preside over HP from 1992 until 1999 was Lew Platt:

> A well-liked engineer who joined HP in 1966, he was an operations expert and a devoted practitioner of the HP Way—perfect qualifications to oversee HP's growth in the mid-1990s. But when PC prices and Asian sales tanked in 1997, HP was not prepared for the next big wave: the Internet.
> —Burrows and Elstrom (1999, p. 84)

During the leadership period of these two CEOs, HP did have one innovation that jumped it to leadership in the personal computer printer market: the inkjet printer, introduced in 1984.

But what was the continuing strategic problem? With HP historically a strong engineering company, why had it not—except for the inkjet printer—been an innovative leader all these years?

> By late 1997, employees were crying out for stronger direction. That December, a poll of the 300 top staffers revealed that HP's workers thought the company needed an infusion of new thinking and more customer focus. . . . By last summer [1999], with revenue growth slowing to low single digits, Platt began to make dramatic changes. . . . Even more important, Platt put his own job on the line: He wanted the board to consider hiring a new CEO. The board took him up on the idea, leading to Fiorina's hiring."
> —Burrows and Elstrom (1999, p. 80)

In late 1999, Platt resigned and the board selected Carly Fiorina as HP's new CEO. In November 1990, at an annual computer show where she was a keynote speaker, Fiorina was interviewed by *InfoWorld's* Editor-in-Chief Michael Vizard and News Editor Katherine Bull, who asked what she saw as the major issues facing Hewlett-Packard. She answered:

> We have to reconnect the people of HP to the fundamental spirit of invention that began with this company 60 years ago. . . . Instead of being slow, we have to be fast. Instead of being indecisive, we have to be focused. We have to lead instead of follow. We have to be bold."
>
> —Vizard and Bull (1999, p. 8)

One can see in the use of these general terms of "fast," "focused," "lead," "bold" that innovation in new products was seen as needed in HP. The reason for this was the impact of e-commerce innovations in which HP had not been a leader.

Fiorina had begun her career at AT&T in its core long-distance business but later moved to its Network Systems group, which manufactured telephone equipment. When the group was spun off from AT&T as Lucent in 1996, Fiorina was one of the new company's top executives. It was from there that she was recruited to head HP. In the *Infoworld* interview, Vizard and Bull next asked what Fiorina viewed as the best of HP, to which she responded, "We have chosen not only to embrace e-services, but to use that as a strategy to drive the entire business." (Vizard and Bull, 1999, p. 8)

Earlier under Platt's leadership, HP had begun to devise a strategy for the company's participation in the growth of the Internet. Two managers met in April 1998 to improve coordination, Ann M. Livermore (head of software and support) and William V. Russell (HP's UNIX computer chief). They began to explore a Net strategy for HP and agreed that their independent groups would cooperate to commercialize a host of Web technologies:

> "This was a big deal—like bringing two armies together," says Nicholas J. Earle, chief marketing officer for HP's Enterprise Computing Solutions Division. "The Net became the great unifier."
>
> —Burrows and Elstrom (1999, p. 82)

HP's strategy for exploiting business opportunities the Internet was making possible was summed up in their term *e-services*. HP meant this term to cover new services on the Web supported by a lot of information infrastructure.

Fiorina made strategic changes. One was to tie top manager's pay closely to the performance of the company's stock. Another was to divest HP's $8 billion test-and-measurement division as Agilent Technologies. This business was large and profitable but not an extraordinarily growing business, compared to the e-commerce businesses. By selling part of Agilent Technologies to the public, HP gained $2.07 billion and retained 84 percent of the new company.

Fiorina's strategic purpose was to change investors' view of HP as one of the new economy firms in information technology and value it as an e-commerce business:

> "The businesses that compose Agilent are basically mature but steady," says Steven Tuen, director of research at IPO Value Monitor, "Now we can value HP against competitors like Gateway."
>
> —(Gustke 2000, p. 42)

Fiorina also altered the organization of HP with a 100-person e-services unit, to provide business customers the hardware and software to use the Internet:

> With Agilent out of the way, HP's main concern is its new e-services business, a catch-all term encompassing anything to do with Internet-related hardware and software. Sun and IBM have grabbed the lead in this business, but Fiorina believes the potential size and growth rate of the market give HP a chance to catch up.
>
> —Gustke (2000, p. 42)

Hewlett-Packard was offering what they called their "e-speak" source code. Fiorina described e-speak as software that provides a building block of e-services, enabling different devices to communicate, brokering different kinds of applications. The e-speak core software provided a universal interface for software runtime that was computer-platform neutral; and e-speak software tools provide developers an ability to create appliances and components that could communicate with each other across the Web.

Even with e-speak, HP was still playing catch-up because it hadn't earlier anticipated the Internet and conducted the innovative research to lead in Internet information technologies:

> Although HP is playing catch-up in some key markets, it dominates others. "The best part of HP," says Salomon Smith Barney analyst John B. Jones Jr., "is its printers and their brand recognition. . . ." In the quarter ended October 31, 1999, total printer sales . . . accounted for abut 50 percent of HP's revenue. . . . Meanwhile HP is rolling out new products. . . . But these products represent incremental improvements rather than the sort of bold breakthroughs for which HP was once famous.
>
> —Gustke (2000, p. 42)

Having missed anticipating the coming of Internet and e-commerce, the giant company had to take the kind of risks involved in any large company's trying a catch-up strategy:

The upside—HP boasts a crackerjack management team, shining financials, quality products and a brand name. . . . The downside—turning a successful, complacent company into a hungry, speed-driven one inevitably involves upheaval. If the process is managed badly, morale could suffer and HP could lose its focus."

—(Gustke 2000, p. 44)

Case Analysis

This case illustrates the importance of strategic planning to formulate and implement new strategy and first having a vision to plan and implement change before other competitors pioneer a new technology. The lack of visionary strategic planning at HP had been part of its corporate culture for a long time. HP had been succeeding by being a technology follower rather than a technology leader. HP had been able to follow a succession of innovations in electronics, computers, and information technology to periodically improve its products.

Sixty years earlier, Hewlett-Packard had begun as a high-tech company in electronic instrumentation, but it had not managed to lead in other major inventions. It had, however, been a quick follower in minicomputers and then in personal computer printers. HP had been known as a good engineering firm with up-to-date products but not as a science firm with advanced, breakthrough products.

With the rise of the Internet, Hewlett-Packard was seeking new products to position itself in the next wave of information technology progress. In HP's case, its information/business strategies were pinned on its e-speak source code.

It is hard to prosper as a purely technology follower, for then a firm needs to find a market niche not covered well by the technology leaders (as HP did with ink-jet printers). Good strategic planning is necessary for a technology follower to jump into the prosperous position of being a technology leader.

By February 2001, Fiorina had strategically focused upon three cross-company initiatives of digital imaging, wireless services, and commercial printing. To implement these strategic initiatives across all product lines of HP, Fiorina reorganized HP into four principal units: Printers, Computers, Corporate Sale, Consumer Sales. The Printer and Computer groups were to create products sold to corporate customers and to consumer customers. This reorganization collapsed 83 HP product units into the two Printer and Computer groups:

Not to tackle one problem at a time, Fiorina is out to transform all aspects of HP at once . . . That means strategy, structure, culture, compensation. . . . Such sweeping change is tough anywhere, and doubly so at tradition-bound HP.

—(Burrows, Peter, 2001, p. 72)

PLANNING

Strategic plans need to be formulated as guided by the direction in a strategic vision. Yet the actual implementation of a strategic plan will require that it be put into action through a sequence of operational plans.

Accordingly, we next need to review the differences between strategic and operational plans:

- Strategic planning is a concern for and laying out of the directions for the long-term future.
- Operational planing is a concern for and laying out of the directions for the short-term future.

The conceptual duality of controlling both long-term and short-term events makes strategic planning *and* operational planning *complementary cognitive functions of management.*

In the case of HP, it had good operational planning capability but not good long-term strategic planning capability. The ink-jet product success was a result of good short-term product development and planning.

Stasis and Change in Operations

Operational planning is aimed at controlling the steadiness of organizational operations, or *stasis*. Stasis in management attention attends to the immediate efficiency of operations, for it is efficiency that in the short term determines profitability in a commercial organization. Efficient repetition of operations, as the production and sales of a large volume of products, creates economies of scale, and upon such economies rests profitability of operations.

In the long-term performance, management practice additionally needs to focus upon making desirable changes for future operations, strategy. Strategy focuses on the long-term, mediate effectiveness of operations—not to the short-term, immediate efficiency of operations.

Efficiency of business operations produces profits in the short term, but effectiveness of business operations creates survivability in the long term.

It is the effectiveness of business strategy in the long term that creates the right kinds of products and services for market share and dominance. It is in market share and dominance that a business survives over time.

Thus both stasis of current operations (which creates profits) and change in future operations (which creates market dominance) are essential to strategic management thinking. Stasis produces the short-term, immediate benefits of organizations, whereas change produces the long-term, benefits mediated through intervening events. Stasis and change are complementary. Few steady-state operations can go on forever without needing change because many aspects of an organization are not static—markets, technology, competition, politics, etc. Thus periodically—at the beginning of an organization, through the growth of an organization, and at subsequent critical periods of an organization—operations need to change for the organization's long-term effectiveness and survival. Planning needs both to continually optimize stasis and periodically change stasis:

- Operational planning focuses upon optimization of stasis in operations in the immediate, short-time horizon.
- Strategic planning focuses upon change in future operations for survival in the mediate, long-term horizon.

Both operational planning and strategic planning become integrated in the annual budgeting activity of organizations. But conceptually they are different. Thus in the planning and implementation procedures of strategic management, the procedures need to facilitate both short-term and long-term planning.

Implementing Operational and Strategic Plans

How is operational planning implemented? How is strategic planning implemented? To answer these questions, we need to remind ourselves of the levels of decision logic in the control of an organization's activities.

Organizations conduct repetitive activities in their operations to add value— such as manufacturing and selling products (e.g., inkjet printers, routers, automobiles) or providing and delivering services (e.g., retailing books, transporting passengers, etc.). Thus the ground logical level of any organization is its repetitive activities that directly transform inputs of resources into value-added outputs of products/services.

The scheme, or order, of how repetitive activities are to be carried out in an organization is called its "operations." Operations are the patterns of order that govern, or control, activities. For example, in automobile manufacturing operations it is the order of the assembly line, where engines are first assembled in parallel with chassis and body assemblies and then engines are mounted onto chassis and then bodies are attached.

In a large organization, precisely how these operations are to be conducted are specified as organizational "procedures." Procedures are the instructions on how to carry out an operation. For example, in automobile manufacturing, there are designs which specify the tooling for production and standards for performing operations. Procedures as designs and standards control operations (as operations control activities).

The next decision-logic level in organizations is "policies," which specify the purpose of procedures. For example, in automobile manufacturing, policies determine the types of autos to be designed, the extent of annual model change, the markets to be targeted for the auto designs, the cost targets for production, and so on. Policies control procedures (as procedures control operations).

The highest decision-logic level in organizations is "strategies," which provide the directions of change for policies. For example, in automobile manufacturing, strategies determine the product lines to be produced (e.g., family sedans, sports cars, SUVs, trucks, vans), acquisition of new brands (e.g., Ford acquiring Volvo and Jaguar), extent of vertical integration of production (e.g., GM selling off Delco), and so on. Strategies control changes in policies (as policies control procedures).

In summary, the hierarchy of decision logic in an organization consists of the following control levels:

- *Activities* that transform inputs to outputs
- *Operations* that control activities
- *Procedures* that control operations
- *Policies* that control procedures
- *Strategies* that control policies

The relationship of operational and strategic planning can now be seen in how their implementation affects differently these levels of decision-control in organizations. Operational planning is implemented by changing the lower-two levels of operations and procedures, whereas strategic planning is implemented by changing the upper two levels of policies and strategy: (1) 0perational planning specifies operations and procedures and (2) strategic planning specifies policies and strategy:

- Operational plans are implemented through targets of operations and changes of procedures.
- Strategic plans are implemented through targets of strategy and changes in polices.

Furthermore, when we look in detail at these two kinds of planning (in later chapters) we will find that their logics and processes are really very different:

- Operations planning uses the logics of known action and specific directions to specify how to achieve specific goals of repetitive types of action.
- Strategic planning uses the logics of unknown action and preparation to launch exploration into action never previously experienced.

The logic of planning is the logic of operations—knowing how to get to someplace we have gone before. The logic of strategy is the logic of exploration—preparing to go someplace we have never gone before. In the logic of an operational plan, one can clearly state the ends and means of action—goals and tactics—since we have performed this operational action before, repetitively, and we understand what it takes to do it. Thus operations plans can be summarized in bullet form, for everyone involved can fill in the story's details—been there, done that—know how to do it again. An operations plan just says how much we are going to do again.

However, in the logic of a strategic plan, one is going exploring, rather that repeating an action previously performed. The logic of strategy consists—not of spelling out the means and ends of known action—but of *refining perception, creating commitments, preparing for action*. Together—perception, commitment, and preparation—constitute the real logic of strategic exploration. (We examine all this more carefully in later chapters.)

Strategy is not planning:

- Strategy is change in long-term direction
- Strategic planning lays out the sequence of steps to implement long-term change
- Operational planning details the immediate steps of implementing long-term change and of continuing stasis

STRATEGY PROCESS

Now we can put these ideas together and depict an effective modern strategy process. We recall that strategic thinking is a process, and a strategic plan is a result of the process. The problem of a strategy process in a large organization is how to have planning procedures that:

- Focuses management thinking on *long-term* prosperity
- Anticipates *relevant* change
- Stimulates *constructive interaction* between top-down and bottom-up perspectives
- Creates *effective* strategic vision
- *Transforms* strategy into action

Strategic Planning Teams

Look at the first step in Figure 1.2, which illustrates that the first two steps in creating a strategic thinking process in an organization is to establish a planning process that encourages the interaction of top-down and bottom-up perspectives by

1. Forming top-down and bottom-up strategic-planning teams
2. Scheduling interactions between teams

The composition of a top-down strategy team needs to consist of a firm-level planning staff and executives of the businesses or divisions that compose the company. A bottom-up strategy team should consist of managers of the businesses in the firm (or of the functional divisions within a firm). Since there are hierarchical differences in the authority positions in these teams, it is important to formally schedule interactive presentations of their planning work to one another as the planning efforts proceed to stimulate appropriate interaction of perspectives.

Planning Scenario and Strategic Business Model

Look again at Figure 1.2, which illustrates that the *next step* is to focus the strategy teams upon creating formal descriptions of the environments and the businesses as outputs of the interactions:

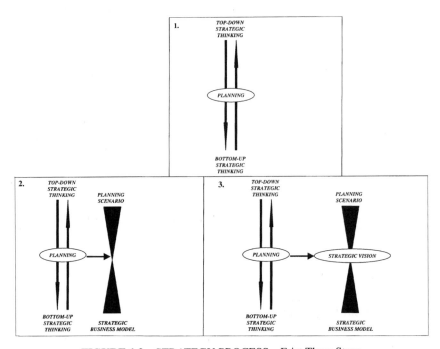

FIGURE 1.2   STRATEGY PROCESS—Frist Three Steps

3. Construct a planning scenario and strategic models.

A planning scenario anticipates from the top-down perspective the future environments of the company. Formulating such a scenario requires anticipating changes in the structures of the societies in which a company operates. In industrialized societies, four general structures exist, economic structures, governmental structures, territorial structures, and cultural structures. The technique of scenario planning provides a systematic way of examining trends and forecasts of possible and likely changes in the future in these structures.

A strategic model of a business summarizes from the bottom-up perspective the intended future policies of the business.

The kind of strategic business model one uses to depict future business policies depends upon which kind of corporate performance one wishes to optimize.

In formulating any strategic business model, important strategic issues are those of markets and innovation competition and structure, operations and control, and information and knowledge. (And these we will address in detail in later chapters.)

Strategic Vision

The procedures for creating a planning scenario and strategic business models facilitate the two perspectives on the future from the general to the particular (top-down) and from the particular to the general (bottom-up); and from this interaction the third step to add in the planning process (as sketched in Figure 1.2) is an integrative picture of the future:

4. Formulate an intuitive strategic vision.

A strategic vision is an intuitive view of the future. For example, in the merger of AOL and Time Warner, both CEOs, Case and Levin, developed a strategic vision of the future merged company, that is, altering boundaries of AOL and Time Warner to become a new firm with the boundaries of content creation and delivery. For AOL, Case's vision was to vertically integrate AOL from media service delivery back into media content creation. Acquiring the media-content creation businesses of Time Warner would change AOL's future business capabilities. For Time Warner, Levins' vision was to merge Time Warner into a major e-commerce business

Vision results from the intuitive cognitive function of the mind—vision is a synthetic view of a totality—a gestalt. How to facilitate intuition in a group setting is a difficult problem, which we will address in a later chapter. For now, the important point is to emphasize that the procedures for strategic plan need to create a strategic vision arising from the team interactions of constructing a planning scenario and a strategic business model.

Strategic and Operational Plan

The final steps in constructing the procedures for a strategy process in a large organization is to translate the strategic vision of the future into strategic plans that can be implemented beginning as near-term operational plans, as illustrated in Figure 1.3, by adding the following procedures:

5. Construct an analytical long-term strategic plan.
6. Construct short-term operational plans in the direction of the strategic plan.

What we have sketched are the key elements in strategic thinking as an *organizational process* in Figure 1.3. The strategic planning process begins with bi-directional views on the future of the company—top-down strategic thinking about changes in the environments of the firm and bottom-up strategic thinking about changes in the businesses of the firm. The interactions (down and up, and up and down) of these two perspectives should create a *strategic vision* about the *directions* the company should go in the future; and the concrete steps to do so constitutes a *strategic plan* for the company.

The top-down view arises from the construction of a *planning scenario,* which

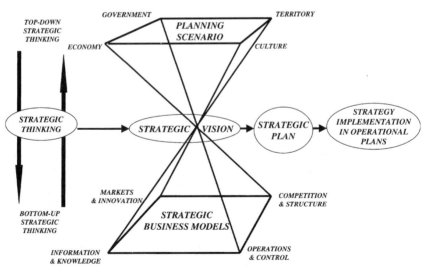

FIGURE 1.3   STRATEGY PROCESS

anticipates the kinds of changes in the environments that will be relevant to the businesses of the firm in the future. The bottom-up view arises from the construction of a *strategic business model* for each business of the firm. This model anticipates the kinds of changes desired and needed to prepare the firm and its businesses for a competitive and successful future.

Multi-Business Firm Strategy

The strategy process depicted in Figure 1.4 assumes a company is a single-business firm, yet most large corporations are multi-business firms. Strategy changes dramatically at the different levels of a multi-business corporation (i.e., the firm level and the business level). For example, Lowell Steele emphasized the perspective differences in strategy between single and multiple business companies:

> One must distinguish between single-business (or closely related business) companies and multi-business companies. Strategic planning at the corporate level for a multi-business enterprise cannot be the same for a company with a single line or closely related product lines."
>
> —Steele (1989, p. 179)

The strategic differences arise from the kinds of competition each kind of company,—single-business or diversified-businesses—faces. The single business

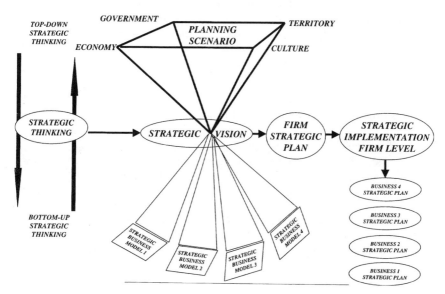

FIGURE 1.4   STRATEGY PROCESS—MULTIBUSINESS FIRM

company finds its principle competition in the marketplace—face-to-face with customers and competitors also directly in contact with customers:

> Competition for a single-business entity is in the market place, offering superior value to customers. If it does that effectively, its performance will be satisfactory—provided that its markets permit an acceptable rate of return.
>
> —Steele (1989, p. 178)

Accordingly, competitive strategy for the single business company must focus primarily upon its products and services—product, production, and marketing strategies. Strategy needs to be focused upon the variables that directly add value to customers, such as product attributes, quality, cost, safety, differentiation, distribution channels, advertising, and so on.

In contrast, the multiple-business company is primarily a financial holder of businesses, so that it performance is not in the customer market but in the financial market:

> Competition for the multi-business enterprise is in the capital markets: Does its present portfolio of businesses and mode of management produce a competitive rate of return . . . ? Multi-business strategy focuses first and foremost on portfolio optimization—what mix of sources of revenue is desired and what allocation of resources will best bring about this preferred mix.
>
> —Steele (1989, pp. 178–179)

Multi-business company strategy must focus principally upon variables that directly impact the rate-of-return of capital, such as business portfolio, position of a business in its industry, business investments, business leadership, business acquisitions and divestitures, and so on. Accordingly, the strategy process for a multi-business firm needs to be modified as shown in Figure 1.4. Therein a common planning scenario is still appropriate for the entire firm and a common strategic vision. But each separate business in the firm (Businesses 1, 2, 3, 4, etc) needs to create a strategic business model appropriate to its business. In a multi-business firm, the strategy process will result in a:

1. Firm strategic plan
2. Strategic business models and strategic plans for each business and strategic business plans and operational plans for each business

SUMMARY: USING THE TECHNIQUE OF STRATEGY PROCESS

Now we summarize the ideas in this chapter as a strategy "how-to-do-it"—how to use the strategy process as depicted in Figure 1.3 as a strategy technique to construct a formal organizational process for strategic planning.

1. Form top-down and bottom-up strategic-planning teams
   - A top-level planning team should be formed and given the task of formulating a planning scenario for the long-range strategic planning exercise.
   - Strategic business unit teams should be formed and given the tasks of constructing strategic company models for each business unit.
2. Schedule interactions between teams
   - Both sets of teams (top-down and bottom-up) should periodically meet during their tasks to present to each other and discuss preliminary versions of the planning scenarios and strategic models as they proceed.
3. Construct a planning scenario and strategic models
   - Express planning scenarios and strategic models in readable forms, emphasizing critical challenges and assumptions about the future.
4. Formulate an intuitive, synthetic strategic vision
   - From consensus in the team interactions, formulate a brief gestalt of the direction for a strategic vision that captures the conceptual totality, principal purposes, and goals and assumptions about strategic change.
5. Construct an analytical long-term strategic plan
   - The strategic vision is then used by a firm-level strategic team to construct a strategic plan for the firm and by strategic-business-unit teams to construct a strategic plan for each business unit. The strategic vision provides the strategic framework and performance measures for integrating the firm-level and business-level strategic plans.
6. Construct short-term operational plans in the direction of the strategic plan
   - Operational business plans for the next-year's budget planning can then be constructed by each business unit within the framework, assumptions, and goals of the strategic plans.

The importance of such a procedural framework for strategic planning is that it specifies the right kind and number of strategic planning teams needed for strategic planning in a diversified firm and/or in a single-business firm. Moreover, it specifies the task of each team and how these tasks are to be coordinated and integrated. Also, it specifies the kinds of outputs of the task and how they are used to create short-term business plans that have long-term strategic directions.

*Essential to proper strategic thinking in an organization is a **proper and logical clarity about the strategy process**—types and tasks of planning teams, construction of consensual top-down/bottom-up strategic vision, and guidance of short-term planning with long-term strategy.*

LOOKING AHEAD

Next we need to look at the complementary process to strategy formulation—strategy implementation. As we noted from one of the CEO's comments on strategy, "And fourth, you want to take action." After all this, we will go more deeply into the subtle ideas in strategy and strategy techniques. To use the strategy process practically, a business team needs to understand the basic techniques of *models, scenarios, vision, and planning.* We will address each of these in detail in subsequent chapters.

Also, we will briefly review the business literature on strategy. The evolution of the theory of strategy has not gone easily because while strategy is one of the deepest and most fundamental process in management, its complexity has made it the most difficult process to fully capture in the rather narrow disciplinary confines of the academic perspectives that study business. Yet we need to review all the schools of strategy in order to be certain that we have indeed captured the important lessons in the literature on the best practices and theory of strategic management.

We will look carefully at the critical strategic issues in formulating strategic models, which include *markets and innovation, competition and structure, and operations and control.* Because of the newness and importance of information and knowledge to strategic management, we will address each in separate chapters on *information strategy* and on *knowledge assets.* Finally because of the differences in strategic models for single-business companies and multi-business companies, we will examine in detail *diversification strategy* for a multi-business company.

For Reflection

Identify several firms that went public in the last ten years. Find their prospectuses, subsequent SEC filings, and trace their stock prices since going public. Have any encountered problems? What were they, and why did they occur?

CHAPTER 2

IMPLEMENTING STRATEGY

PRINCIPLE

Strategy implementation requires managing the paths of innovation.

STRATEGIC TECHNIQUE

1. Choose the correct implementation pathway
2. Form appropriate strategy implementation teams
   - new venture path
   - first mover path
   - dominant-player path
   - diversification path
   - e-commerce path

CASE STUDIES

Welch's Last GE Strategy Conference
Boo Fails and Limited Evolves
Cisco Systems
Industrial Life Cycle of the Auto Industry

INTRODUCTION

Strategy is about change in direction; and implementing strategy is about how to create change in that direction. The kinds of strategic change that occur in business are generally:

1. starting a new business,
2. incremental quality changes to existing businesses,
3. introducing new product models and product lines,
4. acquiring a new business,
5. innovating new technology.

Each of these strategic changes require different kinds of strategic vision and the initiative to implement a vision. For implementation of strategy, a strategic vision can be expressed as a strategic initiative identifying the direction of change.

CASE STUDY: Welch's Last GE Strategy Conference

When Jack Welch was CEO of General Electric, he annually held a strategic meeting for GE's business leaders. In 2001 at retirement, he held his last strategy conference in Boca Raton; and at the beginning of that meeting, Welch introduced his successor. Then at the end of the strategy conference, Welch gave his last exhortation to his GE team, emphasizing the need for integrity in business. He mentioned that GE's criteria for evaluating a manager's performance depended both upon "meeting-the-numbers" and upon "values." He noted that a manager who both met numbers and acted upon values should be rewarded and promoted, and one who neither met numbers nor acted with integrity should be dismissed. However, the manager who had not met numbers but had integrity should be given a second change; while the manager who met numbers but acted without values should be dismissed. Welch was arguing that in the long run, the manager without integrity would harm the company. (And we will see in the case of Sunbeam in the last chapter how a lack of management integrity really destroyed a once valuable company.)

In this meeting, no formal plans were discussed but only strategic directions. Historically this was in contrast to an earlier GE management practice, before Welch became CEO, of holding annual planning meetings. In the late 1970s, GE was famous as a company for its detailed annual plans, which produced an annual plan that would stack high in several volumes upon a GE manager's desk. It was not that GE had abandoned planning, but under Welch, GE put strategy before planning. Strategy at the firm level set the directions, and at the business level, each business planned how to go in the strategic direction.

At Welch's last strategy conference in 2001, GE managers discussed four strategic directions (which they called GE's strategic initiatives): globalization, digitization, e-commerce, and six-sigma.

Globalization was the strategic direction driving GE onwards as a global company with worldwide markets and worldwide production. One business gave an example of the production of a GE product in which its over seven hundred parts were produced in different countries in Europe, Asia, and North America. Each business in GE needed to define its markets, production, and competitors in a global context.

Digitization was a second strategic initiative that emphasized the direction of an increasing use of information technology in all products and production. For example, in a medical products division of GE, its future product lines were to be integrated around a digitized electronic medical record, containing medical information on the patient as both digitized text and diagnostic images.

E-commerce was a third strategic initiative in which the Internet was to be increasingly used to interact with customers, interact internally between managers, interact with suppliers and vendors, and integrate the flows of information throughout GE.

"Six-sigma" was an ongoing strategic initiative of GE as a way of involving all its managers annually in quality improvement projects. Six-sigma was GE's name for an institutionalized program of a methodology for all management to participate in continual improvements in the quality of operations. It had adapted its name and methodology from Motorola; who in turn had adapted the methodology from innovative Japanese management programs of continual quality improvements (whose inspiration had in turn come from the American quality expert Deming). Six-sigma used the terminology from statistical quality control to identify variations in production systems and to reduce variation by improving control of a production variable. At GE, six-sigma was organized into "greenbelt" quality improvement projects, in which each manager was require to complete two greenbelt projects annually as part of the manager's performance. Also some managers would even spend a two-year assignment as a "blackbelt" quality improvement expert to provide help to groups of greenbelt projects. Six-sigma was a method and process for identifying and implementing incremental change; and GE's six-sigma strategic initiative was a way to create a *management culture* of continually improving the quality of products, services, and operations.

Case Analysis

In 2001 at the time of his retirement, Jack Welch, was famous for his strategic leadership, focusing upon change for business growth. While all CEOs at that time desired growth, few had succeeded in managing an older established firm to continual growth. Welch had accomplished growth by creating a management culture for change in GE, which changed primarily through

- incremental quality changes to existing businesses,
- introducing new product models and product lines,
- acquiring new businesses.

Under Welch, GE's businesses had grown by continually improving quality in products and operations, by introducing new products, and by expanding product-lines and markets through acquiring new businesses (and integrating these into existing GE businesses).

The emphasis upon "meeting the numbers" in GE's management performance reviews focused upon meeting numerical targets for growing revenue, earnings, and cash flow and for improving gross margins. The emphasis upon "values" in GE's management performance reviews focused upon providing and improving real value for the customer in GE's products and services. At the firm level, GE's annual strategic initiatives set new directions for change in all GE companies.

Strategy is about change in direction, implemented as strategic initiatives in a company.

Change and Innovation

Any change in an existing business creates something new in that business, an innovation for that business. But that innovation may not necessarily be new to competitors, only catching the company up with competitors. This kind of innovation, new but not unique, helps the business survive but not necessarily ensures its future prosperity. Innovation both new to the business and to its competitors, basic innovation, improves the chances both of survival and of prosperity—basic innovation should be a strategic ambition.

Moreover, historically, basic innovations have been important forces for building new businesses and industries and for long-term economic development. Also historically, patterns of innovation have shown that there are different paths to successful implementation of strategy. What we will do next is to look at how innovation affects the context of strategy implementation. We will find that innovation affects the range of strategic pathways for changes in direction through three kinds of ideas:

- styles of management
- stages of the life of companies,
- dynamics of industrial evolution.

Different styles of management are appropriate to different stages in the life of companies. Both companies and industries have stages of their dynamics as life-cycle stages.

Companies have life stages including starting, growing, stabilizing, changing, stabilizing, changing, stabilizing, and so on—or failing at any stage! Implementing strategic change is a periodic requirement of long-term business prosperity and survival—that requires different modes of implementation depending upon company life stage.

Industries in which businesses operate also have an industrial dynamics, depending upon innovation within the industry, and this has been called an "industrial life-cycle."

The effective ways to implement strategy—strategic implementation pathways—do differ for stages of a company and of the industry in which it operates. One needs to understand

- at what life stage is a company?
- what is state of innovation that affects the company?
- at what stage in the industrial life cycle of the industry of the business of a company?

CASE STUDY: Boo Fails and Limited Evolves

We begin by comparing two cases of companies in different life stages (and in different industrial life cycles, which we will later describe). Boo was a start-up company (in a new industry life cycle of electronic retail commerce). Limited Inc. was in a growth stage, hoping to become a dominant player (in the mature-industry life stage of the clothing retail industry).

Boo.com.   As we saw in the earlier case of Amazon, one important application of the Internet was to establish a new means of retail sales. Following Amazon's lead, several hundred companies began retail sales businesses in the last years of the twentieth century. One among these was Boo.com. This case provides an example of the challenges and dangers of implementing strategy. Business success in times of change depends not just on having a good business strategy but also in how well it is implemented.

Certainly, the last years of the twentieth century was an extraordinary time in stock market history. It was, to quote a famous novel, the best of times and the worst of times. For investors, the best of times is when there is a growing new market that offers staggering business opportunities for rapid business growth (such as AOL and Amazon). The worst of times is when so many imitating new companies arise that it is difficult to pick out which ones have good management and will succeed.

Millions of dollars were invested in new dot.com businesses of the 1990s—some new companies did not even have a business plan or prior management experience or even a clear plan for profitability. Such was the investment market's enthusiasm for the new Internet technology. The initial exponential growth of the e-commerce markets of a new sector of the retail industry was evidence to optimistic investors that almost no new dot.com business could fail. But enthusiastic investors poured money into many that would eventually fail. Early in 2000 the speculation of the stock market peaked and the value of high-tech stocks dropped; and the many not-yet-profitable e-commerce businesses began failing. One of the spectacular failures was a start-up called Boo. Its founder, Ernst Malmsten, had to lay off the people he had hired:

> Malmsten, . . . the shy Swede, who had sold his first Internet start-up for millions, was chief executive of Boo.com Group Ltd., the most hyped and hip of European e-commerce start-ups. . . . When it came time to lay off 130 staffers in a do-or-die cost-cutting drive, Mr. Malmsten wasn't sure how to proceed— "Do you break the news to them one at a time or do it all at once?" he recalls wondering. He finally decided the honorable thing was to call them in one by one.
>
> —(Cooper and Portanger, 2000, p. A1)

Earlier, Mr. Malmsten and Kajsa Leander had sold their first start-up, an Internet book retail business, for several million dollars. Then they thought of starting another Internet retail business. Boo was to sell clothes on a website showing a virtual changing room in 3D graphics that allowed customers to look at clothes from any angle. It was readable in seven languages and calculated prices for eighteen different currencies.

Malmsten and Leander flew to New York and called upon investment banks about their idea for Boo, finally interesting J.P. Morgan & Co. The argument that appealed to the Morgan bankers was that suppliers of fashionable ware would like Boo because it did not intend to sell at cut-rate prices, but would show high fashion to attract customers. The plan promised gross margins of 55% and to be profitable in two years. J.P. Morgan agreed to find early-stage investors for Boo, take its fees in stock, and recruited two of its clients. One was Lucian Benetton, the patriarch of the family that controls the fashionable Italian clothing chain of Benetton:

> Reminded that Ms. Leander had once been a Benetton model, Mr. Benetton invited the founders to supper at the family's Villa Minelli in Treviso, Italy, in the winter of 1998. There, as they ate a simple meal of pasta and red wine in the cavernous dining hall, Ms. Leander explained the finer points of the business. Mr. Benetton put up $5 million in seed money and volunteered his son Alessandro, to take a seat on Boo's board.
>
> —(Cooper and Portanger, 2000, p. A1)

With Morgan's backing and the excitement about the new Internet, many other prestigious investors, such as Bernard Arnault, head of luxury-goods empire LVMH Moet-Hennessy-Louis-Vuitton, joined. With such big-name investors committed, the bankers raised $62 million by the summer of 1999.

By then Malmsten added a third principle officer, Patrik Hedelin, appointing him chief financial officer. Hadlein had been a junior investment banker at HSBC Holdings PLC and had arranged the sale of Malmsten's Internet book-store. However, Hedelin immediately irritated the J.P. Morgan bankers by look-ing for new financial advisers, adding Credit Suisse First Boston and Goldman Sachs. (And later this soured relationship with J.P. Morgan bankers would turn out to be a critical factor in Boo's failure to survive.)

With first-round financing, Boo then spent its capital lavishly, renting office space in Munich, Paris, New York and Amsterdam and hiring hundreds of staffers to service orders. It also spent heavily upon developing complex soft-ware for Boo, with graphics, multiple languages, and automatic currency con-versions. The software development delayed the launch of the website from the middle of 1999 to the end of that year. But the entrepreneurs had not made plans for contingencies:

> Despite the launch troubles, Boo's spending continued apace. The founders trav-eled with an entourage and stayed at hotels like New York's swanky Soho Grand. They set up six offices in cities. "With all those trophy offices, Boo looked more like a 1950s multinational than an Internet start-up," says Marina Galanti, who was marketing director.
>
> —(Cooper and Portanger, 2000, p. A1–A8)

Also Boo continued to spend a large sum in developing their software:

> Boo devised its Internet platform and customer-fulfillment system from scratch, in-house. "It was like they were trying to build a Mercedes-Benz by hand," says a prospective investor who took a pass.
>
> —(Cooper and Portanger, 2000, p. A8)

By September of 1999, the web site was still not up and Boo needed more money. Boo had burned through its $70 million and needed more. J. P. Morgan and investors were unhappy about the lack of financial controls and, particu-larly, Hedelin's numbers: "His figures changed from week to week," one in-vestor said. At a board meeting in October, the founders began another monthly request for more shareholder cash. The Benetton's board designee wanted to take the company public. J. P. Morgan said he couldn't do so until the man-agement agreed to quit tinkering with the Web site and simply launch, and to replace Mr. Hedelin with an experienced CFO. (Cooper and Portanger, 2000, p. A1–A8)

In November of 1999, Boo finally opened its Web site:

to horrible reviews. Customers said that it was very slow for most computers and didn't work at all on Macs, and that it was complex and hard to navigate. . . . Within six weeks, Boo was discounting its clothes 40% in a desperate attempt to move them.

—(Cooper and Portanger, 2000, p. A8)

In January of 2000, Malmsten removed Hedelin from the executive rank and hired an experienced chief financial officer from a German sportware company. And by April, sales had grown to $1.1 million a month, yet Boo's expenses were over $10 million a month. Also in April a major decline in tech stock occurred in the stock market. J.P. Morgan stopped assisting Boo to obtain further financing. The situation for Boo was critical, and the loss of J.P. Morgan as their banker would prove fatal. Mr. Malmsten tried to raise another $50 million through a U.S. private-equity firm, Texas Pacific Group. They examined the numbers carefully and would invest but only if the equity of the previous investors was slashed to nothing. Benetton vetoed that proposal.

Some of the investors talked about raising more money themselves but could not raise enough; and on May 17, 2000, the firm was liquidated.

Later another relatively new e-commerce retailer, Fashionmall.com, bought the trademarks of Boo and in October changed its name to Boo.com:

In one respect the new Boo wants to resemble the original: courting a European customer base . . . Ms. Buggeln (of Fashionmall) became convinced that sticking with the Boo name was the right choice when she learned that 35,000 people visit Boo.com each week even though it has been inactive since spring.

—(Kapner, 2000, p. C8)

Limited Inc. The next part of this case looks at a successful start-up that faced the next life-stage challenge of becoming a dominant player in a mature industry market. Limited Inc. was started by Leslie Wexner in the 1960s, during the shopping mall expansion in the United States. Wexner saw new business opportunities by innovating specialty stores in shopping malls to compete with the former dominance of department stores in the United States in clothing retailing. The new shopping malls in the growing suburbs of U.S. cities provided large enough customer traffic to enable a specialty store to be profitable. Wexner opened his first Limited clothing store in 1963 in an Ohio shopping center and then continued to expand. Within twenty years, Wexner grew Limited Inc. into a corporate empire of 5,000 stores and several major clothing retail brands, including Express, Structure, Victoria's Secret, and Abercrombie and Fitch.

Yet by the early 1990s, growth had stalled in the company and earnings became ragged and its stock value sagged. In particular, Limited's women's apparel stores had lost direction about the changes in fashion and were losing customers. It was then that Wexner tried to find out what had gone wrong:

Mr. Wexner began a long, personal crusade to change both his own management style and the company's internal structure. He met with visionary leaders from inside and outside retailing; consulted management experts on how to reconfigure operations; and eventually began spending less time picking sweaters and more time attending to the company's executive ranks.

—Quick (2000, p. B1)

As a result, Wexner created a centralized organizational structure, which included a corporate team of executives to oversee design, marketing, and distribution across all the company's stores. The nine retail brands of Limited were then encouraged to work together. The executives of these brands hold monthly meetings, sharing information about market trends, fashion, operations, and so on. By 2000, Limited was again making consistent monthly increases in sales in its stores, and its stock was at an all-time high. When interviewing Wexner, Rebecca Quick asked him how his role had changed as chairman and chief executive of Limited; and Wexner responded:

I was an entrepreneur. . . . You start with one store and you do all the jobs in the store, and then you have two stores, and the 10 stores, and then 50 and 100. . . . [Then] I think what went wrong was the . . . entrepreneurial style wasn't working (any longer). The business had outgrown that in terms of complexity.

—Quick (2000, p. B1)

Wexner saw the strategic change in management style going conceptually from managing a collection of specialty stores to managing a family of brands: "The reason I like that word is that I like the association of family, in terms of relationships. I think what it really speaks to is that it is a team. Everyone in the business has to work together as a team." (Quick, 2000, p. B1)

Wexner had learned that the development of a large organization into a dominant competitive organization required building a profession management team for the firm. He himself made the transition from being a successful entrepreneur to becoming a professional manager. He learned the distinction from his talks with well-known professional managers:

It had . . . to do with meeting [General Electric Co. chairman and chief executive] Jack Welch and [former PepsiCo Inc. chairman and chief executive] Wane Calloway. . . . [Wal-Mart Stores Inc. founder] Sam Walton. . . . When I talked with Calloway, I asked him how he spent his time. And he said that he probably spent . . . 40 percent or 50 percent of his time on people. To me, it was startling.

As an entrepreneurial-style manager, Wexner had spent most of his time on operations: "I like people, but I am busy picking sweaters, visiting stores. . . . How do you find that much time? And [Calloway] said, because the talent in the organization is the most important asset that you have."

—Quick (2000, p. B4)

As we will next review, we will see that a classic entrepreneur is a business-starter, high-risk taker, an opportunist, and a do-it-all-yourselfer. And as we will next see a classic professional manager is an organization builder, a risk-minimizer, a planner, a delegator, and a developer of people. Wexner was learning about the idea of a professional manager from other successful CEOs and that an important job of a successful CEO was in developing people to run the large organization:

> I began to see myself as the chief personnel officer. . . . [The first thing I look for in people is] do they know their job? The second thing I look for is, are they whole people? Do they have balanced lives?"
>
> —Quick (2000, p. B4)

In summary, Wexner had transformed himself from an entrepreneurial style to a professional manager style in which things like finding people, motivating people, judging people's performances, seeing that people worked together as teams became the majority of his job. His strategic focus had shifted from starting businesses (he had started 5,000 stores) to running the businesses of the firm (managing a family of brands). In terms of managing brands, the executive team at Limited focused on building and maintaining brand dominance in fashion retailing. When asked what he thought the biggest risk facing the Limited over the next six months was, Wexner answered: "(The) Express (brand) is a top priority. Sales in the first quarter were the best in the brand's history. We are in a fashion cycle that is particularly well-suited for Express." (Quick, 2000, p. B4)

Case Analysis

These cases of Boo and Limited illustrate that strategy implementation faces different challenges for the different life stages of a company. The challenge for Boo was to get the company started with sufficient profits to carry on the business after initial working capital was exhausted. In the case of Limited, Wexner faced a different kind of challenge in learning how to manage a large business. Different management styles are appropriate for implementing strategy in the two different stages of a company between start-up and growth.

ENTREPRENEUR AND PROFESSIONAL MANAGER

We next review what has been learned about different management styles for starting and growing businesses—entrepreneurship versus professional management. Different management styles are appropriate in strategy implementation during the different stages of a company's life—entrepreneurship during the creation of a new company and professional management for its growth into large company.

Entrepreneurial Management

In the business literature on entrepreneurship, three themes emerged:

1. There is an emotional theme. The entrepreneur is a kind of business hero or heroine. They have admirable qualities—initiative, daring, courage, commitment. These virtues are especially admired in turbulent business conditions, when initiative is required for origin of a business or drastic change for survival.
2. In many of the stories of successful entrepreneurs, problems of change of leadership occur, particularly after the organization has grown large and requires rationalization. Then a professional manager is sought to take over after the entrepreneur.
3. Within an organization, some entrepreneurship should always be encouraged, supported, and rewarded if it is to continue to be innovative. Yet balancing rewards for entrepreneurship against rewards for professional management in a large organization is difficult.

For example, Howard Stevenson and David Cumpert compared managerial strategic styles along two dimensions: (1) desire for future change, and (2) perceived ability to create change. They noted that entrepreneurs ask questions, such as

Where is the opportunity?
How do I capitalize on it?
What resources do I need
How do I gain control over them?
What structure is best? (Stevenson and Cumpert, 1985, p. 87)

Stevens and Cumpert also asserted that, in contrast, professional managers more concerned with stability than change adopt a bureaucratic style of strategy that asks different kinds of questions, such as

What resources do I control?
What structure determines our organization's relationship to its market?
How can I minimize the impact of others on my ability to perform?
What opportunity is appropriate? (Stevenson and Cumpert, 1985, p. 86)

Since entrepreneurial vision and risk is a distinctive managerial style, many have studied the psychology of entrepreneurs, hoping to learn why some people are more likely than others to become successful entrepreneurs. Researchers have listed several attitudes and values they found typical of the entrepreneur, such as a desire to dominate and surpass, a need for achievement, a desire to take personal responsibility for decisions, a preference for decisions with some risk, an interest in concrete results from decisions, a tendency to think ahead, and a desire to be their own boss (Vesper, 1980, p. 9).

Others, in attempting to describe entrepreneurial style, have used a sociological perspective. For example, James Quinn viewed the entrepreneur style as a kind of role encouraged by an "individual entrepreneurial system," which is to say a capitalistic system that encourages and supports individual initiative. Quinn (1979) identified several characteristics of an entrepreneurial system that encourages technological innovation:

1. Fanaticism and commitment
2. Chaos acceptance
3. Low early costs
4. No detailed controls
5. Incentives and risks
6. Long time horizons
7. Flexible financial support
8. Multiple competing approaches
9. Need orientation

Quinn saw the single-minded dedication of the entrepreneur as a kind of fanaticism, and an economic or organizational system must tolerate the kind of ruthless, dedicated purpose required of an entrepreneur. The context of such single-mindedness will appear chaotic and disorganized because the entrepreneur is fixed on the goal and will use whatever means or expediency that proceeds toward that goal. The economic and organizational system should tolerate this kind of apparent chaos, which includes little detailed control in the early phase of a new venture. The originators of new ventures operate in an opportunistic, cost-cutting, short-cutting way to a single-minded, clear-cut goal.

Quinn also argued that the economic or organizational system wishing to foster entrepreneurship should provide appropriate rewards for the risks taken in entrepreneurship. Moreover, these rewards must be structured for long-term horizons, since it takes time for anything really new to become a success. At some point when a new thing takes off, observers often think how quickly and rapidly the successful innovation grew, not appreciating the long, painful starts, false starts, and build-up to the take-off stage.

Because of the experimentation and learning that goes into new venture action, it is also important for the system to provide flexibility in financing from many sources and allow for multiple and competing approaches. In the early days of any radical innovation, new ways are being tried out and only down the line will an optimal configuration emerge for a standard design of a new technology. Need orientation should always be the goal of entrepreneurship. Systems that encourage the fulfillment of needs of a marketplace stimulate innovation which lasts and is economically important. Thus for entrepreneurship, the psychological attitudes and the economic and organizational environment are all important the values of the entrepreneur (e.g., risk taking, vision, ambition) and a system that encourages entrepreneurship (e.g., committed, risk-taking. long-term, need-oriented environments).

Professional Management

In recent times, the opposite of the entrepreneur, the professional manager, has gotten a lot of bad press, and is often described as a kind of anti-hero—a bureaucrat. But originally, the idea of a bureaucratic style of management was not bad. A famous sociologist, Max Weber, introduced the idea of a bureaucratic manager as a kind of rational, efficient, honest administrator. And later the famous management theorist, Peter Drucker, argued for the idea of a good manager as a kind of professional manager.

A bias against the idea of a professional manager as a rational bureaucratic management style reveals that the writer is simply ignoring the fact that different kinds of management roles are needed between starting and institutionalizing a new business.

In the early nineteenth century, Max Weber studied the new government agencies that had been emerging in Europe in the late 1800s. He called these organizations 'bureaucracies" and formulated the idea of bureaucratic rationality as a kind of organizational effectiveness and operational efficiency.

The manager of such rationality was a bureaucrat—or what we will call in the later spirit of Drucker's terminology, a professional manager—in the sense that a management style focused on the challenge of institutionalizing rationality and strategy and efficiency in the running of large organizations.

Weber came to this view of the bureaucrat as a kind of rational, professional manager from historically comparing the new forms of government agencies that grew up in the industrializing Europe to the older kinds of governmental administration common to Europe prior to industrialization. This earlier form Weber called a "prebendal" form of office maintained by "feudal" holders of authority. A feudal office-holder exercised authority in the name of a sovereign ruler in order to perform some governmental function (e.g., tax collecting, public order, etc.). The nature of the office and the personal property and interests of the feudal office holder were not at all separated. The first characteristic that Weber noted about a modern bureaucratized office is that the public property and authority of the office should be *separated* from the private property and authority of the office holder.

For example, in the United States government, federal laws forbid office holders from accepting gifts that would create conflict of interest in exercising the responsibilities of the office. As another example in the United States in the private sector, there are federal laws against insider trading in stock held by top-level officials of public corporations.

The second characteristic that Weber noted about a modern bureaucratized organization is that the decision criteria by which decisions are made should be explicitly written down and the procedures by which activities are conducted should be *formalized*. This explicitness of decisions and formalization of procedures introduced a kind of formal order, "rationality," into the operations of a bureaucracy. Moreover, this rational order should be governed by the goal of attaining efficiency and effectiveness in operations.

For example, in the United States the federal agency that has broad responsibility for funding the advance of science is called the National Science Foundation (NSF). It is a bureaucratic policy of the agency to require all NSF science administrators (science bureaucrats) to use peer review procedures in evaluating proposals for research grants. Science administrators send proposals out of the agency to external scientists who are peers of the research proposers. Their peer review evaluation is seen as providing an objective scientific review of the scientific merit of the proposal. Science administrators are then allowed to fund only the proposals that are rated of highest quality by peer review. This rule of peer-review is NSF's rational process for selecting science projects for funding. Peer review by knowledge experts in a research field is held by NSF and its clientele of the scientific community as being the most rational and effective procedure for selecting science-discipline-focused research proposals for grants.

As an example of rational rules for judging corporate performance in the private sector, there are many formal measures of corporate performance, such as profits, profit-margins, return-on-investment, economic-value-added, earnings-per-employee, and so on—each of which formally and partially measure the rational performance of business operations.

Influenced by Weber's studies, the *idea* of rules and rationality in large organizations got to be called by the name of "bureaucracy". And in Weber's view, the concept of bureaucratic management indicated a positive view of the administration of large organizations—bureaucrats were some of the good people in society. However, later students of bureaucracy uncovered a dark side of this idea. Robert Merton began to study the inefficiencies of large organizations. He pointed out that when procedures became formalized to be rational, they also became rigid and inflexible and therefore somewhat irrational. While formality of procedures promote efficiency, it also promotes rigidity and inflexibility. So Merton (and others that followed him) gave the idea of bureaucracy a bad name. This view became popular so that in the second half of the twentieth century, calling a manager a "bureaucrat" came to be viewed as an insult.

So who was right? Was Weber or Merton right? Are professional managers rational or inflexible? Are bureaucracies inherently efficient or inefficient? The answer is that both were right:

- One the one hand, formalization of decision making and procedures in a large organization is essential to providing rationality and efficiency in operating repetitive activities.
- On the other hand, organizational formalization does create rigidity and inflexibility in policies and decision making.

Like the many inherent contradictions of real life in business (such as optimizing both profit and value-added-to-the-customer or minimizing inventory and maximizing sales), there is an inherent contradiction in the style of professional management in large organizations between formalization of processes for rational efficiency and rigidity of processes for irrational inflexibility of operations. All large organizations, business or governmental, do operate as bureaucracies. All large organizations require formalization of decision-making and procedures. All large organizations must become, to some degree, bureaucratized. The concept of Weberian rationality in organizations is the idea of the benefits of bureaucracy, while the concept of Mertonian irrationality in organizations is the idea of the inflexibility in organizations.

In the previous case of Welch's strategic initiative of "six-sigma' in GE, one can see that Welch was using this process and methodology to create a culture of professional managers in GE for which annual incremental changes of quality improvements was an essential part of a professional manager's (bureaucrat's) job.

Strategic initiatives of continual incremental improvements in quality is one strategic way to avoid the professional manager's instinct for excessive stability (and accompanying bureaucratic rigidity).

In summary, the entrepreneur is a business-starter, high-risk taker, an opportunist, and a do-it-all-yourselfer. In contrast, a professional manager is an organization builder, a risk-minimizer, a planner, a delegator, and a developer of people.

In strategy implementation, the style of strategic management is important:

- Entrepreneurs are essential in starting businesses.
- Professional managers are essential to growing businesses toward competitive dominance.

CASE STUDY: Cisco Systems

What is essential in implementing strategy is understanding how the strategy for change fits into the larger societal picture of prosperity through innovative change. To use the idea of the innovation context of industries and businesses for implementing strategy, we must next look at what kind of general competitive challenges that companies must face, as innovation alters the environments of the company. The concepts for this which we will examine are called the "life-stage of businesses" and the "industrial life cycle" of markets.

We will next look at the strategic competitive challenges of one of the successful new information technology companies, Cisco Systems, which was started and grown rapidly during the innovation of the Internet. By 2000, Cisco was one of the most successful new businesses on Earth (and by 2001, Cisco was recovering from a strategic mistake). This case provides an excellent illustration of successful strategy implementation that required changes both in management styles and in strategies during the different early life stages of the new company.

About the same the ARPAnet became NSFnet (which later became Internet), a new company was begun by university personnel at Stanford University to provide a key product, routers, needed to build the Internet system. From the mid-1980s through the 1990s, Cisco prospered on change so much as to become worth $300 billion In the booming stock market of the late 1990s— then one of only three companies in the world at that time to be so highly valued. Even by 1995, Cisco had been recognized by analysts as a very hot stock:

> In its scant six years as a publicly traded issue, Cisco's stock has increased 75-fold . . . It has made millionaires of hundreds of Cisco employees, all of whom have stock options. . . . There is a lot to like about Cisco just on the fundamentals. It is in the exploding business of data networking—that is, it makes both the software and hardware that allow far-flung, otherwise incompatible computer networks to talk with each other and to connect to the Internet. It has a solid management team. It is remarkably lean. Its revenue per employee is among the highest in the technology sector."
> —Nocera (1995, p. 114)

Cisco was founded by Sandy Lerner and Leonard Bosack. The two met and married at Stanford University in 1977. She was a graduate student in statistics and computer science, and he was teaching in Stanford's computer science department (with a master's degree in computer science from Stanford.

Since the middle of the twentieth century, universities have played major roles in the origin of new applied technologies in the information sciences. In the origin of the Internet, the U.S. Department of Defense funded the basic research in networking computers through establishing ARPAnet. By 1979, ARPAnet was having a major impact upon computer science in universities and networking computers (first minicomputers to mainframes and later personal computers to minicomputers and mainframes). Stanford was one of the universities on the cutting edge of innovation in computer networking in the early 1980s. At that time, Stanford had about 5,000 different computers on its campus, and the need to talk to each other was strong. Then computers could talk to each other only by "going outside" through ARPAnet. At the end of 1979, the adjacent Xerox corporate research laboratory, Palo Alto Research Center (PARC) donated to Stanford a copy of its then innovative Altos computer network, along with its Ethernet connectivity.

Using PARC's Ethernet local-area-network (LAN) information technology, Stanford's medical school and computer science department each installed separate networks. Networking engineers were working to connect the different discrete local area networks springing up at Stanford by constructing bridges to extend networks. But the idea of a router to route messages from network to network was a better way to move messages around and through networks, and an engineer, Bill Yeager, working in Stanford's medical school began designing routers for the school's network. In 1980, he developed a prototype of a router using a DEC minicomputer and connected the medical school and computer science department networks. Then from 1980 to 1982, efforts continued at Stanford to construct "intercommunicating" networks across the campus Ethernet connectivity, using workstations running UNIX operating systems. The whole project was named Stanford University Network (SUN). But the project ran only Unix systems and was not effective in connecting everything.

At the time, Sandy Learner Bosack was director of computer facilities at Stanford's business school, and Leonard Bosack was director of Stanford's computer science department. They and some colleagues began their own "bootleg" experiments (without sanction from higher university authorities) with networking using the router concept. Using coaxial cable they ran wires from one building to another across campus and installed routers, servers, and other computers to communicate with each other and ARPAnet. Yeager added code to the Stanford routers to coordinate the network, and others added more code to the routers to provide additional network services: "The project was a success. The router enabled the connection of normally incompatible individual networks. . . . Soon enough, the bootleg system became the official Stanford University Network." (Bunnell, 2000, p. 6)

With this experience, Sandy and Leonard Bosack strategically understood the tremendous importance of routers to connecting computer networks together, going next to Stanford's administration with a proposal to build routers for sale under the school's structure:

... (Stanford's) Office of Technology Licensing was cognizant of the opportunity the couple had offered, ... (but) the decision makers did not give Len and Sandy permission to continue their business on campus or to use school resources for making routers for colleagues at Xerox Labs and Hewlett-Packard. Livid, the couple decided to gather up their technology, quit their jobs, and leave Stanford to start their own business.

—Bunnell (2000, p. 7)

In 1984, they started their own router business, financed with their credit cards and a mortgage on their home. They named the new company Cisco Systems. Sandy Bosack took a daytime job at Schulmberger to support them while the new business was started. For a year and a half, they worked out of their home (along with colleagues including Kirk Lougheed, Greg Satz, and Richard Troiano) to write code, assemble computer hardware, and test new prototypes of routers. They sold routers by word-of-mouth and e-mail to universities and big corporations that knew they needed to hook together their networks:

The building of these early systems was a collaborative effort, with the customers often working side by side with the Cisco engineers. [Sandy] Learner was so intense about keeping customers happy that she gave herself the task of setting up Cisco's customer-support group, labeling it the "customer advocacy" group.

—Nocera (1995, p. 116)

They priced routers between $7,000 and $50,000 dollars. In the fiscal year ending July 1987, they had a profit of $83,000 dollars on $1.5 million sales. Cisco then moved to a business building in Menlo Park, California (also home to Stanford University). The timing of Cisco was fortunate in that by 1985, university and corporate demands for computer networking was exploding, and companies would pay up front for the unique product. Cisco was therefore able to grow on cash flow, without incurring expensive debts for production expansion:

(Cisco was) one of those rare companies that was started at a moment in time where the problem was so vital that customers would pay in advance. . . . Cisco in 1987 filled a desperate need. Customers were tearing the hinges off the door to get the products.

—Valentine in Cringley (1997, p. 306)

LIFE STAGES OF COMPANIES

We will continue the case of Cisco, but its origin illustrates an important theoretical concept in implementing strategy that we need to review here, that there are four distinct stages in the life of a company:

- New venture start-up
- First-mover
- Dominant player
- Mature market

We will start by reviewing the first two stages.

1. New Venture Stage

As we saw in the Cisco case, there are several critical points—milestone goals—that any new business venture must pass along the way to commercial success. Watch for these, and be sure to hit them:

1. Acquisition of start-up capital
2. Development of new product and/or service
3. Establishment of production/delivery capabilities
4. Initial sales and sales growth

Milestone 1. Acquisition of Start-up Capital   Capital is necessary to begin and operate a productive organization with potential profitability until revenues can sustain the operation and provide profits. Start-up capital is required to establish a new organization and hire initial staff, develop and design the product or service, fund production capability and early production inventory, fund initial sales efforts and early operations. Start-up capital can be in the form of

1. The founder's personal wealth, borrowing, sweat equity (e.g., the Bosack's home mortgage, credit cards, and Sandy Learner's job at Schulumberger).
2. Venture capital investments from individuals (called investment "angels") or from venture capital firms (e.g., Sequoia Capital).

Start-up capital is seldom sufficient for rapid growth, and therefore further capital requirements are usually necessary for commercial success.

Milestone 2. Product or Service Development   A new firm is high-tech when its initial competitive advantage is in offering the technology advantage of new functionality, improved performance, or new features over existing products/services. Sometimes new high-tech firms can be started with alternate high-tech production processes for existing types of products or services. But usually a new high-tech product or service provides better competitive advantages with which to start new high-tech firms.

The next event is developing and designing the new product or service. This requires capital and will be a major cost on the start-up venture capital. Ordinarily development and design should be far along before start-up capital can be attracted. However, development problems or design bugs that delay the introduction of a new high-tech product or service can cause serious problems in starting a new firm because such delays also eat into initial capital. Moreover, if the delay is so long that competitors enter the market with a similar new product or service, then the advantage of first entry into the market is lost.

In the case of Cisco, the Bosack's major development work was done at Stanford University. When they went into business, they already had a product prototype developed and ready for final engineering design to produce and sell the new routers. This is why Cisco was a quick financial success, founded in 1985 and profitable by 1986.

Milestone 3. Production or Delivery Capabilities   The third event is to establish the capability to produce the new product or service. In the case of a physical product, parts or materials may be purchased or produced, and the product assembled. The decision to purchase parts or materials or produce them depends upon whether others can produce them and whether or not there is a competitive advantage to in-house production. Establishing in-house production capability of parts or materials will require more initial capital than purchase, but it is necessary when the part or material is the innovative technology in the product.

However, the establishment of any new production capability will also create production problems, problems of quality, scheduling, and on-time delivery. Capital will also be required to debug any new production process.

In the case of Cisco, the early design and production of the routers were financed by the Bosacks' personal savings and by advance payments from customers for their unique and urgently needed product. Later the perceived need for more capital for expansion during rapid growth motivated the Bosacks to find and have Valentine invest in their company.

Milestone 4. Initial Sales and Sales Growth   Initial sales and growth are the next critical event. The larger the initial sales and faster the sales growth, the less room there is for competitors to enter. An important factor influencing initial market size and growth is the application of the new product/service and its pricing. Another marketing problem is establishing a distribution system to reach customers. Distributions systems vary by type, accessibility, and cost to enter. Planning the appropriate distribution system for a new product or service, the investments to use them, and its cost influence on product or service pricing is important for the success of new ventures. Generally, reaching industrial customers costs less than reaching general businesses or consumers. This is one of the reasons a large fraction of successful new high-tech venture are those in which industrial customers provide the initial market. They are usually industrial equipment suppliers or original equipment manufacturers selling to large manufacturing firms. This allows a new firm to get off to a fast start but eventually limits its size and also makes it vulnerable as a part supplier to a commercial customer (who later may choose to integrate vertically downward by producing its own parts). Moreover, a small firm with only a few industrial customers is very sensitive to cancellation of orders from any one of them.

In the general business and consumer markets, distribution system infrastructure will usually consist of wholesale and retail networks. In these access to the customer will depend on wholesaler and retailer willingness to handle the brand offered by a new firm. Establishing brand identity and customer recognition of the brand is then an important problem and a major barrier for a small new firm to overcome. Moreover, in some retailing systems, under-the-counter-practices (such as buying shelf space and/or generous holiday gifts to purchasing agents) may also be barriers to overcome.

As a market grows, the long-term success of a new high-tech venture becomes increasingly dependent on gaining access to and maintaining access to national and international distribution systems.

In the case of Cisco, the Bosacks sold to other university customers like their former employer, Stanford university. Research connections between universities and early adapters of the new network connection technology sent Cisco corporate customers by word-of-mouth. Later when the U.S. Congress made the ARPAnet/ NSFnet public, the rapid growth of the market was anticipated by the Bosacks.

2. First-Mover Stage

After a new venture has successfully started, its next challenge will be to become a large company in its market. Alfred Chandler called the kind of company that succeeds in this a "first mover" (Chandler, 1990). Chandler argued that the early innovator companies in a new industry that continue to grow and survive are those who first move to

1. Continue to advance the new technologies
2. Develop large-scale production capacity
3. Develop a national distribution capability
4. Develop the management talent to grow the new firm

The reason the investment in advancing technology is important is that in the early competition as the technology changes rapidly, a competing company must not lag behind technologically and can gain competitive advantage from being innovative.

Investments in large-scale production capacity and a national distribution system are necessary for an emerging firm to gain a dominant market share in the new national market. Also now in a global economy, a first-mover firm must also move to establish presence in international markets.

Developing management talent to run a growing, large firm is also a necessary investment. For example, a frequent kind of failure of many firms after an initial success has been the failure of the founder of the firm, as an entrepreneur, to build a management team that can succeed the founder.

Accordingly, the next milestones a successful new venture must meet to become a first mover are

5. Production and distribution expansion
6. Meeting competitive challenges
7. Product improvement, production improvement, and product diversification
8. Organizational and management development
9. Capital liquidity

Milestone 5. Production Expansion   As the new market grows and sales are successful, production expansion must be planned and implemented in a timely manner or sales will be lost to competitors because of delivery delays. Production expansion will usually require a second round of capital raising, for the initial capital seldom provides enough for expansion.

The exception is when production can be outsourced. The rapid growth of the market and high margin of Cisco's unique products allowed Cisco to finance production expansion from cash flow. Cisco's hardware products were standard kinds of commodity-type minicomputers, all of whose parts could be outsourced. Cisco used outside vendors to produce their physical product. All of Cisco's proprietary advantage lay in its software and not in its hardware. This meant that Cisco did not need much capital for hard-good production facilities.

Milestone 6. Meeting Competitive Challenges   In a very few areas and rare cases, a patent on a new product or process is basic and inclusive enough to lock out all competitors for the duration of the patent. This is true in the drug industry and occasionally elsewhere. However, most new high-tech ventures are launched with only partial protection from competition by patents, and competitors soon enter with me-too products. The me-too products/services are likely to be introduced with improved performance or features and/or at lower price. The entrance of competitors into the new market is the critical time for new ventures. They must at that time meet the competitive challenges or go into bankruptcy.

Milestone 7. Product Improvement and Diversification   A new firm must upgrade its first-generation products with new products to keep ahead of competition in product performance and features. It also must continually lower its cost of production to meet price challenges by competitors. And it must diversify its product into lines to decrease the risk that a single product problem will kill the firm. The round of capital raised for production expansion also needs to provide for product and production improvement.

As we will see in the next part of the Cisco case, Cisco's major product diversifications occurred through a strategic and aggressive policy of acquisition of potential competitors.

Milestone 8. Organizational Development As an organization grows in size to handle the growth in sales and production, it is important for the firm to develop organizational structures and culture and to train new management. This is an important transition, as the early entrepreneurial style of organization and openness and novelty of culture needs to mature toward a stable but aggressive large organization. In a small firm, coordination is informal and planning casual. In a large firm, both coordination and planning needs formalization.

In the Cisco case, the transition from the management and organizational styles of the founders to the traditional control of experienced strategic managers occurred abruptly (and rather violently) when the venture capitalist took control of Cisco and installed professional and seasoned managers.

Milestone 9. Capital Liquidity The final step for success in a new firm is to know when and how to create liquidity of capital assets and equity. One means is to go public and another is to sell the firm to a larger company. Liquidity of capital enables the founders of the firm and early employees to transform equity into wealth.

Cisco's initial public offering in 1990 was successful and provided the founders with a personal fortune, even though they lost control of the company. The venture capital firm, Sequoia Capital, leveraged its modest $2.5 million investment into billions of dollars by the late 1990s.

DYNAMIC MODELS OF SMALL BUSINESSES

Timing of attaining these new venture and first-mover milestones are the critical factors in the success and failure of a new business venture. Timing is critical when the new product is introduced, when production is expanded, when the product is improved, when competition enters, when production cost is reduced, and when working capital is created. The dynamics of growth or death of new firms center around timing.

Jay Forrester introduced analytical techniques for examining problems that arise from timing of activities in a firm, which he called "systems dynamics" of organizations (Forrester, 1961). Forrester then applied the technique to the startup and growth of new high-tech companies. Figure 2.1 shows the four most common patterns that Forrester found happens to new companies. The first curve, A, is an

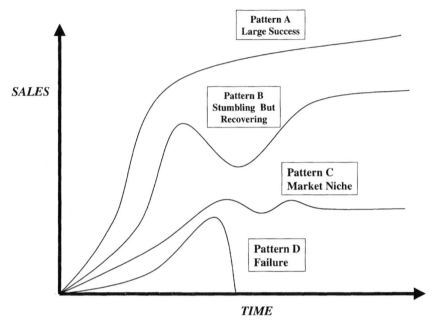

FIGURE 2.1   DYNAMICS OF NEW VENTURES

ideal pattern for which all companies hope. Initial success is rapid and exponential in sales growth and then growth slows but continues to expand as the firm transitions from a small firm to a medium-sized firm to a large firm. Cisco is an example of this pattern in curve A. But curve A is relatively rare. The harsh fact is that few new high-tech firms ever become large firms. Cisco, Intel, Microsoft— these are the exceptions.

Another successful pattern is curve B, in which a problem occurs soon after rapid expansion, but the problems are solved, and growth resumes. Curve B is even rarer than curve A because severe problems early in the growth of a new firm usually kill it, because the working capital of new firms is always thin and fragile.

For new companies encountering troubles, the most common pattern is curve D, when the new firm's capital cannot sustain a period of losses. Boo was an example of curve D.

For many new high-tech new ventures that do survive, pattern C is common. Here growth levels off as competitors enter the market, but the company successfully establishes a market niche for itself and continues on as a small-to-medium sized company or is purchased by a larger firm. Examples of this pattern C are the many smaller firms acquired by Cisco.

The dynamics occurs when all critical activities are not rightly timed. For example, if sales strategy overestimates actual sales, then production will produce too many products, increasing costs from excess parts inventory, product inventory, and production capacity. If on the other hand, sales strategy underestimates actual sales, the production capacity will be too small to meet demand, resulting in delivery delays which in turn result in lost sales to competitors, reducing revenue and lowering profits.

Pattern D is, of course, the pattern of no return and always results from a cash-flow crisis. When the costs overwhelm the revenue long enough to exhaust working capital and ability to generate immediate new working capital, then a firm goes bankrupt.

In the short-term, cash-flow is always the critical variable for survival.

The critical delays in the activities that dramatically impact cash-flow are

- Sales efforts that lag behind sales projections
- Production schedules that significantly exceed product demand
- Delay in delivery of products sufficient to lose sales
- Significant delays and/or failure of market revenue to pay for sales
- Delay in the introduction of a competitive new product model or service

Planning new business ventures is concerned with envisioning business future and anticipating business challenges to meet them on time.

CASE STUDY: Cisco Systems, continued

Now we continue the case of Cisco, starting when the Bosacks sought out venture capital to fund the new venture milestones of production expansion and meeting new competition. The market was getting hot by 1987, and several competitors began entering the router market. A special incentive to the growth of the market that year was a decision by the U.S. Congress to transform the government-sponsored ARPAnet/NSFnet to a commercial Internet.

The Bosacks saw that this commercialization would increase demand for routers exponentially and knew that Cisco Systems needed to grow very rapidly to dominate the market. They decided to seek professional management and capital help to assist in expansion. They pitched their company to about 75 venture capital firms without success until they found Don Valentine, who then was the founder and general partner at Sequoia Capital. But Valentine's terms were tough. He provided $2.5 million for one-third of the company's stock and also stipulated the night to form the company's management team. The Bosack's retained 35% of the stock:

"Did Valentine cut an almost obscene deal for himself? Of course he did; he's a
venture capitalist."

—Nocera (1995, p. 117)

Valentine immediately looked for a new CEO for Cisco and found John P.
Morgridge.

In 1988, John P. Morgridge was 54 years old and president of a failing
personal computer company called Grid Systems. The company was selling a
top-of-the-line and high-priced portable computer without any major proprietary performance advantages and thus was losing out to competitors (and was
then being sold to another company). Morgridge was then 54 years old and
accepted Valentine's offer to run Cisco, receiving a stock option of 6 percent
of the company. Previous to Morgridge's two-year stint at Grid, he had spent
many years as a salesman at Honeywell.

Conflict began right away between the founders and the new management:

Morgridge, as the new boss, found himself sparring with the company's founders
every step of the way. Cisco was becoming a real corporation, and Sandy in
particular was not the corporate type. Sandy had always been a self-proclaimed
rebel and iconoclast.

—Bunnell (2000, p. 17)

One of the things Morgridge did to try to improve cooperation was hiring
a company psychologist to improve relations, but the differences between the
principles about the culture of the organization were unbridgeable. However,
in 1989, they all agreed to take the company public, as that year Cisco had
$4.2 million in profits and was growing. The first quarter of the fiscal year of
1990, profits had already soared to $2.5 million. On February 16 of 1990, the
stock opened at $18 a share and closed that day at $22.50. But tensions continued:

". . . the IPO had made [Sandy Learner Bosack] rich, but with her role dimin-
ished and her company moving further away from its roots, she was deeply
unhappy. Increasingly, she began lashing out. This she largely concedes—"I was
screaming about a lot of things," she says—but her view is that there were a lot
of things going wrong at Cicsco, and her screaming was necessary."

—Nocera (1995, p. 120)

In the summer of 1990, the conflict came to a head: Several executives went
to Valentine and demanded that Learner leave Cisco.

In the contract, the Bosacks had signed with Valentine, there were no pro-
tections on their employment with the company they had founded. They had
agreed to a provision that gave the right to Valentine to purchase the Bosack's
shares.

"It was not my intention to get rich. My intention was to not be poor," said Sandy Learner. . . . "We worked 20 hours per day, saying the check is in the mail over and over to our vendors. In 1987 we finally got money from our seventieth or eightieth venture capitalist. . . . Then I was fired by the venture capitalists in August 1990, and Len walked out in support of me."

—Cringley (1997, p. 306)

In December 1990, they sold their two-thirds shares of Cisco for about $170 million. They were both multimillionaires. Had they hung on to most of their shares until the late 1990s, they could have been billionaires. The Bosacks divorced soon after leaving Cisco. They put much of the money into two charitable funds. Leonard Bosack moved to Redmond, Washington, and started a new company producing network equipment. Sandy purchased a 50-room mansion on 275 acres in Chawton, England that once belonged to Jane Austen's brother. She restored and converted the house into a Jane Austen study center for early English women's literature. In 1995, Sandy had also founded a cosmetics start-up company, Urban Decay.

Thus it came to pass that from 1984 to 2000, Cisco had three chief executives: Sandy Bosack, John Morgridge, and John Chambers. Sandy Learner Bosack was an entrepreneur focused on innovation, new product development, and customer sales and satisfaction. Morgridge was a professional manager experienced in large rationalized, bureaucratic organizations, who was brought in by Valentine. Morgridge later hired Chambers, another professional manager (with a view to possible succession to CEO, which did occur in 1995). Chambers was still CEO of Cisco (at the time of this writing). Both Morgridge and Chambers are examples of a good-kind of bureaucratic manger—a professional manager—those who take successful small start-up companies into first-moving, dominant, large companies.

In 1988 as new CEO of Cisco, Morgridge's first strategy implementation was to instill a culture of tight control capable of building through a period of rapid growth. His earlier experience at Honeywell had taught him that grand schemes were seldom attained and, for tight control, it was the yearly planning that made things attainable. Accordingly, he introduced planning into Cisco but didn't use a long-term basis for planning (such as a five year plan). As he explained: "At Cisco, we build a one-year plan with 80 to 90 percent assurance we'll meet or exceed our goals, so it's not a stretch. Then we modify the plan, because we're conservative." (Bunnell, 2000, p. 32).

Morgridge continued to build on Cisco's culture of working closely with customers, primarily large businesses. His salespeople were technically competent and could fix any customer problem. Morgridge kept expenditures in control, providing modest salaries to employees with stock options.

In 1991, the person Morgridge had hired as senior vice president of worldwide operations was John Chambers, who would later succeed Morgridge.

Morgridge, Chambers, and chief technology officer Ed Kozel reexamined the issues of Cisco strategy. They studied lessons from General Electric, IBM, and Hewlett-Packard. From GE, they would use Jack Welch's stragegy for each business to be dominant as number one or two or not to compete. From IBM, they would use the business strategy to provide complete application solutions to customers but would avoid IBM's rigidity to avoid adaptation to new product lines. From Hewlett-Packward's strategies, they saw the need to periodically reinvent the organization with new products as new technologies emerged. Cisco leadership then devised a strategy: provide a complete solution for businesses, make acquisitions a structured process, define the industry-wide networking software protocols, and form the right strategic alliances. (Bunnell, 2000, p. 33)

The strategic challenge that Cisco leadership saw was the need to continue to ride the wave of Internet expansion as a dominant and fast-moving player. To do this they needed to continue to add new technologies and network products and decided to do this through business acquisitions (externally acquired technology). And they needed to participate in defining the software protocols so that their products continued to fit into the emerging Internet system.

By 1993, an additional growth spurt began in the router market as the Internet continued to expand nationally and globally. Technical advances in networks continued to drive information technology progress. New switches were developed for networks using hardware that did what software in routers was doing but faster. They were also capable of being used to put together larger Ethernet LANs, as switches then could make better LAN hubs than routers. Boeing had been considering a $10 million router order from Cisco but wished to use new switches from a new firm called Crescendo Communications. John Chambers learned from Boeing that unless Cisco cooperated with Crescendo, Cisco would not get the Boeing order. Morgridge decided to buy Crescendo for $97 million. This was the first of Cisco's subsequent strategic policy of acquiring competitors to extend and improve technologies to provide advanced and complete customer solutions.

Also in 1993, it had become apparent that Cisco continued to need better information technology to manage itself. The first thing it needed was a better database system, and they decided to use Oracle, which would cost $15 million to install. The installation required about 100 people in Cisco plus Oracle people and an outside consulting group. There were problems at first as the new system crashed, but it began to run four months after planned date.

In 1995, Morgridge moved up to chairman of the board of directors and made John Chambers the next CEO of Cisco. Chambers had a law degree from West Virginia University and an MBA from Indiana University. After graduation, he took his first job in sales at IBM in 1977. He was there during the

1980s, IBM's decade of strategic errors. IBM failed then to fully exploit the rise of the personal computer and computer networking to successfully reinvent its computer businesses. (And IBM did not begin serious reorientation until the 1990s under new leadership.)

From his IBM days, Chambers learned both of what to do and what not to do. A good to-do lesson was IBM's efforts to make customers satisfied. Another good lesson was a sales person's need to sell information technology at all the levels of the customer organization. A third good lesson was how important software was to IBM's successful mainframe hardware business.

But Chambers also learned what not to do from IBM. One bad practice was IBM's relative neglect of small businesses. From this, Chambers saw the importance of selling not only to big businesses but also to small businesses. Also Chambers learned to avoid the overly restrictive command-and-control structure of IBM, which made it difficult for IBM to make timely and appropriate decisions (it also stifled initiative and entrepreneurship within IBM).

During those years of trouble at IBM, Chambers decided to move to another company, one that unfortunately was to get even deeper in trouble than IBM. In 1983, Chambers joined Wang Laboratories, Inc, which had pioneered word-processing workstations that, however, were soon made obsolete by word-processing software on the personal computer. In 1986, the founder of the business, An Wang retired, just as the company was trying to find new markets using minicomputer technologies. In 1990, Wang came back from retirement to try to save the troubled company. He asked Chambers to become the senior vice president of U.S. sales and field service operations. Unfortunately, Wang died soon after. Chambers then had to try to control the company's continuing decline. He presided over five layoffs of 4,000 people, as sales fell from $3 billion a year to $42 million. Chambers' stock options in Wang became worthless. From both the IBM and Wang experiences, the most important lesson Chambers learned was to adapt to the flow of technology advances in information technology—never to resist them, but to get ahead of, ride on, and exploit progress in new applied knowledge in information technologies. Chambers quit Wang in 1991 and joined Cisco as senior vice president.

With this background, when Chambers was made CEO of Cisco, he first attended to rationalizing a formal means of keeping Cisco advancing on progress in information technology:

> When Cisco's technology started to become dated in the early 1990s, the company saw it coming and adapted . . . Routers were still a hot ticket, but there were at least two new networking technologies . . . switching—(and) asynchronous transfer mode,"

> —Nocera (1995, p. 120)

Chambers formalized the acquisition of new companies as a strategy to continue to get new applied knowledge capabilities and new product lines into Cisco. The information technology challenge was in tying LAN nets into WAN nets. In 1995, fast Ethernet connections were still the preferred LAN technology, but for WAN networking, asynchronous transfer mode (ATM) switches were becoming preferred by customers. ATM was a hardware-based switching technology that transmitted data faster than routers and could be used to connect a finite number of LANs together, with resulting high-speed communication between LANs. Moreover, ATM allowed a digital emulations of traditional switch-based phone networks and could bridge between data communications and telephone communications. Thus Ethernet technology was hooking up computers into LANs, ATM technology hooking together LANs into WANs, and routers were hooking all into Internet.

As this was happening even before Chambers became CEO, we recall he discovered the need for Cisco to move rapidly when just before his promotion, he had visited one of Cisco's largest customers, Boeing:

> One day Chambers . . . was visiting a long-time Cisco customer and discovered to his horror that Cisco was about to lose a $10 million order to a competitor that was manufacturing switches. "What do I have to do to get that order?" Chambers remembers asking the man. "Start making switches," the man replied. So Cisco did. It bought a startup called Crescendo Communications.
>
> —Nocera (1995, p. 120)

When Chambers became CEO, he launched an aggressive continuation of business acquisitions for Cisco. And because the stock market grew throughout the 1990s and Cisco's stock soared with very high price/earnings ratios, Chambers' could use Cisco's highly valued stock to acquire other companies.

Case Analysis

We can see illustrated in this case three of the important lessons of the life stages of new high-tech business ventures.

First, different managerial styles are important to beginning and growing a new venture. The Bossacks displayed the appropriate entrepreneurial styles of being visionary and committing to and working hard to get a new high-tech business started. Later, the professional management experiences and styles of Morgridge and Chambers successfully grew the new business into a large, competitive firm.

Second, the initial success and rapid growth of the market for Cisco's routers came from a new basic innovation of the Internet which was providing a new economic functional capability to the U.S. and the world. New high-tech ventures can only be launched in the opportunties of innovation of new functional capabilities.

Third, Morgridge and Chambers successfully paid attention to all the aspects required to make a new venture into a first mover in a new industry. They simultaneously managed for

1. Continuing to advance the new technologies (through acquisitions)
2. Developing large-scale production capacity
3. Developing a national distribution capability
4. Developing the management talent to grow the new firm

INDUSTRIAL LIFE CYCLES

Now we turn from the topic of life stages of a business to life stages of the industry in which the business exists. The other two life stages of a large business—dominant player and diversified firm—occur because of continuing changes in a new industry as it later matures. The stages and dynamics in the origin, growth and maturation of an industry has been called an "industrial life cycle." In the 1980s, this idea was popularized by David Ford and Chris Ryan (Ford and Ryan, 1981).

An industrial life cycle is the pattern shown by all new industries (whether based upon information technology or any other technology) that originate upon innovation of a new core technology, after which the markets of the new industry grow and then mature as the rates of innovation in industry slow down.

One can see this striking industrial pattern by plotting the size of the industrial market over time, as shown in Figure 2.2. Market volume does not begin to grow until the application launch phase of the new technology of applied knowledge. In this early phase, the rate of growth of the market is exponential as new applications and new customers discover the new products or services of the new industry.

In the case of Cisco, the founders experienced an exponentially growing demand for their new routers as corporate and university customers began connecting their local-area Ethernet networks together and into the Internet. The demand then was so great that Cisco was profitable from the beginning because it was able to charge high prices with high profit margins, without competitors for its routers. Soon competitors began to enter, and that was when Cisco's founders sought venture capital and professional management for growth.

The early phase of any new industry based upon the innovation of new applied

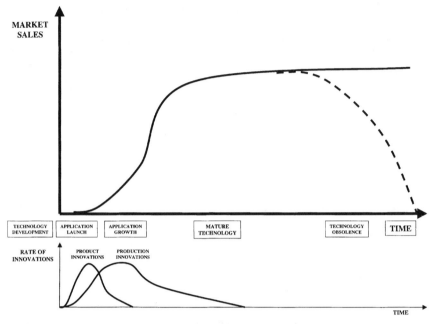

FIGURE 2.2   INDUSTRIAL LIFE CYCLE

knowledge is an exciting time, as new products are developed and rapid improvements to these products continue. Then new companies spring up in the new industry like weeds. They are started by individuals entering the industry and by employees leaving and starting up their own firms. At this time, all the new firms are small and change is swift.

William Abernathy and James Utterback pointed out that early innovations in a new-technology–based industry will usually be product innovations (e.g., improving the performance and safety of the product) and then later innovations shift to improving production process make the product cheaper and with better production quality (Abernathy, 1978; Utterback, 1978).

We have also plotted in Figure 2.2 (underneath the pattern of an industrial life cycle) a second time graph showing how the rates of product innovations and of process innovations in the newly growing industry occur underneath an industrial life cycle. There one sees Abernathy's and Utterback's point that the rate of product innovations peaks at the time of the introduction of a design standard for the new-technology product. Thereafter, the rate of innovations to improve the product declines, and the rate of innovations to improve production increases. Thus applied knowledge in a new industry is first applied in developing the new product and then applied in producing the product.

All technologies eventually mature and innovation ceases. This is why the rate of product innovations stops and eventually the rate of production innovations declines. And this is why eventually the market growth of an industry stops, and a market becomes mature.

Growth of a new market is paced by the rate of innovations in its industry, with any industrial market eventually saturating after industrial innovations substantially cease.

With the product design standard in place, the market growth continues to be rapid but now no longer exponential but linear. In Figure 2.2, is part of the industrial life cycle is noted as the linear growth phase.

In the case of Cisco, the linear growth phase for industrial computer networking began after Congress commercialized the experimental ARPAnet/NSFnet. Standardization of the Internet occurred through national network protocols and name directories, which facilitated a very rapid expansion of the Internet to global proportions. By the mid-1990s, this standardization also facilitated the innovation of commercial business on the Internet and commercial uses of the Internet for business-to-business transactions (e-commerce). When the twenty-first century began, the computer networking industry was still growing rapidly.

Eventually in all industries, the rates of innovation of technologies slow and the market saturates; this results in a mature industry phase of its life cycle. This phase continues indefinitely, unless a new substituting technology is innovated that makes obsolete the existing industry.

Historically, this general form of the core technology life cycle and its impact on the numbers of competitors in an industry has been seen in many industries. For example, James M. Utterback and Fernando F. Suarez charted over time the numbers of competitors in several industries: autos, TV, tubes, typewriters, transistors, supercomputers, calculators, and chips (Utterback and Suarez, 1993). Each of these industries showed the same pattern of a large number of competitors entering as the new technology grew and then the numbers peaking and dramatically declining due to intense competition as the technology progressed and matured.

For a technology-mature industry, market levels remain relatively constant since without new technology, market level is determined only by replacement rates and demographics.

However, if a new technology replaces an older technology (e.g., steam ships replacing sailing ships), the manufacturers of the older industry become obsolete; they must change to the new technology or die. Not all technologies become obsolete, but so me do. For an obsolete technology, the industrial market dies, as illustrated by the dotted downward line in Figure 2.2.

Another important point about an industrial life cycle, which is the number of firms in an industry over time. The number of companies in an industry first expands, then peaks, and then declines to a very few competitors (3 to 5). This is the harsh reality about competitive conditions in a large market.

Eventually only a very few large companies survive as the dominant producers in a mature market.

This is the reason why after industrial markets mature, there is always a rash of mergers of the remaining firms. For example, in the automobile industry at the end of the twentieth century, world automobile makers were merging. Then the German automobile firm Daimler Benz acquired Chrysler (one of the last three of the U.S. automakers). The American Ford acquired Swedish Volvo and British Jaguar.

CASE STUDY: Industrial Life Cycle of the Auto Industry

To see a complete pattern of an industrial life cycle, we need to look at an older industry of the twentieth century that matured during that century. The case of the industrial life cycle of the automobile industry is a good example, and its pattern of market-size over time is charted in the upper graph of Figure 2.3.

The innovation of new applied knowledge in internal combustion engines to horseless carriages created the automobile industry. The first automobile

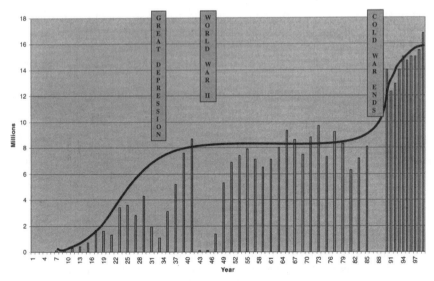

FIGURE 2.3 U.S. DOMESTIC AUTO SALES

was innovated in Europe by Daimler (after the earlier invention of the gasoline combustion engine by Otto).

In the United States, the first automobile manufacturers were from the bicycle industry. Bicycles themselves were an innovation in the 1880s, made possible by three things: (1) the cheapness of steel from the earlier innovations in steel production, (2) the paving of city streets, and (3) the discovery of vulcanized rubber for tires. To invent the automobile, the idea was to put some kind of engine onto a carriage made of bicycle components, for the bicycle provided steering, gearing, and wheels. Three kinds of engines were tried: (1) steam engine and wood fuel, (2) electric motors and battery power, and (3) internal combustion engine and gasoline fuel. It is interesting that carriage manufacturers did not innovate automobiles for they traditionally worked with wood, whereas the new bicycle manufacturers worked with steel.

The year 1896 historically marks the beginning of the U.S. automobile industry because that year more than one auto was produced from the same plan. J. Frank Duryea made and sold thirteen cars in Springfield, Massachusetts. During the next few years, many new automobile firms were founded and a variety of auto configurations were offered (Abernathy, 1978). Races were held between the three principle configurations of automobiles as steam, electric, or gasoline powered. In 1902, a gasoline-powered car defeated electric and steam cars at a racetrack in Chicago, establishing the dominance of the gasoline engine. Thereafter, this engine was to become the core technology for the automobile.

In 1902, the Olds Motor Works constructed and sold 2,500 small two cylinder gasoline cars priced at $650. The next six years in the United States saw the growth of many small automobile firms selling different versions of the gasoline engine machine. The next key event in the history of the U.S. auto industry was Henry Ford's introduction of the famous Model T. Henry Ford was producing automobiles and racing them to establish a reputation for performance for his automobiles. His cars were expensive as were all other cars, principally for the well-to-do to purchase. But Ford had in mind a large untapped market—a car for people living on farms. Around 1900 over half the United States population lived on farms. A car for farmers had to be cheap, rugged, and dependable. One of the major technical problems that Ford was facing then was the heaviness and expense of the steel chassis of the automobile.

While attending a road race in 1905, Henry Ford saw a French-made automobile crash. Since Ford was very interested in what kinds of cars competitors were making, he went over to investigate the wreckage after the race. Examining the broken engine of the auto, he picked out a valve stem that seemed unusually light. He took this back to his factory to learn its composition and found it was made of vanadium steel (steel with the element vanadium added as an alloy). On measuring the strength of the vanadium steel sample,

Ford learned that it was more than twice as strong as American steel. This was the innovative breakthrough Ford was seeking! Building a car chassis from vanadium steel could weigh less and be stronger than current autos!

The strength of the chassis was critical. As an early car bumped and bounced over rough roads, its chassis could crack, wrecking the car. Also at the time motors were bolted directly to the chassis, and sometimes in the violence of a pothole, a motor could be literally twisted in half. Ford said to Charles Sorensen, who helped him design the Model T, "Charlie, this means entirely new design requirements, and we can get a better, lighter, and cheaper car as a result of it" (Abernathy, 1978, p. 31).

Ford contracted with a small steel company in Canton, Ohio, to research how to produce vanadium steel. With vanadium steel for a new strong, light-weight chassis Ford began the design of his Model T. He decided to put the engine in front and suspend it on rubber mounts, rather than bolting it directly to the chassis. Ford also chose the best ideas of the time—a magneto ignition (no batteries), drive-shaft powering of the rear wheels (no bicycle chains), and so on. Ford designed the basic form of the automobile that was to dominate for the next fifty years. The Model T became the design standard of the auto industry.

Ford's Model T was the right product at the right time for the right market at the right price. Performance, timing, market, price—these are the four factors for commercial success in innovation. Ford captured the auto market from 1908 through 1923, selling the majority of automobiles in the United States in those years.

In the early stages of any new industry of new products and processes created by new applied knowledge, the dynamics of competition is strongly determined by competing applications of the new knowledge. Ford used new knowledge about a new kind of steel steel to build an advanced, strong, light-weight automobile. We recall that this pattern of product and process innovations of a new industry impacting market growth is a very general pattern across all industries the "industrial life cycle." In the case of the automobile, the auto industrial life cycle began when Duryea made and sold the first 13 cars from the same design. in 1896.

We recall that the first phase of an industry will be one of rapid development of the new product during the applications growth phase, which for the automobile lasted from 1896 to 1902 as experiments in steam, electric, and gasoline powered cars were tried.

We also recall that when a standard design for the product occurs, rapid growth of the market continues, which for the automobile occurred with Ford's introduction of the Model T design. Industrial standards ensure minimal performance, system compatibility, safety, repairability, and so on. Sometimes these standards are set through an industrial consortium and/or government assistance (e.g., safety standards). But usually in a new technology a performance standard emerges from a market leader.

After 1918, the final form of the automobile was set by the Model T, and subsequent improvements in the automobile occurred infrequently over the next seventy years, such as the all-steel body, all-wheel drive, automatic transmission, fuel injection, and so on. Most of the innovations in automobiles during that time have been in production processes.

In summary, one can see that the pattern of the industrial life cycle fits the graph of the U.S. auto market over time:

1. An applications launch phase as the new automobiles were marketed from 1903 to 1918
2. An industrial standard design when the Henry Ford innovated the Model T in 1918
3. A linear-growth phase as the market was expanded with every family wishing for a family car
4. After World War II, an eventual saturation of the market with family multiple ownership of automobiles
5. A continuing mature technology market as long as the internal combustion automobile is not technically obsoleted

Also one sees that there were four exceptions to the simple theoretical pattern of the industrial life cycle:

1. The steep decline in auto sales due to the Great Depression
2. The ceasing of domestic auto production during the second world war (1942–1946), when auto factories were converted to weapons production
3. Business expansion-recession cycles on the tops of the pattern (1950–1985)
4. No technology obsolescence of the automobile, as no substituting basic technology for the internal-combustion-engine-and-petroleum-fueled automobile system has yet commercially succeeded

As an industrial life cycle progresses, the number of firms in the industry first peaked and then declined. In 1909 in the new U.S. auto industry, there were sixty-nine auto firms, but only half of these survived until 1916 (Abernathy, 1978). In 1918, Ford's new model T began putting many of these out of business. Competitors had to quickly redesign their product offerings to meet the quality of the Model T and its price. By 1923 in the United States, only eight firms succeeded in doing this and remained—with about twenty-six firms failing in the four years from 1918 to 1923. The eight remaining firms then were General Motors, Ford Chrysler, American Motors, Studebaker, Hudson, Packard, and Nash.

The depression of the 1930s and the second world war interfered with the normal growth of the auto industry in the U.S, but after the war, the market growth of the U.S. auto industry resumed. The average annual sales of cars in the United States peaked around 1955 at about 55 million units sold per year. By then General Motors had attained close to 50 percent market share.

In 1960, the number of domestic auto firms remaining were four: General Motors, Ford, Chrysler, and American Motors.

The 1970s were the beginning of significant U.S. market share being taken by foreign auto producers. During that decade, gasoline prices jumped due to the formation of a global oil cartel, and American producers did not meet the demand for fuel-efficient cars. In 1980, U.S. auto producers faced desperate times with obsolete models, high production costs, and low production quality.

During the 1980s, the foreign share of the U.S. market climbed to one third, and there were three remaining U.S. based auto firms: General Motors, Ford, and Chrysler. In the 1990s, the German automobile firm Daimler acquired Chrysler. When the twenty-first century began, only two indigenous U.S. auto firms survived: GM and Ford. Further consolidation in the world's auto industry has continued. For example, Ford acquired the British Jaguar, Swedish Volvo, the Korean Daewoo, and the Japanese Mazda.

IMPLEMENTATION STRATEGIES DURING AN INDUSTRIAL LIFE CYCLE

The innovation context of companies and industries affect their ability implement strategy due to the competitive challenges that arise in the different life stages of industries and companies.

Looking back to Figure 2.2, we recall that the industrial life cycle showed the general pattern of any industry as to:

- **The early market growth and eventual saturation of any market over time**, which explains phenomena such as the rapid growth of new markets and saturation of older markets, such as Cisco's extraordinary market opportunities and the saturation of the U.S. automobile market.
- **The peaking of rates of product and production innovations over time**, which describes that all technologies mature and explains why all industries (based on the technologies) eventually mature of become obsolete, such as the automobile industry (mature) or the sailing industry (now recreational and no longer commercial).

- **The peaking and decline of the numbers of companies in an industry**, which describes the initial spurt of new companies in an industry and the harsh and continual weeding out of businesses to a handful of survivors of very large firms in a mature industry, such as the continuing concentration of firms in the world's automobile industry.

We can see that this strategic idea about the impact of innovation upon industrial dynamics is a very powerful descriptive and explanatory concept for describing why business opportunities change in an industry and what challenges business's face over time in an industry. Moreover, it is a very useful concept for strategy implementation. It can provide a theoretical ground for considering what kind of needs strategy implementation must address in the different conditions of life stages. We will call the different strategic directions in these different life-stage conditions "strategy implementation pathways."

Strategy implementation depends upon the stages of the life cycle of the industry in which the company operates.

As we have seen major force for change and opportunity for business has been innovation. Now we can see just how innovation conditions affect the implementation of strategy. To do this, we will define an implementation pathway for putting strategic change into action.

An implementation path is a pattern of strategic challenges that need to be met in implementing strategy due to the innovation context of an industry and business.

As laid out in Figure 2.4, we can classify the general patterns of implementation pathways according to the challenges of

- Starting a new business
- Rapidly growing a new business
- Moving a business into market dominance
- Diversifying a large business
- Meeting the new challenges of e-commerce

For new business ventures, in implementing the strategy of starting a new business, management styles should be entrepreneurial, and the early milestones of new business ventures should be met with the challenge of becoming profitable before working capital is exhausted. This did not happen in the case of Boo.com, as it failed to meet the critical milestone of sales generating sufficient revenue

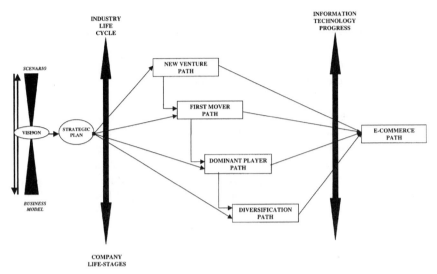

FIGURE 2.4   STRATEGY IMPLEMENTATION PATHS

before working capital was expended. However, Cisco under the Bosacks met all these milestones successfully.

To grow a large business, the next stage in a successful start-up of a new business (e.g., as Limited Inc.), is to grow from a small business into a large, well-run competitive business. The challenges needed to do this have been called a "first mover" implementation strategy. This was the strategy challenge Wexner faced in changing the management style of the Limited from entrepreneurial to professional in order to develop people to run a large organization. Morgridge successfully moved Cisco onto a first-mover path that enabled it to grow to dominate the market.

However, becoming a major player and then the dominate player is a path that requires not only being an early first mover but later adopting a redirecting strategy. Chambers took Cisco into a strategy of appropriate acquisitions that made Cisco dominant. Repositioning a large business after achieving success through start-up and into a large, competitive organization, is always periodically necessary to meet new competitive challenges. All large organizations periodically require strategic redirection to meet new competition and to adapt to new innovation and market changes. The challenges here are to attain and retain a dominant market position, through new strategic policies and redirecting business practices.

Diversification of a large firm—all dominant players in an industry will find the control of the firm under challenge when growth of the market ceases and business diversification can be sought to create new growth. How strategy is implemented for successful diversification and to meet the challenges of managing a diversified firm is the final stage in the life of large businesses.

A strategic plan may involve one, two, or all of these styles and challenges. And finally, because of the continuing pace of progress in information technology, modern strategy implementation also needs to address the challenges of integrating the business trade-offs between physical facilities to serve customers and electronic means of serving customers as an e-commerce path.

New Venture Path

The first application of a new applied knowledge as technology is important. This is what makes forecasting the conditions of competition in the applications launch phase difficult. New technology will create new applications and new markets. That is easy to forecast. But which new applications and which new markets? That is what is difficult. We do know that in the applications launch phase, the more radical the functional capability provided by a new technology the more radical and different will be the applications created for the technology.

> *The best implementation strategy in the applications launch phase will be an early innovator strategy that fosters a particular application and quickly gets to market a new product focused upon an application.*

In the case of Cisco, the Bosacks were early innovators of network routers focused on academic and corporate networking applications. In the case of the U.S. auto industry, Ford was an early innovator of the Model T, focused upon a rural market for automobiles.

In a new industry, many competitors fight to enter the market and for market share as the applications of the industrial product evolve:

1. A clear performance/price advantage is necessary to enable product entry through product substitution.
2. When competitors enter with new high-tech products, the competitor's product that facilitates new applications of the products will gain the competitive advantage.
3. Applications generation is encouraged most rapidly by software or standards compatibility that enables the immediate and easy transfer of the new technology to existing applications.
4. Applications generation is also encouraged by customer service which adapts or generates new applications of the technology.

Thus in the applications launch phase of an industrial life cycle, early-innovator strategies need to provide either unique function or dramatic increases in performance/price ratio over existing products (based on older technologies). Competitors that provide a better applications focus will have an advantage over other early competitors.

An application focus on creating a new market, although slower to build, may over the long term prove to be beneficial and profitable by creating new market niches. Only after an innovative product is completed with the availability to the customer of the needed supplies and peripherals for an application system will the product succeed in the marketplace as a new application. An innovative product without proprietary information in the design of the product or in its production will eventually have no technology competitive advantage against competitors with me-too products.

The pace of development of applications for a new-technology product is critical to the rate of market growth. Several techniques for strategy implementation can facilitate the development of applications, including:

1. Research performed by the product manufacturers for user applications development (e.g., when aluminum producers developed aluminum beverage cans for its beverage-producing customers)

2. Assisting the formation of technical-user groups and communicating with them and assisting them (e.g., when Apple helped hobbyist groups share information on new personal computer applications)

3. Use of selected customers as beta test sites for first use of product prototypes (e.g., when a software company releases new product upgrades to a few clients for early feedback)

4. Provision of industrial standards or open architectures to facilitate third-party development of application software (e.g., when Cisco participated in the formalization of Internet standards).

First-Mover Path

The first rapid winnowing out of the early innovators always occurs during the product design-standardization phase. This happens even as the market grows dramatically, with only a few firms capturing this growth. Survival depends on gaining market share.

Which companies of the many companies begun in a new industry finally succeed in emerging and surviving as dominant firms? We recall that Alfred Chandler suggested that a successful firm that is the first among its competitors to make necessary investments in (1) advancing the new technologies, (2) in large-scale production capacity, (3) in developing a national distribution capability, and (4) in developing the management talent to grow the new firm.

The reason the investment in advancing technology is important is that in the early competition as the technology changes rapidly, a competing company must not lag behind technologically and can gain competitive advantage from being innovative.

Investments in large-scale production capacity and a national distribution system are necessary for an emerging firm to gain a dominant market share in the new national market. In a global economy, a first-mover firm must also move to establish presence in international markets.

Developing management talent to run a growing, large firm is also a necessary investment. For example, many firms fail after an initial success because the entrepreneural founder of the firm does not build a management team that can succeed the founder.

Therefore, the concept of a first mover in a new industry is the strategic leadership in a company that establishes strategic policies covering essential functions of the business—innovation, production, organization.

The best implementation strategy in the design/standard phase will be a first-mover strategy that emphasizes building market share and brand recognition and develops efficient national/global production and distribution capabilities.

Cisco under the entrepreneurial Bosacks is a good example of a successful early-innovator implementation strategy, while Cisco under Morgridge and Chambers is a good example of a successful first-mover implementation strategy. In the automobile industry, Ford was not only a successful early innovator in the design of the Model T but a first mover in the development of mass production systems in automobile manufacturing.

First-mover strategies are also important to the later dominance of the firm, as they become business conventions. Lowell Steele emphasized the point that all successful firms embody their strategic policies into what then becomes the business conventions of successful firms:

> To a remarkable extent, a management team manages from a base of shared beliefs and conventions. These beliefs and conventions are not so much taught or inculcated as they are absorbed. Many of them so deep in the bones that they are not even evident to those who live by them. They may persist for decades and literally go back to the foundation of a company.
>
> —Steele (1989, p. 71)

In Steele's view, the shared beliefs and conventions are the sets of policies, explicit or implicit, that strategically guide the operations of a firm:

What are these shared beliefs and conventions, and why are they so important? I am not referring to the values and practices that characterize leadership, interpersonal behavior, or organizational development, important though they are. I am talking about the way you run a business and determine what it will become in the future.

—Steele (1989, p. 71)

Included in the these policies are informal management assumptions about the nature of the business:

- Common beliefs and rationalizations about the nature of the business of the company
- Shared assumptions about competitive advantage is gained
- Shared stories about how the company successfully grew
- Shared conventions about what is appropriate quality of operations and products
- Assumptions about what are appropriate procedures for decision making
- Conventions about appropriate kinds of control and measures for evaluating performance

Shared beliefs and conventions that act as implicit policies for a management team arise in the history of the firm from policies that first led to the success of the firm. As Steele commented:

I have always been struck by the time and energy people in management spend in reassuring themselves about the rationale that glues their business together. People in GE talked of the "electrical ring," the self-reinforcing sequence of generating electricity, transmitting, and distributing it, and then using and controlling it that has traditionally provided the unifying themes of the company. Even the apparent departures from electrical power were rationalized: engineered materials grew out of a need for insulating materials with unusual properties; GE Credit grew out of service to dealers; Apparatus Service grew out of service to industrial customers; jet engines and gas turbines had their roots in high-speed rotating machinery, and Medical Systems, in high-voltage engineering and electronics."

—Steele (1989, p. 73)

Dominant-Player Path

After an industrial product standard is established, the business and consumer markets grow around the design standard as the technology continues to improve with successive improved design standard models. These markets then partition into luxury, middle, and economy markets with corresponding differences in performance and features in their product focus. Firms establishing reputation for quality prosper, and survival in the economy markets become fierce with price cutting margins to where many firms fail.

Technology innovations will have shifted from a focus on product improvement to production improvement. Increasingly the competitive conditions turn from performance (and product differentiated by performance) to similar products competing primarily upon price and quality.

Excess production capacity can also appear in the mature phase, when the market has saturated but competitors have continued to expand capacity. Also during the course of this market growth, if technology discontinuities in product performance do occur, new product design standards will also occur. These discontinuities may also partition the market into different industrial segments for high-end performance market to low-end performance market. When technology discontinuities occur, there are opportunities for new firms to began. Existing firms that are slow to change over lose market share rapidly and may fail.

As we noted all successful businesses create conventions in the shared experiences of their management teams that are transmitted as company folklore and beliefs to younger members. For a dominant player, these conventions arise from a successful first-mover strategy. If however times change, these strategic policies, conventions, and shared beliefs create practices that no longer lead to success, and then a strategic reformulation is necessary:

> If these fundamental and often hidden . . . considerations become mismatched with the requirements for competitive advantage, then heroic efforts to improve product-specific or process-specific aspects of technology are unlikely to succeed. Many of these conventions were adopted, at least in part, because of what was . . . practical at an earlier point in time. But if they have not been revisited to assay their utility in the light of present technology (and competitive situations), they may be dysfunctional.
>
> —Steele (1989, p. 72)

Diversification Path

Finally as technology progress in the product slows or stops and the market enters a product-technology-maturity phase, products by all producers look alike in performance, differentiated only by fashion or luxury features. In the so-called technology maturity phase of the technology life cycle, products are of a "commodity-type," that is, functionally undifferentiated. Then innovation continues primarily in production processes, and competition is principally in price.

Here competitive conditions become very fierce, with low profit margins.

High productivity, and low-production costs, and high-quality production capability are critical. Brand recognition, market-share and shelf-space determine the survivors. The surviving competitors will shrink to a handful, each large. Only those with at least a 20 percent share of the market have much chance for long term survival. Business cycles provide critical conditions for eliminating marginal competitors.

Moreover, if a superior new technology is then innovated, the existing product technology will become obsolete and a process of product substitution in the market will begin. Large firms in the existing technology will likely be slow to move because of their major investments in plant and equipment and are then likely to fail. New firms in the substituting technology will be founded and grow.

If, however, no substituting technology occurs, the market for the technology will continue, and market volume becomes a function of product replacement and demographic growth (with business cycles superimposed). If production improvements continue to lower the cost of the product dramatically, the market may grow even if market demographics do not change due to multiple copies of products purchased by the market. International competition in mature technology markets will become the dominant competitive conditions with whole industries in different countries struggling to survive (if cartels are not allowed).

In the industrial maturity stage, implementation strategy should focus on four principal factors:

1. Cost/quality leadership in production,
2. Market share and brand recognition
3. Product variation and market niching
4. Fashion and demographic and life-style and environment/regulatory changes

Of these, improvements in product performance and functionality will likely play only a small role (due to the maturity of the product technology) in competitive strategy. Product differentiation will be mostly limited to market niching and fashion. Product improvement will, however, still continue, but mostly in the adding of features of luxury models down into economy models, improving product dependability and reducing product maintenance, and improving product safety. Products may also have to be improved to meet increasingly higher standards of public safety and environmental quality. But technological change will play primarily the important role principally in production technology through improving production quality and lowering production costs.

Comparing the maturity-stage of a industrial life cycle to its early-applications stage, the major difference in competitive conditions is the later dominance of production cost/quality competitive factors over the early dominance of performance/applications competitive factors.

However, even when successful in maintaining a survival and dominant player position in a mature-market industry, corporate growth cannot be found in this market. Accordingly, most mature market companies turn to diversification in businesses in different markets to find new corporate growth.

E-Commerce Path

Because of the pervasive innovation impact of the Internet upon all business, now a new strategy implementation path must be added to the traditional paths. This is an e-commerce path. As we saw in the case of Amazon versus Barnes and Noble in Chapter 1, Barnes and Noble had to add a book retail Web site to compete with Amazon. However, because of Amazon's early lead barnes and noble.com had a very small share of the e-commerce book retail market in 2000.

The applications of the Internet to commerce created several kinds of e-commerce businesses, including:

- Internet portal services (e.g., AOL)
- Retail consumer businesses (e.g., Amazon)
- Commercial supply businesses (B2B)
- Auctions (e.g., eBay)
- Materials trading markets (commodity products)
- Financial trading markets (stocks and bonds)
- Financial services (banking, mortgages, etc.)
- Information and reservations services
- Entertainment services
- Educational and training services (e.g., distance education)

For each kind of business, the kind of information strategy and business strategy required for success differed, and the paths to successful implementation of e-commerce modes differed.

SUMMARY: USING THE TECHNIQUE OF STRATEGY IMPLEMENTATION PROCEDURE

We can now use the concepts of strategy implementation to construct a practical technique for guiding successful implementation of strategic plans in a large organization. Look again at Figure 2.4.

1. Choose the correct implementation pathway. Successful implementation of strategy requires using the right kind of strategic pathway, depending upon the life stage of an industry and of the business in the firm.

2. Form appropriate strategy implementation teams. Each kind of strategic pathway requires different kinds of organization and procedures of implementation.

   A. New venture path
      - If the strategic plan (or one of its parts) is directed toward exploiting the technical progress of a basic innovation, a new-venture team should be formed to launch an early-innovator new business in the new industry.
      - The milestones of a new venture can be used to guide the implementation of the new business plan.

   B. First-mover path
      - If the strategic plan (or one of its parts) is directed toward taking a successful new start-up business into a growth mode, then a first-mover strategic team should be formed to implement the plan in all its first-mover components of:

         Continuing to advance the new technologies

         Developing large-scale production capacity

         Developing a national distribution capability

         Developing the management talent to grow the new firm
      - The milestones of an expanding first-mover strategy can be used to guide the implementation of the new business plan.

   C. Dominant player path
      - If the strategic plan (or one of its parts) is directed toward reengineering a large firm to maintain and strengthen its dominant competitor challenges through:

         Maintaining leadership as low-cost, high-quality producers,

         Maintaining national and international distribution capability and brand recognition.

   D. Diversification path
      - If the strategic plan (or one of its parts) is directed toward diversification of businesses for corporate growth, then implementation of strategy should take the form of systematically searching for, acquiring, and integrating new growth businesses (and divesting inappropriate businesses. (We will discuss this approach in greater detail in a later chapter.

   E. E-Commerce Path
      - For the type of e-commerce business, devise an appropriate information strategy and strategic business model to implement.

For Reflection

Find a biography of a person who started a successful new company? What were the conditions for the initial success? As the company grew, what were the strategic challenges in meeting competition? What was the eventual fate of the company under successors?

CHAPTER 3

STRATEGIC BUSINESS MODELS

PRINCIPLE

Strategic business models focus strategic change to optimize specific performance measures.

STRATEGIC TECHNIQUE

1. Formulate alternative strategic models
2. Construct appropriate strategy policy matrices
3. Select the most robust strategic model
4. Formulate a strategic policy matrix

CASE STUDIES

General Motors and Ford's Strategic Rivalry
Red Envelope Begins in 1999

INTRODUCTION

In traditional approaches to strategy, there have been a variety of techniques used to help think about formulating strategy. Some of these include mission and vision statements, internal assessments of weaknesses and strengths. While all these have some use in helping strategic thinking, the real bottom-line of all strategic thinking should be the business strategy. This is the changed set of policies for the future operations of the business, the future business model. Accordingly, change in strategic direction of any enterprise should be summarized in the sets of changes in business policies controlling the future of its business functions. A modern strategic management formulates strategic change as the firm's strategic business model.

In this chapter we will examine in detail the forms of strategic business models. Strategic models result from a strategic vision and provide the conceptual basis for a strategic plan. A systematic technique for constructing strategic models is the tool of a strategic policies matrix. This technique facilitates the identification of and interaction between a complete set of strategic policies, including information strategy, that are needed to construct a strategic plan.

CASE STUDY: Elf Atochem Installs Enterprise System Software

As we earlier noted, there are two strategic "totalities" in strategic thinking, the future environment and the future business. We first look at a case of how information technology in the 1990s was implementing control for the totality of a business, its "enterprise system." This case of Elf Atochem nicely illustrates both the importance of information technology and the totality of a business in its 1997 implementation of software to integrate all the operations of the company. This kind of software was called enterprise resource process (ERP) software, and it was intended to coordinate sales with production.

Elf Atochem was only one of many companies then successfully installing and operating with this piece of information technology. This case is interesting because this kind of software goes far beyond the problem of simply installing a complex piece of software in a company's computer network; it raises very important strategic issues of how a company does business and competes. For example, Thomas H. Davenport studied some of the early adopters, in addition to Elf Atochem, and reported:

> Enterprise systems appear to be a dream come true. These commercial software packages promise the seamless integration of all the information flowing through a company—financial and accounting information, human resource information, supply chain information, customer information. For managers who have struggled, at great expense and with great frustration, with incompatible information systems and inconsistent operating practices, the promise of an off-the-shelf solution to the problem of business integration is enticing . . ."
>
> —(Davenport, 1999, p. 121)

By 1998, the market for enterprise integration software had grown to about $10 billion a year. The largest competitor was the German firm SAP, whose sales had risen from $500 million in 1992 to over $3 billion in 1997. Other competitors were Baan, Oracle and PeopleSoft. In the 1990s, enterprise integration was one of the fastest growing software industries in the world.

One of the firms that successfully installed enterprise integration software was Elf Atochem North America, which was then a subsidiary of the Elf Aquitaine and a $2 billion regional chemicals firm. It had twelve business units, each with different information systems. Moreover, the different functions could not seamlessly communicate, so that ordering systems were not integrated with production systems, and sales forecasts were not linked to budgeting systems or to performance-measurement systems. And all units were independently monitoring and tracking financial data. Because of the many incompatible computerized information systems in the firm and its business units, operating data in the firm were not flowing smoothly through the organization, nor was proper information flowing to top management for sound and timely business decisions. The executives decided to install SAP's integration software, seeing an opportunity to review the company's strategy and organization.

In review, Elf Atochem's executives recognized that not only were its many information systems shared by the same customers, but each unit was managed autonomously. This resulted in the need of customers to place many different phone calls to many different units and to pay by processing a series of invoices. Elf Atochem's order-processing time was four days and required seven information hand-offs between departments. All units independently managed inventory and scheduled production, making it impossible for the company to consolidate inventory or coordinate manufacturing at the corporate level. Because of this lack of corporate control, each year about $6 million in inventory was always written off. Moreover, plants had to be shut down frequently for unplanned production-line changes.

Sales representatives couldn't promise firm delivery dates because ordering and production systems were not linked, and this lost some customers. This lack of coordination and speed of responsiveness was serious because the petrochemicals business of Elf Atochems was in an industry where many products are commodities and the company that could offer the best customer service often won the order.

Accordingly, the goal of management's information strategy in implementing enterprise integration software was to dramatically improve its service capability and make Elf Atochem into an industry leader in service and efficiency. To do this, management focused on four key processes for information integration: materials management, production planning, order management, and financial reporting. These were the control processes most inefficient in the fragmented organizational structure. Also, improvement in these would have the most impact on the company's ability to manage its customer relationships to enhance customer satisfaction and improve corporate profitability. Each of these control processes were then redesigned to take advantage of the new integration software capabilities, particularly to simplify the flow of information.

The installation of the modules was also accompanied by selected organizational changes. For example, in the financial installation, all of the company's accounts-receivable and credit departments were combined into a single corporate function, enabling the company to consolidate all of a customer's orders into a single account and issue a single invoice. This also allowed the company to monitor and manage overall profitability of serving a given customer, something previously not possible when orders were fragmented across business units and disconnected information systems. Elf Atochem also combined all of its units' customer-service departments into one department, providing a single contact point for each customer to check on orders and get problems resolved.

The new information technology provided Elf Atochem with real-time information needed to connect sales and production planning—demand and supply. When orders were entered or changed, the system automatically updated forecasts and factory schedules. This provided a core competency in the company of the capability to quickly alter its production runs in response to customer needs (at the time only one other company in the industry had this capability, providing Elf Atochem an edge over the remaining competitors).

Elf Atochem's management strategy to integrate information technology into corporate-level control systems was a basic understanding that data must be usable to decision making.

The effort required to implement and enterprise integration scheme was enormous. Elf Atochem's project was led by a sixty-person core implementation team, which reported to a member of the company's executive committee. That team included business analysts and information technologists. They installed the system in one business unit at a time, with each unit implementing the same system configuration and set of procedures for order processing, supplier management, and financial reporting. This implementation team and unit-by-unit implementation allowed Elf Atochem to staff the effort mainly with its own people, engaging only nine outside consultants. The principle reliance on internal resources not only reduced costs but also ensured that the company employees would understand how the system works.

While Elf Atochem's case of installing enterprise integration software in the late 1990s was successful, not all installations were as successful. As in any major new innovation in information technology, some users gained benefits and some were not. Davenport commented in 1999:

The growing number of horror stories about failed or out-of-control projects should certainly give managers pause. FoxMeyer Drug argues that its system helped drive it into bankruptcy. Mobil Europe spent hundreds of millions of dollars on its system only to abandon it when its merger partner objected. Dell Computer found that its system would not fit its new, decentralized management model. . . . Dow Chemical spent seven years and close to half a billion dollars implementing a mainframe-based enterprise system; now it has decided to start over again on a client-server version.

—Davenport (1999, p. 121)

The challenges of using this new information strategy arising from new information technology came from three sources: (1) the complexity of the software, (2) the complexity of installing the software into practice, and (3) the fit of information strategy to business strategy. Davenport commented upon these problems:

Some of the blame for such debacles lies with the enormous technical challenges of rolling out enterprise systems—these systems are profoundly complex pieces of software, and installing them requires large investments of money, time, and expertise. But the technical challenges, however great, are not the main reason enterprise systems fail. The biggest problems are business problems. Companies fail to reconcile the technological imperatives of the enterprise system with the business needs of the enterprise itself.

—Davenport (1999, p. 130)

This is nicely illustrates the heart of the strategic issue between any information technology and business operation, reconciling the imperatives of information technologies to business needs. As Davenport further emphasized in enterprise integration software:

An enterprise system, by its very nature, imposes its own logic on a company's strategy, organization, and culture. It pushes a company toward full integration even when a certain degree of business-unit segregation may be in its best interests. And it pushes a company toward generic processes even when customized processes may be a source of competitive advantage. . . . If a company rushes to install an enterprise system without first having a clear understanding of the business implications, the dream of integration can quickly turn into a nightmare."

—Davenport (1999, p. 131)

Case Analysis

When the twentieth century ended, using ERP software to integrate operations of an entire business enterprise was one approach to using information technology to control a total model of a business. Later with the advance of the Internet, other approaches to integrating a total business enterprise began evolving (which we will examine in a later chapter). For now, this case emphasizes two general points about any implementation of any information technology into any business strategy:

- There are always major costs and risks in strategic implementation of information technology into the ongoing operations and processes of a firm.
- To be effective, the installation of major new information technology needs always to be accompanied by process and policy reengineering in the firm's operations and competitive strategies.

In modern business strategy, the component of information strategy is how information technology is used to gain economic benefit. The management practices of the enterprise often need to be changed to implement the economic benefit. In other words, strategic management—change in the firm as a whole—is usually needed to provide the potential competitive advantages of new information technologies.

ENTERPRISE SYSTEM

Organizations are goal directed and create productive transformations to reach these goals. All businesses make profits by directly adding value to purchased resources by transforming them into manufactured goods or delivered services. The production or delivery system of the firm is the coordinated set of activities (system), which directly adds value. For a business, this goal-seeking has been called the "concept of the enterprise," and its productive transformations constitute the "enterprise system."

We recall that we indicated that all models of organizations involve the concept of an "open system" in which the totality of the organization is described as receiving inputs from its environment and transforming these into outputs into its environment. A famous example of this is Michael Porter's value-adding model of a business (Porter 1981). Figure 3.1 compares an open system model to Porter's direct value-adding activities (which he described in the form of an arrow). In both schemes, resources as inputs are transformed into outputs production. Porter's version adds overhead functions to the direct production (transformation) center of the open system model. This particular model of a business is one fundamental way to look at any business.

A model of a business's "enterprise system" can be constructed as (1) overhead activities above a (2) transforming open-system; and the open-

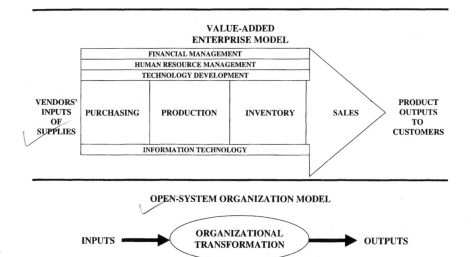

FIGURE 3.1 VALUE-ADDED & OPEN-SYSTEMS MODELS OF A BUSINESS ENTERPRISE

system portion acquires material, capital, and personnel resources from the economy, transforms these into goods and/or services, and sells the goods/services into the markets of the economy.

Models of a business express the core ideas of a business, as Lowell Steele emphasized:

> Every business is based ultimately on a few simple ideas, principles, or even assumptions. They address the fundamentals of the business: What products or services do we provide? Who are our customers? How do we compete? How do we define success? How do we behave toward each other? In the aggregate these fundamental features could be termed *the concept of the enterprise.*"
>
> —(Steele 1988, p. 69)

We recall that Steele also pointed out that to a great extent the answers to the fundamental questions of the business enterprise is implicit in the shared beliefs and conventions in the culture of the firm. From past experience in successful business operations, managers develop a culture of shared beliefs and conventions about how the firm should operate, including assumptions about:

- The nature of the business
- The way competitive advantages are gained
- A sense of how and why the company became what it is
- Conventions about the guidance and operational control of the enterprise

Also we recall that policies control procedures which control operations which control activities. Strategic planning is also concerned with changing policies over the long term. Therefore, a strategic business model formulates the policies that will guide business activities in the future.

> *A strategic business model makes explicit the strategic policies of future operations.*

CASE STUDY: General Motors and Ford Strategic Rivalry

In the case study of the U.S. automobile life cycle in Chapter 2, we looked at the industrial life cycle of this industry and noted that competition was so fierce that, by the end of the twentieth century, only two U.S. producers were still independent companies, General Motors and Ford Motor Company. Also we saw that Henry Ford's strategic policies (1) to build the Model T established a dominant industrial standard and (2) to build these cars in an assembly line provided Ford with a first mover capability that quickly leaped the company to dominance of the car industry in the United States in the 1920s. In contrast, Durant's strategic policies at General Motors were to assemble a conglomeration of automobile manufacturers and parts supplies, intending to create a strategic robustness for GM. In this case, we examine in detail the competitive rivalry between GM and Ford. We shall see that it required Sloan (GM's first CEO after Durant) to rationalize Durant's conglomeration into a set of coherent strategic business policies before GM could overtake Ford in the marketplace—a better strategic business model.

After Ford's initial dominance of the U.S. automobile market in 1918, GM went on to overtake and pass Ford to become the dominant U.S. automobile manufacturer until the 1970s. Then GM's market share began to decline under the impact of foreign competition. In the 1990s, Ford was beginning to overtake GM.

Arthur Kuhn summarized the jockeying for competitive dominance between GM and Ford:

> By 1923 Ford Motor Company had become the envied giant of American industry. In 1924, however, the automobile market began to change. Over the next two years a metamorphosis occurred that rendered Ford Motor's Model T machinery and the accompanying historical information worthless. Even worse,

(Henry) Ford had long held that Ford Motor's non-factory organization should be minimized. Without a central headquarters to guide it, (the company) foundered in 1927.

—(Kuhn, 1986, p. 5)

Henry Ford had innovated the Model T and built efficient mass production lines, which quickly made Ford the largest automobile manufacturer in the world. But while Ford was doing this, another automotive entrepreneur, William Durant, was putting together a big automobile company by acquiring many smaller auto companies and auto parts companies. He called this empire of auto manufacturers General Motors. Durant's strategy was that since the auto market was fickle, he would assemble as many different makes as possible to cover the market. In 1908, GM was founded by William C. Durant who traded stock in his successful Buick Motor Company to acquire additional auto firms. But in 1910, a group of bankers took control away from Durant, critizing his expansion as too rapid. Durant then founded Chevrolet and used its stock to regain control of GM.

Yet in only four years GM was again in trouble and need rescuing once more. Inventories were high and cash ran short. GM needed to borrow $83 million in October 1920 to pay salaries and supplies. Durant was forced resign, and Pierre duPont was elected president as the duPont family held the controlling interest in GM.

Pierre S. duPont had led the growing U.S. chemical firm of duPont but had retired. As the chemical firm was producing great profits for the duPont family, it had invested in Durant's building of GM. Pierre duPont soon appointed Alfred P. Sloan to succeed him, stating: "I greatly admire Mr. Sloan and his business methods and look upon him as one of the most able partners in the management of General Motors Corporation." (Forbes, 1924, p. 760)

Sloan had been trained as an electrical engineer at the Massachusetts Institute of Technology and afterwards built and operated Hyatt Roller Bearing Company. He sold bearings to both Ford Motor and Cadillac. In 1916, he sold Hyatt to Durant, continuing as president of Hyatt, which operated as a GM subsidiary. When duPont replaced Durant, Sloan was promoted to president of GM. Sloan's challenge was to revamp GM to make it into a profitable competitor and overtake Ford Motors. Although Henry Ford had genius in the development of auto design and production, he had an autocratic management style and neglected to groom management talent to succeed him. Sloan had experience in managing production and running a business and, it would turn out, genius in building an efficient and effective very large organization. (Ford is an classic example of a successful entrepreneur, while Sloan is a classic professional manager.)

Sloan took taken over a confused and cumbersome organization from Durant's previous strategic management. As a part of Durant's earlier team, Sloan intimately knew how much GM sorely needed better strategic management. This Sloan set out to emphatically improve the organization. Also Sloan had the assistance of another talented manager, Donaldson Brown, whom Pierre duPont had brought into GM from duPont. Brown served as GM's vice president of finance. Together, Sloan and Brown began to redesign GM, starting with its system of financial control.

In Sloan's extensive redesigning of GM's business practices, he would change several areas of GM's policies about information, product, innovation, marketing, production, organization, finance, and diversification. Altogether, this new set of policies would transform GM's previous model of a business under Durant as a conglomerate into a tightly coordinated enterprise system for automobile production and market dominance.

Information Strategy The first strategic policy that Sloan and Brown implemented in GM was a systematic, detailed, and uniform performance reporting system across all the divisions of the firm. Durant had been primarily concerned for the stock price of the firm, since a high stock price had enabled him to acquire companies for assembling GM. However, the duPonts were long-term investors and not stock speculators, so they were primarily concerned with dividends. Sloan also held significant holdings from his sale of Hyatt to GM, and he thought: "Naturally, I like to see General Motors stock register a good price on the market, but that is just a matter of pride . . . What has counted with me is the true value of the property as a business return on investment." (Sloan, 1941, p. 103)

And this view of the true value of a business as the return on the investment in the business is one of the great eternal strategic truths of business—applicable as it was then in that "old economy" of hard good production of the twentieth century, such as automobiles, and later even to twenty-first's century "new economy" of information services. From duPont, Donaldson Brown had brought with him a financial model he had developed there for measuring return of investment: $R = T \times P$.

This reads that the way to calculate the (R) rate of return on invested capital is as the *multiplication product* of the (T) the rate of turnover of invested capital multiplied by (P) the percent of profit on sales. This formula emphasizes that management should monitor how *quickly capital was returned from investments* in the production of cars and how large was the *percentage of profit* being made on sales of cars.

Brown measured the capital turnover variable T in terms of fixed-capital variables such as plant and equipment and working-capital items such as cash balances, in-process and finished inventory, and accounts receivable. These were reported in terms of a ratio to sales, whose inverse-ratio expressed the rate of turnover T. The analytical model allowed management to see: ". . . a specific disclosure of causes and effects for the return on investment. . . . Effective control, or lack of it, for any item on either side of the equation (T or P) could be identified, thus making possible efforts to improve conditions." (Brown, 1957, p. 27)

These two quantities, of production-investments-return and production profits, became Sloan's measures of internal performance on GM's production operations (and Sloan's bureaucratic rationalization of GM).

Sloan put into place procedures of an information system that reported the performance of divisional operations in variables that directly affected return on investment. Sloan's information policy required that all divisions in GM adopt and report information in this standardized and systematic financial analysis model. Such information was used not only to control but also to plan operations:

> By means of our accounting system, we can look forward . . . and can alter our procedures or policies to the end that a better operation may result. In one case we are in principle, looking backward—in the other case, forward. We are able to forecast our operations four months ahead with a certainty that would hardly seem possible.
>
> —(Sloan, 1929, p 92–96)

Brown and Sloan separated planning into short-term and long-term horizons. Short-term planning included events such as sales stimulus, more advertising, temporary pricing discounts, and so on. Long-term planning included factors relating to consumer appeal in style, functioning, serviceability, and so on. In addition, information for planning included economic analyses in terms of seasonal variation of demand, long-term growth of demand, and business cycles. For these, GM collected many kinds of consumer-demand related information, such as upon country-by-country personal income, agricultural conditions, employment, commodity price trends, and so on. GM analysts also developed leading barometer indicators of economic patterns and studied past business cycles. GM also watched their competitors closely: "At the proving grounds near Midland, Michigan, 42 miles from Detroit, every competitive model (was) tested under the most exacting road conditions." (Rukeyser, 1927, p. 376)

Another information strategic problem had to do with the evaluation of the performance of the parts divisions of GM. Sloan's concern with controlling costs in mass production was to create as much as possible interchangeable parts for all its automobile divisions—interchangeable engine components, chassis parts, automotive accessories, and automobile bodies. Sloan retained all the component and parts suppliers which Durant had acquired. However, to control costs, Sloan instituted a transfer-pricing scheme to evaluate the performance of parts suppliers.

Sloan and Brown instituted a pricing market, within and without GM and between divisions, to arrive at interdivisional prices. The prices one GM division charged another GM division for parts had to be priced competitively similar to what an external parts supplier would charge.

Sloan's information strategy aimed at control and planning of operations proved its great value in the onset of the depression in 1930. In the late spring and summer of 1929, GM had been charting the decline in car sales, responding with reduced production to keep inventories low: "dependable forecasting and planning were of outstanding importance during those difficult years. Production by all divisions was held in reasonable bounds." (Brown, 1957, p. 71).

In the decades from 1900 to 1930, more than a hundred auto companies had been started, but less than a dozen survived the great depression of the 1930s. Sloan's information strategy had been aimed at creating a firm responsive to change and market needs and conditions. And about GM's survival of the great depression of the 1930s, Sloan wrote: "We had simply learned how to react quickly. This was perhaps the greatest payoff of our system of financial and operating controls" (Sloan, 1964, p. 199).

Product Strategy   In addition to Sloan's imposition of uniform information and reporting requirements geared to controlling the performance of GM's return on investment to stockholders, he also had to rationalize the products of GM. Sloan's predecessor, Durant, had strategically acquired a variety of automobile companies to minimize sales risk: as Durant expressed his basic strategy:

> The business of an individual manufacturer was hazardous because the model on which he staked his chances of sales might prove to have some mechanical defect or the body design might fail to strike the fancy of the buying public. . . . I was for getting every kind of thing in sight, playing safe all along the line.
> —(Crow, 1945, p. 74; Epstein, 1928, p. 182)

However, Durant's diversification strategy had not yielded financial safety, for the automobiles from his acquired divisions overlapped in features and price:

Chevrolet	(4 cylinder)	$775–$2075
Sheridan	(4 cylinder)	$1685
Olds	(4 cylinder)	$1445–$2145
Olds	(6 cylinder)	$1395–$2065
Oakland	(6 cylinder)	$1395–$2065

Scripps-Booth	(6 cylinder)	$1545–$2245
Buick	(6 cylinder)	$1795–$3295
Cadillac	(8 cylinder)	$3790–$5690

Durant's GM products had been competing as much between themselves as with other competitors. This was another strategic issue Sloan began to address. In 1924, Sloan had reorganized the product line:

Chevrolet	(4 cylinder)	$ 510
Olds	(6 cylinder)	$ 750
Oakland	(6 cylinder)	$ 945
Buick	(4 cylinder)	$ 965
Buick	(6 cylinder)	$1295
Cadillac	(8 cylinder)	$2985

1924 was a recessionary year for car sales. Although Ford's Model T was out-of-date, it was well-made, reliable and cheap, and it sold. However, the larger cars of GM sold well, and Sloan thought the rationalization of GM's product-lines was moving in the right direction. Later, Sloan would continue to refine the GM divisions to Chevrolet, Pontiac, Olds, Buick, and Cadillac spaced over prices and features to cover the U.S. passenger automobile market up until the 1960s.

Innovation Strategy   Another strategy of Sloan's was to innovate new features and performance improvements that would provide the customer a higher quality product. Ford's Model T had an open body and wooden frame on top of a steel chassis. Sloan had all GM autos constructed as closed bodies entirely metal. Other improvements were to be added, such as brakes on all four wheels. With this strategy of innovating product improvements, Sloan could price the GM autos at the top of their price brackets as quality products. He then reduced price down toward that of the Model T as soon as production volumes lowered unit-costs-of-production.

Sloan established a research laboratory under Kettering to develop improvements to the automobile system: "We are searching for the facts that we may . . . add value to the performance and effectiveness of our products" (Sloan, 1927, p. 550). At the time of founding the laboratory, GM was being innovative, which "compared to competitors like Ford, whose engineering departments were dominated by self-taught mechanics, GM's research unit was quite progressive" (Leslie, 1983, p. 98).

In addition to innovation of product improvements, Sloan also innovated a policy of styling. In 1927, Sloan had Cadillac introduce a lower priced car between the Cadillac and the Buick, which was called the La Salle. The La Salle had a body styled by Harley Earl, a Los Angeles custom-body designer hired by Fisher Body; and as Sloan wrote: "The car made a sensational debut in March 1927 . . . The first stylists' car to achieve success in mass production" (Sloan, 1964, p. 269).

Earl's intention in styling was to lengthen and lower the American automobile in appearance. Also a variety of colors could be painted on cars in 1927, thanks to advances in paint technology that decreased the drying time of colored paints. The combination of price bracketing of automobiles, product improvements, styling, and colors devastated Ford's share of the market. Sloan's product strategy was to provide the appearance of product differentiation in a standardized market. Sloan knew that GM's future would depend upon fashion and luxury.

Marketing Strategy   GM's strategy was to sell automobiles through independent but franchised dealers. The GM Finance Committee established by duPont and continued by Sloan decided not to establish company-owned dealerships. The amount of capital needed for 20,000 retail outlets would have been astronomical; and the number of employees in the outlets would have been too large to manage. Accordingly, franchising dealerships avoided the large capital and management problems of retailing the automobiles to customers. Finally, some control over retailing by GM could still be established though the financing agreement.

However, to GM there were still market challenges in financing both the auto inventory and the purchase of automobiles. Banks were reluctant to extend loans to dealers for inventory and to customers for automobile purchase. John J. Raskob, then chairman of GM's finance committee suggested to start a finance company, limited to financing the paper of those who dealt with GM cars, and this was established as the General Motors Acceptance Corporation (GMAC).

Finally, a third element of GM's marketing strategy under Sloan was to introduce the deliberate obsolescence of product models through annual introduction of new product models. GM designers created annual product model variety through annual model changes. This annual model change policy routinized change for GM, giving customer's reason to buy new cars before while their older cars were still functional.

This combination of product and marketing strategy matching GM's policies to customer preferences proved devastating to the older market dominance of Ford:

> Sloan did not vanquish Ford's popular Model T by cajoling consumers to buy upgraded automobiles and annual models. Rather he monitored changing tastes and pushed GM's divisions to keep their offerings synchronized with market demands.
>
> —(Kuhn, 1986, p. 7)

As fleet sales increased in importance, GM in the 1930 established the General Motors Sales Corporation. To improve fleet sales performance, GM staff:

compiled comprehensive records to show where our market was, who was getting the business, and where effort was most needed. Now we keep these records constantly to date and they are to us what a compass is to a sailor. We know who our prospects are, how many units they operate, the make and age of each unit and many other facts. If one of our accounts begin to slip, we know it promptly and are governed accordingly.

—(Baird, 1935, p. 188)

Production Strategy   Sloan's mission was to optimize return on investment to shareholders. Therefore a strategy about production was necessary to make money from the product variety for the auto market. Sloan's strategy for lowering production costs was to standardize as much as possible parts and production across all operating divisions. It was from Sloan's prior experience with mass production in running Hyatt that his production strategy was to strive for the largest possible production runs with lowest product costs. This would provide the largest margins.

Because GM bought parts from outside firms, the strategy for part standards required GM to foster standardization across parts industry. For example, Sloan crated GM's General Purchasing Committee, which published a Book of Standard Parts "containing 196 pages descriptive of standard parts, 100 pages on materials, and about 50 pages of miscellaneous information" (Baird, 1923, p. 336).

The internal production of standardized parts and external purchase of standardized parts lowered GM's production costs by fostering a competitive market in its suppliers.

Organization Strategy   Sloan's organizational strategy was to have a decentralized product and production capability while yet tightly coordinating these from a central authority:

> Sloan started this trend toward centralization simply because he had inherited from his predecessor . . . a group of division managers almost totally ignorant of financial matters. Durant's "automobile men" . . . often accumulated dangerously excessive inventories through inattention, poor forecasting, or even inflationary speculation. The Sloan-Brown team imposed strict financial controls over divisional operations.
>
> —(Kuhn, 1986, p. 97)

This organizational strategy began when Pierre duPont took control of GM, and he and Sloan spent many of their early days at GM visiting and evaluating the different divisions, which Durant had assembled as General Motors. After Sloan became president, he continued his inspection trips and visited dealerships to gain first hand impressions of operations.

Then Sloan and Brown institutionalized yearly and monthly formalized performance reviews of divisional operations. The central control over divisions used the detailed performance information from divisions to review past performance and to compare current performance against planned performance.

Sloan reorganized GM's Executive Committee to exclude most of the division managers in order to carry out this centralized oversight of distributed divisional operations. Sloan then established an Operations Committee to which all general managers of the divisions belonged along with all the general officers on the Executive Committee. The purpose of the Operations Committee was to coordinate implementation of centralized policies of the Executive Committee.

Financial Strategy   As emphasized, Sloan's overall strategy was to promote growth in market share and optimize shareholder return on investment. Accordingly, an external Finance Committee (which had been established by duPont) focused investment upon operations which promoted the highest return. From this committee came several strategic investment decisions, which focused GM investment principally upon production assembly and some parts production. Auto assembly was the highest value-adding operation that created the largest profit margin. Parts production was not as profitable but critical, and so an investment decision was to finance limited part production capability and purchase remaining requirements from outside suppliers.

Diversification Strategy   Accordingly, GM's diversification strategy under Sloan remained focused upon the production of land transportation products and their parts. GM produced autos, trucks, and tractors. The first acquisition Sloan made was to acquire its principal body supplier, Fisher Body Corporation. Sloan secured control of Fisher by purchasing 60 percent of the stock and thereafter purchased the remaining 40 percent. Sloan moved the Fisher brothers into corporate and divisional positions.

For the basic commodity industries that provided materials for the parts producers, a deliberate decision was made by the Finance Committee to not invest in these basic commodities (such as steel and chemicals). The financial investment in basic commodities would be high and the margins lower than from the automobile business and the several competitive suppliers already kept commodity prices low.

Also as earlier noted, a decision was also made not to invest in the auto retail industry; instead using dealer franchises to avoid the extremely large investments required to establish an auto retail network.

However, as also earlier noted, an investment strategy was to facilitate the growth of credit available to automobile retailing by establishing a GM credit company.

Competitive Strategy   The strategic policies that Sloan formulated and implemented in GM provided GM with a competitive strategy to overtake Ford and take market away from Ford and other competitors. In terms of Ford, GM priced the Chevrolet just on top of the Ford price but provided improved features and performance to justify the price and more fashionable models. In terms of other competitors, GM covered all the market niches for cars with similar strategies of pricing and advanced performance and fashionable style.

Sloan's competitive strategy proved successful because the automobile market was changing in America, and Sloan's new strategies adapted GM to exploit these changes. The market changes that Sloan saw and exploited in his new strategies included market factors such as installment selling, used-car trade-ins, closed body auto models, improved roads, increasing urbanization, rising prosperity, changing fashions. Sloan's business model was to adjust GM's operations and control to better match the product to the changing demands of the market.

Strategy is about change, changing the business model to meet new demands of the market. Sloan's strategy for GM to overtake Ford's market leadership required Sloan to formulate strategy in several different areas of the complex management problem of running a large firm—the strategic areas of information, product, innovation, marketing, production, organization, finance, and diversification. These strategies together altered how GM did business, reformulating the business model from Durant's aggregation of companies to Sloan's streamlined, rationalized, and controlled business model of an integrated set of automobile businesses. Sloan's new strategic business model was an enormous success for a very long period of time. GM's overwhelmed the U.S. automobile market from 1926 to 1976—a competitive dominance lasting fifty years. In 1976, GM produced 55% of all passenger cars sold in North America.

But just at the peak of GM's success in the 1960s, its serious decline had begun. GM was battered by a new style of competition and changing world conditions and demographic life styles. From 1976 over the next twenty years, GM's market share slid from a high of 55% to 27% in 1999. Then Ford with a 22% share was in position to finally again overtake GM. What happened? Why had the strategies Sloan implemented successfully in the 1920s and 1930s begun to fail in the 1970s?

The reason for GM's decline were trends in Sloan's strategy that were inappropriately emphasized by his successors, as Kuhn summarized:

> Sloan kept GM's management narrowly focused on stockholders' interest at the expense of labor, consumers, and the general public. Over the years, he placed too much stress on styling and on larger mid-price automobiles . . . Sloan's emphasis on rate-of-return performance would eventually thrust too many financially oriented executives into GM's top corporate decision-making positions, men without knowledge of automobile design, production and marketing.
>
> —(Kuhn, 1986, p. 318)

It was the last trend in successive leaderships' preoccupation only with financial strategy that exacerbated the other trends into a poor overall business strategy. GM's successive leaders' antagonistic attitude toward labor fostered continuing labor troubles for GM (that greatly contributed to the deterioration of GM's capability for efficient, high-quality, and low-cost production). GM's leaders actively resisted providing leadership in automobile safety and environmental quality (eventually fostering U.S. government intervention to mandate safety and environmental standards for the auto industry). GM's successive leaders continued centralizing control in GM until GM's automobiles lost distinctiveness, quality, competitiveness, and performance—compared to imported foreign-produced automobiles.

For example, by the 1970s, GM had failed to compete effectively in the low-end automobile market and also had lost quality leadership in the middle and high end. German automobile makers dominated high-end sector quality; whereas Japanese automobilte makers dominated lower and middle sector quality.

Case Analysis

Sloan used a complete set of new strategies to alter the way GM did business in order to overtake and supplant Ford as the leader in the U.S. automobile market. This set of strategies taken together provided a new model for the way GM operated and controlled its businesses—a new business model.

However, strategy requires periodic readjustment to meet new challenges, competition and market place changes. The GM CEO successors to Sloan failed to bring GM strategies up to date in a proper way as the world changed. Competition requires both proper strategies and proper balance between strategies in the different areas of a company's operations. The failure of GM's leaders after Sloan was due to *unbalanced strategy,* in which financial strategy was over emphasized at the expense of product strategy as well as quality and market strategy.

This problem of balanced strategies is a common one. For example, in 1978, Lee Iacocca left Ford to run Chrysler Corporation, which then was on the verge of bankruptcy. He found poor management with excessive inventories and no overall system of financial controls, later commenting: "If the bean counters are too weak, the company will spend itself into bankruptcy. But if they're too strong, the company won't meet the market or stay competitive" (Iacocca and Novak, 1984, pp. 43).

When strategic policies get out of proper balance, particularly between finance and market, then business can be lost to competitors.

STRATEGIC BUSINESS POLICIES

All businesses consist of activity and control of that activity. Accordingly, strategy must be formulated for carrying on that activity and controlling it.

ACTIVITY		CONTROL	
PRODUCT STRATEGY	PRODUCTION STRATEGY	FINANCE STRATEGY	DIVERSIFICATION STRATEGY
MARKETING STRATEGY	ORGANIZATION STRATEGY	INFORMATION STRATEGY	INNOVATION STRATEGY
			COMPETITIVE STRATEGY

FIGURE 3.2   STATEGIC BUSINESS POLICIES

A strategic business model systematically lists and describes the set of strategic policies that guide the operations of a business.

This list of strategies can be generated from the general functions of any business and its control, as summarized in Figure 3.2. The four basic activities of any business are product, production, marketing, and organization.

- All businesses must design a product and produce it. (The product can be a hard good, software, or service.)
- All products must be produced in volume. (The production can be manufacturing, production, or service delivery.)
- All products must be marketed and sold for the business to obtain revenue.
- All business must construct an organization to carry out the activities of product design, production and marketing.

The basic activities must also be controlled by management of the organization; and the four modes are finance, diversification, information, and innovation.

- All businesses control activities through finance, buying resources and selling products and measuring performance as profits.
- All businesses control the nature of the business activity through diversification, defining the set of businesses carried on by the firm.
- All businesses control operations through information, structuring the planning, scheduling, coordination, and access to activities.
- All businesses control change in operations through innovation, introducing new means to perform activities through improved processes and technologies.

Accordingly, a systematic list of strategies required to construct a model of how a business operates consists of a description of the management strategies for product, production, marketing, organization, finance, diversification, information, and innovation. The strategic concept of the business model applies to all businesses and all times. A strategic business model is a ubiquitous and enduring technique.

CASE STUDY: RedEnvelope Begins in 1999

We can illustrate the generality and timelessness of strategy formulation as a business model by examining a case of a new company RedEnvelope. At the next turn of the century (from 1900 for the automobile industry to 2000 for the e-commerce industries), many new e-commerce businesses were struggling with the age-old strategic problem of creating a successful business model. An illuminating case of this was the experience of a new Internet business for gifts, 911 Gifts which was later renamed RedEnvelope:

> Noon, Christmas Eve (1999), San Francisco. In stores on Union Square and on Market Street, registers will ring up sales for another six hours. But in a tower near the western end of the Bay Bridge, in the offices of RedEnvelope, an online gift company that has been up and running for all of 60 days, the holiday is over. The staff gathers for toasts and sips champagne from plastic glasses. "Next year it will be crystal," laughs Hilary Billings, the 36-year-old C.E.O.
> —(Bekke, 2000, p. 30)

In last week of the annual winter holiday season, RedEnvelope had been mailing packages at the rate of 20,000 per week. It sold millions of dollars of its merchandise which included 500 kinds of products-for-sale: 750 electric full-body-massage mats, 620 Zen tranquility fountains, 1400 Japanese body-and-soul bath kits, 1600 boxes of pistachios and caramels, 1600 amber bead necklaces, 600 antique thermometers, and 1100 chocolate body-paint kits.

During that season, online retailers like RedEnvelope had sold $9 billion in merchandise in the last two months of 1999. Still none were profitable. RedEnvelope paid $4 in marketing for every dollar of sales. It paid AOL a great deal to get into its commerce area (which brought "steady but unspectacular sales") and paid $15,000 to set up on Yahoo (which brought in 600 orders a day at maximum). Of the people who visited RedEnvelope's Website, 5.8 percent purchased (compared with the industry conversion rate of 2 percent). Of its orders, RedEnvelope filled 98 percent accurately and on-time, with only 2 percent wanting to return gifts in January (compared with the then 8 percent industry average return rate).

Thus setting up as a business, they viewed their first season as a success, except they had only $31 million in venture capital left (which would last them only six more months). RedEnvelope needed to get more venture capital or go public. In the heady Internet-stock market boom of the late 1990s, venture capital had been readily available to e-commerce start-ups; many had pursued this market strategy to buy market share at the expense of working capital.

The company was begun by Scott Galloway and Ian Chaplin, who in 1992 had graduated from the University of California Berkeley's business school and started consulting. When the Internet began growing, they started a online pet product business and then sold it. Next they tried selling gifts online, founding 911Gifts. In 1998, its site did about $1 million business, but able to move only 1,000 packages a day. They realized they needed both venture financing and better strategic experience in gift retailing. Galloway called one of his investors, Pat Connolly, who was a Williams-Sonoma executive vice president, and asked him about who were the top merchandisers in the United States. Connolly advised him to call Hilary Billings, who had worked for Connolly at Williams-Sonoma. Williams-Sonoma was a kitchen and home furnishings retailer. In 1991, Billings had developed its Pottery Barn catalog business.

After Galloway called Bllings, she looked at 911Gifts Website but liked neither its name nor appearance. She thought the products being advertised were ordinary and poorly photographed. She thought the site also was poorly designed without inspiration. But she did like the idea that one could navigate the site to find a gift by recipient or by occasion or by department. Then Billings and Galloway talked, and she signed on as director of merchandising in May of 1999.

Immediately Billings began making rounds of Silicon Valley venture capital firms, and in six weeks secured $21 million from Sequoia Capital (the same venture capital firm that had funded Cisco) and $10 million from Weston Presidio. Together, the venture firms would own between 30 and 40 percent of the firm. Billings became CEO and her strategy was to make the company not just seasonal for gifts but continuous, with a substantial margin of 50 percent markup. Billings also proposed to change the business's name. Her team thought about it and settled upon RedEnvelope from an idea of Krisine Dang, director of merchandising. Dang's family had left Saigon for the United States, and she recounted an Asian custom of marking special occasions by giving cash or a message enclosed in a red envelope

The next strategic focus was to think about who should be the customers and what kinds of products they would desire. Billings focused upon the types of customers she knew from her work at Pottery Barn, high income, well-educated professionals of both sexes. Then advertising was the another strategic issue to attract customers to the site. Billings considered TV campaign but thought the expense too high and instead chose ads in newspapers and magazines, which targeted the desired customer demographics. She also paid $2 million for access on web portals such as AOL, Yahoo, Excite, and so on.

The next strategic issue was organization, and Billings decided to own its inventory, merchandising, marketing, systems management, and customer service systems but not the physical handling of products and order filling. The company had been using a order-filling company called ComAlliance in Wilmington, Ohio, and this arrangement was expanded with 32,000 square feet of work space leased for a year. This warehouse was located at the edge of an airport so that overnight merchandise deliveries were possible.

Thus with new brand, product, customer, marketing, and delivery strategies in place, the revamped 911Gifts was in business on October 1999 as RedEnvelope, and by early December sales were doubling every week. Control of operations also involved new kinds of strategy:

> One way a Web site differs from a paper catalog is that it gives a store the ability to track what's selling and then to reformulate the mix of featured merchandise at will . . . when the big wave of buying broke over the site, she (Billings) was checking sales every hour, and she was ruthless about replacing featured products that weren't bringing in dollars fast enough.
>
> —(Bekke, 2000, p. 33)

Case Analysis

This case illustrates that the logic of the strategic business model applies to new economy of e-commerce businesses at the end of the twentieth century, just as it did to old-economy businesses in the early part of the twentieth century. Businesses change and innovations in business change, but the logic of the strategic business model is universal and always timely. All businesses require sets of assumptions and policies that guide the procedures and practices of the business. This is the theoretical concept of the strategic business model.

In the case of RedEnvelope, one can clearly see that the strategy for building a gift business as a niche in online businesses required the formulation and implementation of a set of new strategic policies.

The decisions to change the name of the firm, as a brand issue, and to focus on gifts for upscale customers at holidays and special occasions provided a refocusing of *marketing strategy*. There was also a policy to reexamine all products to be appropriate for this market (200 were dropped and 300 added), which became the new *product strategy*.

Organization strategy was the decision by Connelly to recruit a person with experience in gift retailing to run the company, and that was person Billings. *Production strategy* was the policy to continue out source production through ComAlliance's capabilities. *Innovation strategy* was to become the first and thereby dominant focused online gift shop. *Financial strategy* was to continue to raise venture capital money or public money to fund the expansion of sales to a dominant position on the Internet. Not yet addressed was a financial strategy to eventually become profitable.

Information strategy was to use the Internet and Web tools to establish a business presence. While Billings did not like the Web appearance and products of 911Gifts, she did like its search strategy.

Competitive strategy was to use the novelty of the Internet to start a business opportunity and obtain capital to pursue the opportunity.

What we can see in this case is that while an overall strategy to start an online gift boutique by Galloway and Chaplin did focus the business strategy, it did not in itself construct a successful *strategic business model*. It took the complete set of strategies, reformulated policies for 911Gifts to reposition itself as Red-Envelope.

STRATEGIC POLICIES MATRIX

In strategy, the concept of a business model is fundamental because of completeness, implementation, and interaction. Business strategy is, in reality, a set of strategies because all businesses are complex, coordinated sets of activities—a business system. Therefore, the strategic business model is a technique for accounting for all necessary strategies for a business system.

The implementation of all strategy in any organization is in the policies that govern the procedures and operations of the organization. A strategic business model allows the explicit identification and examination of how strategy is implemented. Do the business policies align to the strategic intent of the business mission?

Since all businesses are systems, the interactions between different aspects of the system are important to the whole system performance. In any business model, it is the interaction between strategies for the different activities and controls that provides the competitive power of the firm. Accordingly, the set of strategies (strategic policies) within a business model should be examined as to their interactions. A formal technique for doing this is to construct matrices of strategy interactions. Figure 3.3 shows a two-interaction matrix model of strategic policies, that is, strategic policies taken in pairs.

Each strategy category is listed both vertically and horizontally in the matrix. The way to read strategic matrix is to think of the item on the vertical list as a

FIGURE 3.3 STRATEGIC POLICIES MATRIX

	PRODUCT	PRODUCTION	MARKETING	ORGANIZATION	FINANCE	DIVERSIFY	INFORMATION	INNOVATION	COMPETITION
PRODUCT									
PRODUCTION									
MARKETING									
ORGANIZATION									
FINANCE									
DIVERSIFY									
INFORMATION									
INNOVATION									
COMPETITION									

FIGURE 3.3   STRATEGIC POLICIES MATRIX

kind of strategic impact and the item on the horizontal list as a kind of strategic effect.

Thus, for example, Sloan's strategy for facilitating credit would be an entry along the finance row of the strategy matrix impacting the diversification column, so that the entry in the box of (finance-row and diversification-column) would be GMAC, the strategic move of establishing the General Motors Assurance Corporation.

Also we saw how GM's successors to Sloan failed to change strategies to meet changing environments and failed to *understand the interactions* of their strategic categories—all of which opened opportunities to new business competitors that took away GM's former strategic market dominance.

The strategic policies matrix provides a technique for systematically identifying and formulating the interactions of all strategies in operating and control business activities and to understand these policy interactions.

It is particularly useful for examining the strategic impact upon a given business of progress in information technology.

If one examines the strategy of any business along the information row of the strategic policies matrix, one can systematically formulate and list all of the potential impacts of a forecasted progress in information technology.

For example in the case of GM's strategies under Sloan, we saw how the kinds of strategies Sloan devised in the different business model strategic categories interacted to provide successful business practices which grew GM's dominance in the U.S. auto industry. Finance strategies of Sloan to help the credit situation in the automobile market, led GM to establish GMAC as a business diversification of GM. As another example, Sloan's market strategy for covering the price segments of the automobile market led to the strategic rationalization of GM's product lines by price segment. Moreover, Sloan's strategy for optimizing financial return on investment led to Sloan's strategy to standardization of parts and production to minimize costs.

TYPES OF STRATEGIC BUSINESS MODELS

As we noted earlier in this chapter, value-added, open-systems models of an enterprise organization receive inputs to transform to outputs. In constructing a strategic company model of an enterprise, one can use the four operational issues of any enterprise for inputs or outputs:

- Resources
- Sales
- Profits
- Capital

Resources and sales provide the issues for the direct production transformations of a business operation. For example, in manufacturing operations, material resources are manufactured into physical products for sale. In service operations, Requests for services are transacted into service sales.

Profits and capital measure the value of the business operations. Profits is a measure of business efficiency, the difference between prices and costs of sold products or services. Capital is a measure of the asset value of the business, the difference between investment and current stock value.

How many different types of business models can one use to construct strategic business models in formulating strategy? We can list all the logically possible types by taking all possible combinations of the four categories (resources, sales, profits, capital) two-at-a-time as inputs and as outputs. Doing this, one can construct six different models for to describe a business as illustrated in Figure 3.4.

The strategic business, strategic enterprise, strategic market, or strategic learning models are appropriate for looking at a company consisting of a single business. The strategic firm model is appropriate for looking at a company that con-

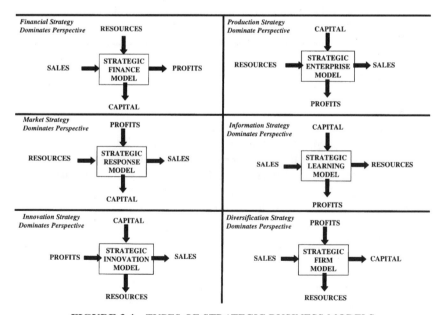

FIGURE 3.4 TYPES OF STRATEGIC BUSINESS MODELS

sists of several different businesses. The strategic innovation model may be used only over the very short term of a company but never over the long term because it fails to focus upon optimizing financial performance.

These types are useful in providing different models of a business for future operations depending upon what kind of performance one wants to optimize in business strategy.

STRATEGIC FINANCE MODEL

In the case of Sloan's strategic restructuring of General Motors, we saw that Sloan's strategic emphasis was upon maximizing profits and optimizing capital worth. Accordingly, a strategic finance model characterizes how Sloan strategically viewed GM.

We recall that both Sloan and the DuPont family were major shareholders in General Motors and desired short-term returns in the form of dividends and long-term returns in the form of stock appreciation. Thus in Sloan's mind, the appropriate strategic model for GM was a business model in which sales and resources were inputs to be transformed into profits and capital—a focus on optimizing financial strategy. To do this, Sloan's first strategic policies were

1. To rationalize the product-line sales, automobile brands, to cover the auto market in the U.S. without competition between brands (e.g., Chevrolet, Pontiac, Oldsmoble, Buick, Cadillac—in ascending order of price and quality)
2. To establish uniform accounting practices across all automobile divisions and uniform transfer pricing of components purchased from GM parts divisions

From Sloan's corporate perspective, the strategic policies were aimed at controlling the impact of all the independent manufacturing divisions of GM upon GM Corporation as a strategic whole—in a rational (i.e., bureaucratic) set of uniform policies—that all together the operations of the GM manufacturing divisions would optimize GM corporate profits and capital appreciation. Sloan continued his strategic perspective on GM with further policies that emphasized styling and annual model change to continue to use sales as an input to GM's overall competitive position.

Of course, this strategic perspective of Sloan for the whole of GM would not have been an appropriate business model for each of GM's divisional managers. Strategic management of the GM divisions required a different strategic model for effective strategic focus of a division, and this would be a strategic enterprise model (which we will describe next). For example, the annual model style changes imposed an annual investment cost upon the divisions that their presidents had to deal with to fit into their business environment as a GM division. Accordingly, a GM division needed to focus upon the strategic use of capital as an input, as well as resources, in order to optimize both sales and profits of the division. The important point here is:

> *The strategic model appropriate to a unit of a corporation depends upon the performance needed to be optimized for its business totality.*

Thus as illustrated by the case of Sloan's strategy for GM, the type of strategic model useful for viewing a corporation with an intention to maximize both profits and capital can be called a "strategic finance model."

> *A strategic finance model provides a perspective for optimizing both short-term profits and long-term capital appreciation by rationalizing sales and resource utilization.*

Because this perspective on the totality of the corporation is focused upon optimizing profits and capital, financial strategy dominates in the strategic policies set.

> *A strategic finance model of a company is a useful perspective to use in viewing the totality of a single-business firm when financial strategy is to be the dominant strategic policy.*

STRATEGIC ENTERPRISE MODEL

The production function of a business needs to view its business in a model that can show how to optimize production efficiency. For this a strategic enterprise model is useful because it treats sales and profits as outputs and capital and resources as inputs.

> *A strategic enterprise model provides a perspective for optimizing both short-term sales and long-term profits by rationalizing capital and resource utilization.*

Because this perspective on the totality of the corporation is focused on optimizing sales and profits, product and production strategies dominate in the strategic policies set.

A strategic enterprise model of a company is a useful perspective for viewing the totality of a single-business firm when product/production strategies are to be the dominant strategic policies.

This kind of model is very useful to think strategically about how to change and improve production operations of a business, since the heart of the model's description of transformation processes is as production. The model is strategically important to use for manufacturing operations and hard-good production. It is also important for hard-good logistics, as in retail or physical services.

An illustration of the use of an enterprise model to focues upon strategic issues of operations was in a case of a manufacturing problem that occured in the 1980s between two strategically allied auto companies, Ford and Mazda:

> Consider the case of Ford versus Mazda . . . which unfolded just a few years ago. Ford owns about 25 percent of Mazda and asked the Japanese company to build transmission for a car it was selling in the United States. Both Ford and Mazda were supposed to build to identical specifications. Ford adopted Zero Defects as its standard. Yet after the cars had been on the road for a while, it became clear that Ford's transmissions were generating far higher warranty costs and many more customer complaints about noise.
>
> —(Taguchi and Clausing, 1990, p. 67)

The difference in warranty costs was traced back to a problem of "tolerance stack-up" in the manufacturing assembly of the transmissions. Every physical part in a hard-good product is produced within a specified tolerance of the geometrical dimensions of the part design. This tolerance provides a target mean value for a physical dimension so that some parts will end with specs near the mean and others with specs near the tolerance limit. The tolerance is necessary because no physical manufacturing process ever makes one physical part exactly identical to another part because of uncontrollable variations in the production process (e.g., variation in process speed, temperature, vibration, masking, tool wear, etc.) These uncontrollable geometric variations will be random in dimensional variation. Therefore, the part features will all vary a tiny bit, within the tolerances specified for and controlled for in the physical formation of the part.

At the time, Ford manufacturing was using a strategic manufacturing policy called the Zero Defects quality standard, which meant that all parts would be within tolerance limits. Yet this policy did not take into account an effect that sometimes occurred when assembling parts into a product called tolerance stack-up, when all parts in a particular assembly were all near the tolerance limit. Ford was using this policy to control for having all parts for an automobile transmission produced within tolerances. Ford would not use a defective part that would be out of tolerance limits, and therefore Ford assumed all its transmissions would have no defects due to defective parts. So using this policy left Ford engineers very puzzled why Ford transmissions were having more warranty costs and complaints than Mazda transmissions.

Both transmissions were identically designed in a strategic partnership then between Ford and Mazda. With identical product designs, why were products differing in quality in use? Ford engineers disassembled and measured all the parts in samples of transmissions made by both companies. What they found was that the parts in transmissions made by Fords were all within specifications. But the parts in the Mazda gearboxes were all on the *mean targets* of the specification! This meant that the differences in the accuracy of the production processes at Ford and at Mazda was the reason that Mazda's identically designed transmissions worked better over time than did Ford's transmissions. Moreover, in producing these transmissions with more accurately machined gears, Mazda transmissions were also being produced with less production scrap, reworking as well as having lower warranty costs.

Although the Ford gearbox components were all within tolerances, many were near the outer limits of the specified tolerances. These many small deviations tended to accumulate, or "stack-up." For example, in the Ford transmission a slightly larger diameter (from its target diameter) of one gear might come up against another gear that also has a slightly larger diameter from its target, and then when the two gears are pressed together in operation, they were pressed slightly harder than designed (and harder than the equivalent two gears in Mazda's transmissions). The Ford-made transmissions were wearing out sooner than the Mazda-made transmissions because the former's transmissions with more out-of-target-but-within-spec gears were wearing out faster than the latter's transmissions.

The point is that strategic operations policies matter, and an enterprise model encourages the strategic attention to operations policies.

STRATEGIC RESPONSE MODEL

A strategic perspective appropriate to entering an established market dominated by a large competitor may require use of resources and nonoptimization of profits in order to optimize short-term sales to acquire a significant share of a market. Over time, establishing one as a dominant player in a market will optimize long-term capital value.

For example, market share began to dominate strategic thinking in manufacturing competition of the 1980s because hard-good manufacturing in things like automobiles had entered a mature-market industrial life stage, and in that stage competitors were being weeded out. The only strategic way to survive over the long term was to establish and maintain a significant market share (usually at least 20 percent of a market). Thus the strategic emphasis on time in responding to market changes as a competitive factor became then possible because of progress in information technology in the productive capabilities of hard good manufacturers. Through computer-aided design and computer-controlled flexible manufacturing, manufacturing firms could move faster in product development.

To emphasize the goals of optimizing market share in mature markets, it is useful to see the totality of a single-business company from the perspective of a strategic market model.

A strategic response model provides a perspective for optimizing short-term sales and long-term capital appreciation by rationalizing profits and resource utilization.

Because this perspective on the totality of the corporation is focused on optimizing sales and profits, marketing strategy dominates in the strategic policies set.

A strategic response model of a company is a useful perspective to use in viewing the totality of a single-business firm when market strategy is to be the dominant strategic policy.

In the late 1980s, many observers of the new impact of information technologies in competing in markets through new products, emphasized the importance of the "response time" capabilities in a company to aggressively create market share. An illustration of this at the time was the responsive operations of Toyota. As Bower and Hout described Toyota:

> Let's look at Toyota, a classic fast-cycle company. . . . [The] heart of the auto business consists of four interrelated cycles: product development, ordering, plant scheduling, and production. Over the years, Toyota has designed its organization to speed information, decisions, and materials through each of these critical operating cycles, individually and as parts of the whole. The result is better organizational performance on the dimensions that matter to customers—cost, quality, responsiveness, innovation.
>
> —(Bower and Hout, 1988, p. 111)

For Toyota's product development, self-organizing and multifunctional teams focused upon a particular model series. They accepted full responsibility for the whole cycle of product development—making the style, performance and cost decisions and establishing schedules and reviews. In addition, the product-development teams selected and managed the supplier input, bringing suppliers early into the design process. As a result in 1988 Toyota was capable of a three-year product development cycle. In comparison, the average car development cycle of U.S. automobile manufacturers at that time was five years.

A fast response-time was used not only for product development but for production control in Toyota. Toyota dealers in Japan were connected online to Toyota's factory scheduling system. As soon as an order was taken, the information on the selected model and options were entered immediately into the scheduling information. Toyota's purpose of integrating real-time sales information with production scheduling was to minimize sharp fluctuations in the daily volume of production, while also minimizing inventories. Toyota could produce on each production line a full mix of models in the same assembly system, using flexible manufacturing cells.

Bower and Hout saw that Toyota's attention to responsiveness pervaded its organization:

> Much of Toyota's competitive success is directly attributable to fast-cycle capability it has built into its product development, ordering scheduling, and production processes. By coming up with new products faster than competitors do, it puts other manufacturers on the marketing defensive. . . . By continuously bringing out a variety of fresh products and observing what consumers buy or don't buy, it stays current with their changing needs and gives product development an edge market research cannot match.
>
> —(Bower and Hout, 1988, p. 112)

Time could be strategically used as a sustainable competitive advantage, and in the 1980s such policies were called "fast cycle companies." Speeding up the response time of companies to changes in customer needs and the economic environment required more than simply working faster. It required working differently—thinking about why it takes time to respond, whether responses are correct, and how to respond more quickly and correctly. The sustainable competitive advantage from attending to "time" was through better and more quickly satisfying customers. Fast-cycle companies developed new products sooner than competitors, processed customer orders into deliveries faster than competitors, were more sensitive to customer needs than competitors and made decisions faster than competitors on how to add value in their products/services to the customer. A strategic response model is useful to strategically think about the policies that facilitate business reponsiveness.

The competitive importance of a strategic response model lies in a firm's ability to respond to market changes faster than competitors:

- *Correcting product mistakes*
- *Refining product successes*
- *Emulating competitors' product successes*

STRATEGIC LEARNING MODEL

This form of a business model that emphasized resource growth and profits as a business outputs has been called a "learning organization model."

> *The learning model emphasizes the optimization of resource strategy for increasing market share in new markets.*

The primary resource of an organization are the knowledge and skills of its employees to satisfy customers and build market share. As the same time as increasing resources for successful market growth, it views profits as another measure of success in satisfying markets. Strategically, it views as an input the valuable knowledge gained in the organization of the sales information in meeting customers needs and desires for products and services. It also strategically views capital as an input to improve the company's ability to be responsive to market changes.

> *A strategic learning model provides a perspective for optimizing both short-term resources and long-term profit potential by rationalizing sales growth and capital utilization.*

Because this perspective on the totality of the corporation is focused upon optimizing resource and profit capabilities, information strategy dominates in the strategic policies set.

> *A strategic learning model of a company is a useful perspective to use in viewing the totality of a single-business firm when information strategy is to be the dominant strategic policy.*

The early growth of America Online (AOL) in the 1990s illustrated a strategic model of business that optimizes both profits and resource growth as outputs. We saw in a previous case how AOL's high market value was used by its CEO, Steve Case, to acquire Time Warner in 2000, thus increasing AOL's resources by adding to its capabilities the media content giant and its cable assets. Case fostered AOL's enormous stock value by growing the largest audience then reachable by the new Internet from a single portal. AOL had about 11 million paying subscribers in 1998, reaching about as many homes as did the cable businesses of Time Warner. Case's strategy had been to build audience, understanding from the beginning that this would be AOL's most important resource and then it could be turned into profits. At the time AOL's stock climbed to a high of $115 a share, over 100 times earnings, when it acquired Time Warner.

We recall that a first mover in a new industry was the first to strategically make four kinds of moves:

1. continue to advance the new technologies,
2. develop large-scale production capacity,
3. develop a national distribution capability, and
4. develop the management talent to grow the new firm.

Case focused on (1) and (3), developing AOL's user interfaces and developing a national market. In 1996, Case's next challenges were in (2) and (4), production capacity and management talent. In 1996, membership soared when AOL replaced is $2.95-per-hour usage charge with a flat rate of $19.95 per month. Then AOL's capacity could not handle the suddenly larger traffic, and AOL had to respond by expanding capacity.

As the AOL organization and operations grew, Case needed to build an efficient organization and control costs, and to do this he hired good professional manager, Bob Pittman. Pittman had quit college to work in radio and had become a star programmer at NBC in his twenties. He then co-founded MTV, which he left in 1986.

Pittman attacked costs, reducing personnel in parts of AOL that were losing money. Pittman used AOL's size to bargain connect time for AOL with Internet backbone providers from 95 cents per hour to 50 cents per hour. He spent less on marketing. But he also knew that cost cutting alone never creates substantial profits but also needs increased revenues.

He found a new way from working with another business person, Dan Borislow, who was building a long-distance telephone company called Tel-Save. Borislow's competitors were bundling services on the same customer bill—services for local, long-distance, cellular telephoning, and Internet access. Borislow's idea for approaching AOL was to add the billing itself into the same bundled services. This would allow Tel-Save to sign up new customers, bill them, and charge their credit cards—all online. Borislow could see that this could cut his costs dramatically. Then it cost one dollar to send out a bill and thirty-five cents to cash a check. Borislow found that he saved 50% by doing business through AOL's online services and could underprice his competition.

Borislow paid $100 million to AOL for exclusive access for three years for phone services sold on AOL to customers. Also AOL obtained a share in future profits from the long-distance business. AOL then began a flurry of business deals, for using AOL as access to customers. For example, Preview Travel paid $32 million to become AOL's travel agent service, N2K paid $18 million to become the sole AOL music retailer.

We see in this example of the early growth of AOL, a strategic business model was first used that first focused strategy upon building a large market of customers as an asset, a corporate resource, and then AOL next focused upon transforming this asset into profits through deals with businesses that used AOL for access to customers.

STRATEGIC INNOVATION MODEL

One can use a strategic innovation model to build early growth of of a new business—but not for too long. This model uses capital to finance early market growth.

A strategic innovation model provides a perspective for optimizing both short-term resources and long-term sales by rationalizing the use of profits and capital to implement innovation.

Because this perspective on the totality of the corporation is focused upon optimizing resources and sales, innovation strategy dominates in the strategic policies set.

A strategic innovation model of a company is a useful perspective to use in viewing the totality of a single-business firm (or a new division of a firm) when innovation strategy is to be temporarily the dominant strategic policy.

For a viable strategic corporate model over the long term, either profit or capital must be one of the outputs of the model to be optimized or the enterprise will eventually run out of working capital. Therefore, the strategic innovation model cannot provide a viable long-term strategic model.

An example of using this strategic model for a time occured in the middle 1990s, when Jeff Bezos built early market dominance by Amazon. Recall from Chapter 1 that Bezos was a pioneer of retail electronic businesses, establishing Amazon as the first dot.com business to sell books over the Internet. The excitement of that kind of idea and the rapid growth of Internet new marketing intrigued investors in the last years of the 1990s; and large sums of capital were invested in start-up electronic businesses. Amazon's early strategy was to grow without regard to profitability and, of course, such strategy was bound eventually to end:

[June 24, 2000) Shares of Amazon.com, the Internet industry's bellwether stock, plummeted 19 percent in heavy trading yesterday as investors grew nervous about the company's financial health and its prospects for profitability.

—(Morgenson, 2000, p. B1)

Amazon went public in May 1997, and in 1998, its stock began a sharp climb from the teens to more than $100 dollars a share in 1999. In 2000, it declined to $34 after the market began to discount its earlier highly inflated prices of the new dot.com companies started in the last three years of the twentieth century. Amazon's stock performed on the basis of rapid growth of customers, so that by 2000 it had more than a 20 million base of customers who had purchased books from Amazon. Amazon's strategy had been to use profits and capital as inputs to optimize sales and resources growing its market share of electronic book sales.

In June 2000, a credit analyst at Lehman Brothers, Ravi Suria, wrote a report on Amazon's $2 billion in bonds issued during the last few years, after he had scrutinized Amazon's first quarter results of 2000. Suria said:

> What we think truly pushed a weak credit off the cliff was the inept working capital management during the last holiday season. Because the company does not generate positive cash flow on each piece of merchandise that it sells, Amazon has had to rely on obliging investors to finance its operations. But as Internet companies have begun to fail, the market has become a lot more selective.
>
> —(Morgenson, 2000, p. B1)

What was really evident was that over the long-term any company has to be profitable: "For Amazon to be a successful business," Mr. Suria wrote, "it must be able to generate the cash operating profile of a successful retailer. It is essentially this yardstick that we use to analyze the company and as the rest of this report shows, we find it woefully lacking." (Morgenson, 2000, p. B1).

Then the general perception that e-commerce businesses had to make money eventually had created the sharp decline in e-commerce stocks, with Amazon's share decline of 55 percent in the spring of 2000. On the day Suria issued his report, Amazon's bonds values fell by 10 percent. In July of that same year Amazon posted continuing losses in its operations: "Amazon.com said yesterday (July 26) that it had lost $317 million in the second quarter of the year. . . . The company, the largest online retailer, posted sales of $578 million, up 84 percent from a year earlier. (Hansell, 2000a, p. C1)

In 2000, Amazon's combined book, music, and video business in the United States had $385 million sales in that last quarter, with a modest profit of $10 million on those sales (but with the company still losing $317 million overall for that quarter). Amazon's continuing losses was still from its rapid businesses' expansions. Its German and British units lost $25 million on sales of $73 million. New businesses in selling toys and electronics had lost $40 million on sales of $31 million. Amazon's total losses for the year 2000 was $1.4 billion dollars, compared to losses in 1999 of $0.7 billion and in 1998 of $0.1 billion (Norris, 2001).

Thus Bezos' early strategy for Amazon to rapidly build a market had succeeded in part. Amazon had business with a total of 22.5 million customers, perhaps about 8 percent of U.S. households, from 1997 to 2000. Still in the long run all companies, even dot.com businesses, had to pay attention to the fundamentals of any business—profits. Amazon had lost money because it had focused on strategies of building market share without focusing on strategies for building efficient operations and profits. It had not used a complete kind of strategy set (e.g., as Sloan had done in building GM). For this reason, Amazon needed to change its strategic focus from market share to return on investment. Amazon then needed to change its strategic business model from its early strategic innovation model to a strategic finance model (as Sloan had used at GM).

STRATEGIC FIRM MODEL

A diversified firm can use profits from sales by its businesses to grow capital and provide corporate level resources. The diversified firm has responsibility for seeing that its businesses are well managed for profits from their businesses to continue to fuel corporate prosperity.

> *A strategic firm model provides a perspective for optimizing both short-term resources and long-term capital appreciation by rationalizing sales and profit utilization.*

Because this perspective on the totality of the corporation is focused on optimizing resources and capital, diversification strategy dominates in the strategic policies set.

> *A strategic firm model of a company is a useful perspective to use in viewing the totality of a multi-business firm when diversification strategy is to be the dominant strategic policy.*

The earlier strategic business, enterprise, learning, and innovation models are appropriate only for describing a single-business company. If a company has more than one business, then a firm-level strategic model is necessary to capture a firm's totality, as opposed to the totalities of each of its businesses. A strategic firm model needs to conceive of a firm as a strategic unity, even though the firm may be composed of a portfolio of businesses. And a strategic firm model must emphasize the optimization of financial valuation strategy in order to increase the stock-market valuation of the firm and return-on-investment of shareholders.

An example of a successful use of a strategic firm model in the 1980s was in the Japanese firm NEC. C. K. Prahalad and Gary Hamel compared NEC in 1990 to a similar U.S. firm of the time, GTE:

Consider the last ten years of GTE and NEC. In the early 1980s, GTE was well positioned to become a major player in the evolving information technology industry. . . . In 1980, GTE's sales were $9.98 billion . . . NEC, in contrast, was much smaller, at $3.8 billion. . . . Yet look at the positions of GTE and NEC in 1988. GTE's 1988 sales were $16.46 billion, and NEC's sales were considerable higher at $21.89 billion.

—(Prahalad and Hamel, 1990, p. 84)

But it was more than the relative levels of their sales that changed. NEC had emerged as a world leader in the information technologies, whereas GTE had not:

GTE has, in effect, become a telephone operating company with a position in defense and lighting products. . . . NEC has emerged as the world leader in semiconductors and as a first-tier player in telecommunications products and computers. It has consolidated its position in mainframe computers. . . . NEC is the only company in the world to be in the top five in revenue in telecommunications, semiconductors, and mainframes. Why did these two companies, starting with comparable business portfolios, perform so differently? Largely because NEC conceived of itself in terms of "core competencies" and GTE did not.

—(Prahalad and Hamel, 1990, p. 80)

Prahalad and Hamel viewed the differences as arising from NEC's superior strategic capability. NEC created a corporate level committee to plan core corporate technical competencies and to oversee the development of core products for the businesses of the firm. This committee established groups across the individual businesses to coordinate the research and development efforts for core products. This committee, which NEC called the C&C committee (Computers and Communications) identified three directions of technologies in computers, components, and communications. NEC's management saw computing evolve from mainframes to distributed processing, while components evolved, and communications evolved.

The strategic vision of the C&C committee foresaw the convergence of technologies in computing, communications, and components businesses. They judged that there would be great opportunities for any company to serve all three markets. They were anticipating that the two previously different industries, of computing and communications, were going to come together into a single and more complex industry. The components for the two businesses were to become increasingly more complex, common to the two and interrelated. With this technological vision, NEC positioned itself for a restructuring of the industries, and GTE did not. In this restructuring, NEC saw that the competitive factors lay not only in the systems-integrator sectors but in the major-devices sectors and in the components-and-parts sectors. So NEC strategically positioned itself in all three sectors.

In contrast, GTE had been managed only as a conglomerate, a diversified set of businesses without core corporate competencies:

> No such clarity of strategic intent and strategic architecture (as at NEC) appeared to exist at GTE. Although senior executives discussed the implications of the evolving information technology industry, no commonly accepted view of which competencies would be required to compete in that industry were communicated widely. Decentralization made it difficult to focus on core competencies. Instead, individual businesses became increasingly dependent on outsiders for critical skills.
> —(Prahalad and Hamel, 1990, p. 81)

Senior management of GTE had failed to develop the corporate strategic insight to fully exploit the evolving market and competitive opportunities in information technology. They failed to develop strategic technical competencies *common* to its several and independently run businesses.

A strategic firm model views sales and profits of its business portfolio as inputs to the firm. As strategic outputs, it provides centralized resources (such uniform management practices, executive education, performance measures, corporate research, etc.) to each and all of its businesses of the firm. It also provides investment capital to its businesses for improving operations and for acquiring new businesses or launching new business ventures.

SUMMARY: USING THE STRATEGY TECHNIQUE OF STRATEGIC MODELLING

We recall from chapter 1 that in thinking about the future capabilities of an organization, it is important to think strategically about the *resources, processes,* and *values* in an existing organization as compared to the *challenge of needed change.* Also we recall that the "value" management of an organization (sometimes called "corporate values") consists of the standards by which management and other employees set priorities and judge the importance of activities and results. Corporate values are standards about how resources are used and how processes are run, as Christensen and Overdorf commented:

> A company's values reflect its cost structure or its business model because those define the rules its employees must follow for the company to prosper."
> —(Christensen and Overdorf, 2000, p. 69)

Therefore the values of an existing organization are essential to strategic direction in comparing the present values to the challenge of needed change. The first decision in strategy is about what should be the primary corporate values in the direction of change. And we have seen that there are six different types of strategic models for constructing a corporate strategic model in any given strategic process depending upon the strategy policy to be dominant in change. The model one chooses in a formulating a strategic plan will describe the totality of the future company through the perspective of the dominant value of the strategic plan. Any strategic plan must first choose a dominant strategic value because in real life one can never optimize all values simultaneously in any given situation. The nature of life is always a trade-off in desirable values—security versus risk, profitability versus growth, and so on.

> *The first decision in modeling the totality of the future of a company is to choose which dominant value is to be optimized for strategic survival and prosperity.*

The choices of a strategic model of a company provides different ways to think about optimizing a company, according to what kind of performance one strategically desires. Accordingly, the types of strategic corporate models of Figure 3.4 can be used to help select dominant strategic values for a strategic plan and to examine the interaction of strategic policies of a total company within the plan.

SUMMARY: USING THE TECHNIQUE OF STRATEGIC BUSINESS MODELS

1. Formulate alternative strategic models

   A top-level corporate planning team (top-down strategic perspective) and a divisional-level team(s) (bottom-up strategic perspectives) should meet to formulate different appropriate perspectives of the totality of the company in the future by selecting one or more strategic model(s) that would meet the challenges anticipated by the planning scenario.
2. Construct appropriate strategy policy matrices

   In each appropriate strategic model, the divisional team should construct a strategy policy matrix to determine what kinds of policies would optimize the dominate strategy in the strategic perspective of the model.
3. Select the most robust strategic model

   Next, the top-level strategy team and division-level strategy team(s) should together select the most desirable strategic corporate model to guide the strategic plan of the company that provides the strongest basis for survival and prosperity of the company's future in the face of the strategic challenges and opportunities of the planning scenario.

4. Formulate a strategic policy matrix

>   The two teams should then formulate a set of business policies for this strategic corporate model, along with a strategic policy matrix that emphasizes the assumptions and the desirable and anticipated interactions between the policies to be attended to in the implementation of the strategic plan.

For Reflection

Find books on the automobile industry and particularly look up the history of the Ford Motor company. Also search Fortune magazine for articles on the automobile industry since 1975. What were the strategic challenges Ford faced over the company's lifetime in the twentieth century? How did it meet them? Why have GM and Ford been the only U.S. auto firms that survived independently? What happened in the consolidation of the automobile in the world when the twenty-first century began?

CHAPTER 4

PLANNING SCENARIO

PRINCIPLE

Planning scenarios provide a useful technique to systematically anticipate future change in the environments of business.

STRATEGIC TECHNIQUE

1. Form a planning scenario team.
2. Divide the scenario team into societal-groups scenario teams.
3. Use appropriate external experts to assist in societal groups.
4. Prepare appropriate forecasts.
5. Present summary of planning scenario to company executives.
6. Modify planning scenario by strategic modeling teams.
7. Extract strategic issues from planning scenario.
8. Use the planning scenario to construct appropriate strategic models.

CASE STUDIES

Energy Forecasts
3M's Strategic Stories
House of Mitsui

INTRODUCTION

We recall that the two important strategic totalities to think about are those (1) of the company and (2) of its environments. In Chapter 3, we addressed the totality of a company as a strategic business model. In this chapter we turn to the totality of the environments of a company, and we review the strategic technique of planning scenarios as a useful way to capture this second totality. A planning scenario systematically and intuitively explores and summarizes the picture of the future environments in which a company expects to operate and within which a strategic corporate model is formulated.

Also we recall that the top-down strategic perspective has a logic that begins with the general and moves to the specific, such as:

1. Scanning the environments of a firm to identify major future trends and changes
2. Interpreting the changes as threats and opportunities to the businesses of the firm
3. Analyzing the present firm's activities in terms of strengths and weaknesses to face such threats or seize such opportunities
4. Redefining the missions of the firm's businesses to match the future operations to future threats and opportunities
5. Setting goals and targets for businesses to meet in a time horizon

But we also recall that while this kind of linear thinking in strategic planning is correct in summarizing the results of scenario planning the real process of formulating strategy is not merely linear and analytical but also interactive and intuitive. And this is why process of strategy must be depicted in a nonlinear and interactive form, such as we symbolized in Figure 1.4. The top-down strategic perspective has as a base the technique of the planning scenario, and the bottom-up strategic perspective has as its base a strategic business model.

Now we will focus on the planning scenario. We will see that to adequately capture the totality of future environments of a business, the planning scenario needs to use the method of scenario narratives and societal models. Scenario narratives provide a method for describing and thinking about the possible impacts of the future upon a current business. Societal models provide a format for describing and thinking about the totality of future environments a business will likely face.

CASE STUDY: Energy Forecasts

When the twentieth century ended in the United States, one area of major economic change was the deregulation of the energy sector of the nation.

Deregulating the electrical power industry changed the structure of power operations from a state regulated monopoly structure toward an unregulated "free-market" structure. It was important to be able to forecast energy prices as the structure altered, and at this forecast many failed. This case examines how scenario forecasting and underlying societal structures interacts in California, Enron, and Duke Energy.

California   At the time, Mark Gimein described the energy crisis of the state of California:

> When it comes to big-time economic disasters, it can be hard to specify when "the situation" turned in to "the crisis" . . . For the citizens of California, that (came) when their governor . . . threatened to use the power of eminent domain to take over the power plants owned by big, out-state energy companies. For Wall Street, it . . . came . . . when the state's second-largest utility simply didn't pay the half billion dollars it owed to bondholders and power suppliers. The origins of the crisis . . . lie in a filed deregulation scheme under which the state's utilities old many of their power plants to national power companies, including Duke Energy . . . which now sell back the power generated from those plants.
>
> —(Gimein, 2001, p. 111)

Earlier in the 1990s, the California state legislature got together with the two big regulated electrical power companies in California, Pacific Gas & Electric and Southern California, and decided that deregulation would be a good thing for the citizens and companies of California. The two companies agreed to sell off their power generation capabilities in exchange for being deregulated. The state agreed to hold the customer's electricity prices fixed for a number of years, and then let the free market take over. Everybody forecast that the price of electricity would go down. Everybody was wrong. Instead, consumer demand for electricity rose, no new generating plants were built, the price of gas and oil jumped, and the only thing really deregulated was the out-of-state power generation companies that now generated power. The average price for electricity which PG&E and Southern California paid jumped from $20 dollars per megawatt hour in 1999 to $300 per megawatt hour in 2001. PG&E went bankrupt and Southern California nearly so in 2001. The companies that benefitted from the deregulation change were power generation companies, such as Enron and Duke Energy.

Enron   In the 1980s in the United States, there had been a widespread movement to deregulate the energy industry. The first industrial sector to undergo deregulation by tight government controls was the natural gas transportation industry. The interstate gas-pipeline industry had been regulated by federal law as a kind of "point-to-point" system requiring gas pipeline companies to sell gas only to a few designated gas and electric utilities along a pipeline route. Furthermore, pipelines were required to buy from well producers and sell to utilities the natural gas at the same price plus only an added transportation and storage cost. If gas pipelines were deregulated, they could sell gas at any price, which balanced demand against supply. So potentially a deregulation of the gas pipeline industry could change the whole economics of the industry. With such a deregulation coming in the middle 1980s, a particular company, Enron, moved first strategically.

In 1985, Enron's Board of Directors first strategic move was in 1985 was to hire a new CEO, Ken Lay, who had been in government arguing for deregulation. Lay had obtained a doctorate in economics and worked in the 1970s at the United State's government agency of the Federal Energy Regulatory Commission (FERC). There he had argued for deregulation of the natural gas markets: "As an economist, I look at how markets ought to operate. I spent a lot of time at FERC arguing for new ways to price gas and got people thinking differently about markets" (O'Reilly, 2000, p. 149).

As new CEO Lay retained a McKinsey consultant, Jeff Skilling, to identify business opportunities in the early deregulation of the gas business and later hired Skilling as COO (Chief Operating Officer). Together, they began to create a new kind of trading company of Enron:

> Once a medium-sized player in the stupefyingly soporific gas-pipeline business, Enron in the past decade (1990s) has become far and away the most vigorous agent of change in its industry, fundamentally altering how billions of dollars' worth of power—both gas and electric—is bought, moved, and sold, everywhere in the nation.
>
> —(O'Reilly, 2000, p. 148)

Enron buys and sells both gas and electricity according to demand of customers and sources of supply—Enron has created a flexible market for moving energy. For example if an electrical power company needs more electricity to meet a seasonal demand, the Enron finds another power company with excess production capacity for the same period and arranges transport from source to destination, buying and selling power:

> But saying the Enron trades electricity and gas is like saying the Thomas Edison made records. In most cases, Enron executives didn't just start dabbling in the natural gas and power trading business; they invented the entire concept. Never before had gas and power been traded like commodities.
>
> —(O'Reilly, 2000, p. 150)

In 1990, Enron had earned $226 million on revenue of $4.6 billion from owning 30,000 miles of regulated gas pipeline. In 1999, Enron earned $893 million on revenues of $40.1 billion, with 75% of the earnings coming from trading.

One can see that trading had become Enron's major business and therefore the ability to forecast the conditions of supply and demand in energy is an important strategic need for Enron's planning.

Duke Energy   Another expanding company in the deregulation scenario then was Duke Energy:

> As the U.S. market deregulates, Duke is increasingly free to be an "energy merchant," a wholesaler in many regions of the country. It operates "merchant" plants from New England to California-producing not just for a regulated local market, but also for the open market—and it is building more in Mississippi, the Midwest and elsewhere.
>
> —(Wysocki, 2000, p A1)

Duke's strategy was to be a national trader and marketer of natural gas and electricity in the increasingly open markets of the United States. In 1997, it merged with PanEnergy Corp. of Houston to acquire capability as a marketer, having then a large energy-trading operation. It also used financial derivatives to hedge and manage positions. In 2000, it's annual revenue was running at $22 billion in annual revenue, of which half came from its trading and marketing activities.

In the year 2000, strategic planning at Duke Energy Corp focused upon trying to forecast the U.S. economy over the next decade.

> At Duke Energy Corp., engineers-turned-executives are at work on what they call a 'wind tunnel for testing strategy.' They have taken the company's ambitious growth plans and tried to test them against various economic winds that might blow either in their favor or against them.
>
> —(Wysocki, 2000, p. A1)

The accuracy of the forecast was important because the strategic plan called for major investments in expanding production capacity, whose rate-of-return depended upon the health of the economy: "The forecasting was It's a timely exercise, amid rising fears that the longest economic boom in' U.S. history is losing its force. 'There are some big risks in the U.S." (Wysocki, 2000, p. A1).

The planning staff had constructed three scenarios, for the senior executives to consider at a two-day strategy meeting in Houston in the last week of June 2000. One scenario projected a rapid slowing of the U.S. Economy with growth declining to only 1% a year—which they called the "economic treadmill" scenario (or Big Slowdown in the economy). This forecast would result in a difficult future for Duke's financials, since Duke would then have built new power plants coming on line just when too much energy capacity would weaken prices. A second scenario assumed a strong U.S. Economic growth rate of 3% annually. But in this scenario was also the assumption that the deregulation of the energy industry in the U.S. would proceed unevenly; and they called this scenario the 'flawed competition' forecast. A third scenario focused less upon the economy and more upon the impact of the Internet, providing an electronic market for buying and selling electricity and natural gas. In this scenario, buyers had more influence on prices than did sellers.

The scenarios were important for Duke's strategy to consider the range of possible impacts upon Duke's future financial performance:

> The scenarios are just a part of what goes into the making of strategy at Duke, but they point to the large economic uncertainties facing every business in the summer of 2000. Is the U.S. economy headed for a so-called soft landing or for' a hard landing? Will interest rates keep rising? At Duke, with more than $9 billion in debt on the balance sheet, every one-percentage-point rise in rates could reduce pretax income $24 million this year.
>
> —(Wysocki, 2000, p. A1).

In strategy, timing is important and forecasts address the issues of timing in strategy: " 'If we get the cycles right, we're successful. If we get the cycles wrong, we're less successful or unsuccessful,' says Mr. Priory, the CEO." (Wysocki, 2000, p A1).

In economic forecasting, several indicators were being used to extrapolate trends. Although forecasting always results in uncertainties about direction and timing, judging the robustness of strategy within the uncertainties is important.

The other aspect of the scenario planning at Duke that had executive focus was not only the economic forecast but also the projected impact of the Internet; and Duke Energy's strategy was beginning to prepare for this scenario by starting an e-commerce unit in 1999.

Case Analysis

None of the energy forecasts had foreseen the re-emergence of a political cartel among nations producing oil, OPEC. In the middle of 2000, OPEC agreements were reached to limit production and the world price of oil shot up from $10 a barrel to $35 dollars a barrel. It was structurally a new ball game for energy in the world. Forecasts are primarily extrapolations from the present into the future. But when underlying structures of the present change, the forecasts are always wrong.

FORECASTS AND EXTRAPOLATION TECHNIQUES

Forecasts and trends are essential techniques for scenario planning because they start with patterns of the present and project these into the future. All forecasts and trends are therefore *extrapolations* of the present, and all extrapolations depend on the structures underlying the trends and forecasts.

The rawest sort of extrapolation consists of simply fitting near-term event data to recent event data, when the underlying form of the curve, the generic pattern of the class of events, is unknown. This can be done by arbitrarily fitting straight lines to series that appear to be monotonically increasing or decreasing. If the series appears to be periodic, one can arbitrarily fit sinusoidal curves. If the series appears to have no underlying pattern, one can take the extrapolate using the average of the last three points. One can use even more sophisticated running average methods, such as the Box-Jenkins methods.

However, no matter how clever one is in fitting curves or averaging points, one is still left with the basic weakness in forecasting about the structures underlying the extrapolated patterns. For example, in 1990, E. Mahmoud, J. Motwani, and G. Rice compared forecasts for U.S. exports using two different extrapolation techniques, time series and econometric models. They concluded about the accuracy of each of the these methods of extrapolating the pattern of exports:

> Exports have usually been forecast using econometric methods. Nevertheless, some studies have shown that time series methods can also predict exports . . . and the methods studied have been sophisticated ones such as Box-Jenkins . . . Research has indicated the simpler forecasting techniques can be more accurate than Box-Jenkins techniques. . . . (And) our findings suggest that time series methods can provide as accurate if not more accurate forecasts than an econometric approach.
> —(Mahmoud, Motwani, and Rice, 1990, (p. 375)

The point Mahmoud et al. are making is that in practice the exact technique for forecasting societal activities (such as exports, market, and economic patterns) makes little difference between techniques. This is because a more powerful underlying feature of markets and economic patterns dominates major changes in forecasts—changes in the underlying structures of societal activities.

Forecasts that presume a fixed underlying structure for events (e.g., econometric models) or forecasts which use no underlying structure or form (e.g., time-series methods) are about equally accurate and equally inaccurate in the realm of economics, when the underlying economic structures do not change.

In an extrapolation (trend or forecast) of economic activities, such the accuracy of extrapolation depends up the underlying forms or structures of the events one is extrapolating.

Any forecast (whatever the extrapolation technique) will be much improved by understanding the underlying forms and structures of the extrapolation.

In any forecast, four general classes of structural features will underpin events:

1. Structures of current technological capabilities,
2. Structures of economic activities and markets,
3. Structures of nature and natural potential,
4. Structures of demographics and cultures.

Whenever underlying structures alter, forecasts based principally upon extrapolation will be in error.

Forecasts extrapolate present trends but should be used with the identification of critical structural variables, that if changed would invalidate the forecasts extrapolation.

Since trends are only identifiable patterns of change and forecasts are only attempts to quantitatively anticipate the direction of change in the trend, it is important to understand in strategic scenario planning that any forecasting attempt to anticipate the future can proceed with different levels of sophistication:

1. Extrapolation
2. Generic patterns
3. Structural factors
4. Planning agenda

When a forecaster has almost no knowledge about the events except historical data on past occurrence, then the forecaster can do little more than extrapolate the direction of future events from past events. Extrapolation forecasting consists of fitting a trend line to historical data.

When a forecaster has some knowledge about the general pattern of a class of events but little knowledge about the specific exemplar of that class at hand, then the forecaster can use the generic pattern to fit the extrapolation of the specific exemplar case. Fitting a generic pattern to an extrapolation has more knowledge than mere extrapolation because one knows before hand the form of the curve to be extrapolated.

In addition to knowing the generic pattern of an event, knowing something about the kinds of factors that influence the directions and pace of the events provides the basis for even better anticipation. Extrapolations from past data always assume that the structure of the future events is similar to the structure of the past events. Changes in structural factors will render extrapolation meaningless and create the most fundamental errors of forecasting.

The deepest level of forecasting requires understanding not only the generic pattern of the class of events to be anticipated but the structure of the events. It then proceeds to intervene in the future by planning to bring about a desired event. A research agenda provides an anticipatory document required to bring about a technological future.

Accordingly, experts should know about the underlying structures in forecasts and be sensitive to factors that alter structures, but experts do not necessarily have quantitative models of structures. Consequently, some experts will be accurate sometimes and sometimes not. The trouble with using only experts to forecast is that there is no way in anticipation of an event to calibrate the reliability and accuracy of any given experts. The accuracy of experts in forecasting can only be judged in hindsight, and even past accurate performance is no guarantee of accurate future performance.

Thus scenario planning needs to use not only forecasting (e.g., the economic indicators in the Duke Energy scenario case) but an understanding of the structural change that may alter an extrapolated forecast (e.g., the impact of the Internet upon energy markets). A knowledge of structures is the real importance for having experts involved in forecasting and not merely extrapolating past patterns.

CASE STUDY: 3M's Strategic Stories

In 1998, Gordon Shaw (then executive director of planning at 3M in st Paul, Minnesota) and his colleagues, Robert Brown and Phillip Bromely, reported on an approach to strategic planning that emphasized the importance of telling "stories" to communicate, so that the stories get at the structures of future events at 3M:

> At 3M, we tell . . . stories about how we failed with our first abrasive products and stories about how we invented masking tape and Wetordry sandpaper. . . . We train our sales representatives to paint stories through word pictures so that customers will see how using a 3M product can help them succeed. . . . Stories are a habit of mind at 3M, and it's through them—through the way they make us see ourselves and our business operations in complex, multidimensional forms—that we're able to discover opportunities for strategic change. Stories give us ways to form ideas about winning.
>
> —(Shaw et al., 1998, pp. 42)

Shaw et al. had come to the conclusion that stories also might be a better why to think out and present strategy:

> over the course of several years overseeing strategic planning at 3M, Gordon Shaw became uncomfortably aware that 3M's business plans failed to reflect deep thought or to inspire commitment. They were usually just lists of "good things to do" that made 3M functionally stronger but failed to explain the logic

or rationale of winning in the marketplace. He began to suspect that the familiar, bullet-list format of the plans was a big part of the problem.

—(Shaw et al., 1998, pp. 42)

Many companies have used the format of lists of bullets in writing and presenting planning information. Bullet lists help reduce the complexity of business situations to a few points and help to focus discussion. But Shaw et al. thought that strategy presented in the apparent simplicity of a list of bullets also lost many of the subtle issues of strategy, as issues neither presented nor discussed. The form of the language in the presentation of a planning report expresses the depth of thinking underlying the plan. A bullet format does not show whether the strategic thinking going into the plan was shallow or profound, because the bullets themselves do not tell the whole story of the strategy underlying the plan: "Bullets allow us to skip the thinking step, genially tricking ourselves into supposing that we have planned when, in fact, we've only listed some good things to do." (Shaw et al., 1998, p. 42).

Shaw et al. concluded that it was the format of the strategic plan summarized as a list of bullet-sized points that in itself encourages intellectual laziness in strategy. First, they judged that a strategy expressed as a list of bullets only results in presenting the strategy as issues that are too generic, only summarizing a list of things to do that would apply to any business.

A bullet-list format can result in a plan that really fails to focus on specifics, specifics of how the business will win in its selected markets. Shaw et al. gave an example of a selection from a planning document submitted by a 3M business unit, in which planners had proposed major strategies, listed in bullets as to both reduce costs and increase customer choice:

- Reduce high delivered costs:
  - Reduce international parent head count by three
  - Explore sales cost reductions
  - Determine vision for traditional products and appropriately staff
  - Continue to reduce factory costs
  - Refine unit cost management system
  - Reduce process and product costs
- Accelerate development and introduction of new products
- Increase responsiveness.

What would have been important in presenting this plan would be the specifics of how these strategic directions were to be accomplished. Shaw et al. emphasized that these bullets were so generic as to be applicable to any manufacturing business. The managers presenting in this format fail to discuss the important issues of planning, which are specifically how to accomplish these things.

Moreover, in addition to facilitating a too generic level of strategic thinking, Shaw et al. judged that a bullet-list strategy encouraged a kind of one-dimensional thinking—one dimensional in terms of the real complexity of strategic relationships. Since a list can only logically present the membership, sequence, or priority of a set of things, a list will fail to present the interactions between the factors of the list or of the structure of the business activities underlying the list. As a planning format, a bullet list will fail to examine the interrelationships of factors in a business.

To illustrate this lack of sophistication about strategic interrelationships, they offered an example of typical kinds of major objectives in a standard five-year strategic plan:

- Increase market share by x percent.
- Increase profits by y percent.
- Increase new-product introductions to a larger number z per year.

Shaw et al. pointed out that the trouble with this strategy list is that it neglects to discuss how these objectives tie together. For example, is it the case that improved marketing by itself can increase market share (from which improved profits will follow and from which funds for increasing new product introductions will be available). This one of the possible sequential causal assumptions implicit in this list. Or alternatively, will it take both increased new-product introductions and increased market share together to increase profits? A bullet list does not make explicit the relationships between points of the list.

In summary, Shaw et al's criticism of the bullet-list format for presenting strategic plans is *the illusion it may create that strategy really has been thought out when in reality it hasn't.* A plan expressed only as a list of bullets will leave unstated the critical assumptions about how the business does or should work. Consequently, a bullet plan can give an illusion of clarity, when in fact the future remains obscure. Shaw et al. emphasized that thinking one is clear about the future, when the future is still obscure can be a very expensive illusion in business.

Shaw et al. then gave an illustration of a strategic scenario in narrative form for division called Global Fleet Graphics. Their scenario first set a scenario stage by describing the business of Global Fleet Graphics, which makes high quality and durable graphic-marking systems for buildings, signs, vehicles, and heavy equipment. Then the scenario described a major strategic challenge to the business in facing more demanding customers and more aggressive competitors. The customers desired greater flexibility in design, larger graphics but low cost; and they also varied, as some customers wanting graphics products that were easy to remove and others wanting durable graphics. The overall sales of graphic materials were increasing, which sales of traditional, painted graphics were declining because of high cost.

3M had 40% of the graphics materials market and had been the technological leader. The three major competitors for 3M were the companies of AmenGraphics, GraphDesign, and FleetGlobal. (The narrative next described how these competitors did business.) AmenGraphics expanded its product line using 3M technologies after patents expired, with its market share of graphic materials growing from 10% to 16%. Another competitor, GraphDesign, competed on price, using direct distribution and new manufacturing capability (but recently had quality problems, with market share dropping from 18% to 15%). The third competitor, FleetGlobal had comparable quality products to 3M but with lower prices, and its market share had grown from 24% to 28%. The conclusion was that while 3M was losing patent advantages in materials, they faced three competitors that were competing with low cost/price strategies.

Next the narrative story of the strategic plan described a dramatic conflict that was then the challenge, 3M's Fleet Graphics would no longer be profitable in the short term future. Nor would strategies of only incremental product and/or process improvements met the challenge, since these same strategies were accessible to competitors.

Finally, the strategic narrative proposed a resolution to these challenges by planning a dramatic move from analog to digital printing-and-storage technology. In addition, the final printed product will be improved through new film and adhesive technologies. The story narrative then summarized the new strategy as three thrusts. One thrust was to make a dramatic change in the production system to deliver products more quickly and more cheaply, at a competitive price. The second thrust was to develop a new generation of patented technologies and products to differentiate the company's products in the future from competitors' products. A third thrust would upgrade sales and marketing staffs' skills to match with technology-driven strategy.

Shaw et al. suggested telling the strategic planning scenario story in three stages: (1) setting the stage, (2) introducing the dramatic conflict, and (3) providing resolution. They argued that the format of a strategic scenario needed to first set the strategy stage, defining the current situation in an insightful and coherent way. This setting the stage should include analyzing an industry's dynamics, the forces that drive change, and the factors providing competitive success in the industry. Next in the planning scenario story, a strategic planner should introduce the dramatic conflict, as to what challenges a company must face in that situation? What will be the obstacles to success and threats of failure? Then the story should conclude by proposing a resolution of the challenges in a convincing way. The planning scenario should indicate the directions of how the company can overcome obstacles and win.

Case Analysis

This case illustrated how the planning format itself, in the presentation of strategic planning, can either impede or facilitate thinking about strategy. The technique for thinking about and capturing the uncertainty and ambiguity about the future impacts upon business is what Shaw et al called the concept of the strategic story, more commonly called the strategy scenario.

SCENARIOS

Scenario planning uses the forecasts of trends within strategic stories to envision *adventures* of the business in the future. The future will be an adventure because it will be a time in which no one has yet experienced. Experience is always of the present, with memories and stories of the past. The future consists of anticipations and/or surprises and plans for the future conceived in the present. All existence is always in the present. It is in intelligence that the past and future exist.

Scenario planning is about planning a future adventure, an exploration into the future.

Scenarios depict the trends that provide opportunities for success and threats to survival. The plot of any good adventure is the depiction of a fortune or success to be won by a hero or heroine and the challenges and opposition to their course of pursuing the fortune, with an eventual successful conclusion though courage, skill, and luck. So too is facing an uncertain future a kind of adventure for any business. In the short-term time horizon, there is usually a greater certainty about the nature of the market, efforts of competitors, success of products and services, profitability and balance of finances. It is in the long-term time horizon, where much change is possible and uncertainty can be about everything. Thus short-term planning, operational planning, can be detailed as a recurrence of current operations—a projection of the present. But long-term planning, strategic planning, cannot be so detailed for it may not be a recurrence of current conditions—but change, major change—an adventure of the future.

Scenarios provide a strategic technique for anticipating changes in the environment of a business or firm. All businesses operate in a complex set of environments, including the environments of industrial and commercial structures, markets, government regulation, financial and economic systems, international competition and environmental systems. What strategy needs to do is recognize trends and anticipate patterns of changes in these environments, for they can influence both the kinds of businesses that maintain future viability and the kinds of conditions such businesses will encounter in the future.

Strategy scenarios should be presented a in narrative format to help thinking through and capture the logical issues of strategy in their specificity, complexity, and interrelationships. Planning scenarios are business stories, but particular kinds of stories, stories of future strategy:

- Stories of the business environments of the future
- Stories of the market opportunities and perils of competition in the unknown adventure of the future
- Stories of change and survival in the future time toward where business has not yet gone

What then should the strategy story try to tell? Strategic stories provide a view into the future, which is relevant to planning. This view encompasses the changes likely to occur in the business environments of the firm, in the competitors to the firm's businesses, in the markets of the firm's businesses, and in the technologies the firm uses. A common form of describing future change is called a "forecast." Forecasts are extrapolations of past and current trends into the future. Forecasts are based upon structures of activities, whose pattern is being extrapolated.

Thus in the strategic story what one is looking for in a scenario about the future are the trends, forces, patterns of change, opportunities, threats of the future. The future has not yet occurred, and therefore one cannot predict what has not yet occurred when human intervention can alter the future. The future is where meets the determinism of mechanical mechanisms and the freedom of human wills in cooperation and in conflict with each other.

CASE STUDY: House of Mitsui

As a technique to facilitate strategic thinking about the totality of the business environments, we will illustrate the concept of the narrative planning scenario by looking at a famous case in business history, that of the long-lived firm of Mitsui in Japan. The case provides an excellent illustration of the complexities involved in very long-term issues of business survival, particularly when the society in which the firm exists undergoes very great change.

Mitsui is one of the world's oldest and continuously operational modern firm; but it's form has changed over time, in pace with and just as dramatically as has changed the history of modern Japan, in transforming from a feudal society into an industrial nation. All countries, or territories, of the modern world have made (or are still making) this important societal transition from feudal/tribal societies to industrialized societies. In Japan, this change occurred quickly and successfully and dramatically in only 100 years—from the second half of the nineteenth century through the first half of the twentieth century. During this time, one particular merchant clan, the House of Mitsui, became a giant, global, commercially powerful modern corporation, Mitsui Gumi Inc.

We will summarize this case as a narrative story, a scenario (but one looking backward rather than forward—a historical scenario). This narrative provides a very dramatic illustration of long term changes over time in a business's environment that a business must face. Changing environments create business opportunities and challenges and threats to success and survival. The historian John Roberts has nicely summarized the scenario drama for Mitsui in the middle of the twentieth century:

> At the end of World War II in 1945, Japan was a shambles . . . (and yet) no time was lost on self-pity, regret over the mistake of waging a hopeless war, or hatred of the conquerors. With resilience, determination, and accommodation, the nation quickly lifted itself from the ashes of defeat . . . she became the world's third great industrial power. . . .
>
> —(Roberts, 1989, p. vii)

The firm of Mitsui traced its history back to a founding samurai family in 1600s and survived as a family-controlled enterprise through the eighteenth, nineteenth, and twentieth centuries. Its history mirrored the history of the economic and social system of modern Japan, transforming in the 1800s and 1900s from a feudal, preindustrial society into a modern nation of military might and economic strength. During this time, Mitsui emerged as a major corporate entity in Japan, becoming the prototype of the powerful combines, zaibatsu, which emerged in Japan as the economic forms for industrializing Japan. These zaibatsu, with government encouragement, also served as instruments of national policy in the building of a modern, industrialized Japan.

The founder of the house of Mitsui, Sokubei, gave up his status as a samurai in 1616 to become a become a merchant, a chonin. He began a small brewery to make sake and soy sauce. Then his wife and children added a small drapers shop and money exchange. From this, the following generations continued to build, creating in the early 1900s, a giant zaibatsu, a huge conglomerate running most types of commerce and industry, including banking, insurance, shipping, foreign trade, retail merchandising, construction, engineering, mining, brewing, textiles, chemicals, paper, glass, electronics, optics, and real estate.

Sokubei's action of changing from samurai to chonin was due to his perception that the times were changing—after the ascendancy of Tokugawa Ieyseu as shogun in 1616, after a century of civil war. Military stability in feudal times promoted commerce and economic prosperity. During the previous century, there was constant war among the feudal lords of Japan, but gradually local warlords increased their areas of control and feudal unity was strengthened which also fostered the growth of internal commerce. The amounts and quality of exchanged goods and services increased. With this the traditional bartering rice for handicrafts and other commodities was increasingly supplanted by the use of money. The feudal lords, daimyo, had also to borrow money to meet the expenses of warfare. Lending the money to the daimyo were a group of commoners prosperous from trade. These formed a new merchant class, chonin, growing in numbers and given official status in the feudal society.

In 1568 a warlord, Oda Nobunaga, subdued most of Japan, and upon his death, two of his best generals, Toyotomi Hideyoshi and Tokugawa Ieyasu, faced off as rivals. In 1600, a final battle occurred at Skeigharaa, as 160,000 samurai took the field in the opposing armies of Toyotomi and Tokugawa. Tokugawa's army won, taking 40,000 heads as trophy. Then Ieyasu was the undisputed warlord, shogun, over all Japan and established a government that was to last until 1863, the Tokugawa shogunate.

Sokubei had been a lower ranking samurai, who could not partake in the success or rewards of the new regime. There was no powerful lord, daimyo, to which the Mitsui family owed allegiance and from whom, conversely, fortune would come to guarantee the family's continuing existence as samurai. Sokubei had to think about his family's future without proper feudal ties to the new regime. He traveled to Edo, the capital of Tokugawa's new government. There he saw prosperity and decided to become a merchant. Since his wife was from a wealthy merchant family, merchants were familiar to him. Upon his return, he gathered his household together—his wife Shuho, his children, his retainers and servants—and he told them he was giving up the family's traditional status as samurai: "A great peace is at hand. The shogun rules firmly and with justice at Edo. No more shall we have to live by the sword. I have seen that great profit can be made honorably. I shall brew sake and soy sauce, and we shall prosper." (Russell, 1939, pp 67–68).

Sokubei's family house of Mitsui began brewing, and people called his sake shop, Lord Echigo's sake shop (because Sokubei's father had been called Lord Echigo). It was unusual to them that a former samurai had become a shopkeeper, a chonin. At first, business was slow, and Sokubei was not a good shopkeeper, but his wife Shuho, the daughter of a successful merchant, grew the business. Without the feudal pretensions of aristocratic class, Shuho could converse with her peasant customers and would gain the favor of servants as customers by offering tea or tobacco went they came on errands. Sometimes, customers spent more than they had cash, and Shuho would loan money, accepting some valuable as a security. In this way, Shuho began the first expansion of the family business, from sake and soy sauce to pawn brokering, with interest on the loans soon becoming more profitable than brewing. Sokubein died in 1633, and Shuho continued to run the family business. She sent her eldest son to Edo with capital to open a draper's shop, called Echigoya, which prospered. (And even today, its descendent in Tokyo, Mitsukoshi Department Store, still stands near that original location in the central Nihombashi district.) The youngest and the third son, Hachirobei, was sent to Edo to help the eldest son. After training, he opened a second shop in Edo. Then Hachirobei took over managing the draper shops, while the eldest son began a cloth purchasing system. (The middle son had returned home to help his mother.) Thus the second generation of the house of Mitsui was established, with a growing cloth merchant business in the capital of Japan.

The principal customers for cloth were the aristocracy, and Hachirobei had his oldest son serve the Tokugawa government (called the bakufu). In 1689, Mitsui was assigned to be purveyors of apparel to the shogun of the time, Tsunayohi. However, the business of selling to the aristocracy required capital, since aristocracy paid when they pleased. Earlier In 1683, Hachirobei had established a money exchange, ryogaeya, and expanded these to locations not only in Edo but also in Kyoto. Thus the second and third generations of Sokubei, beginning with Hachirobei and his sons, established a trading and financial house with ties to the government. This trade and finance was to be the basic foundation of the future firm of Mitsui.

Sokubei and Shuho's son, Hachirobei, had inherited his mother's business ability and began building a superhouse of Mitsui, as a clan business. Harchirobei kept the houses of his sons in this larger clan establishment, forming the Mitsui clan into a comprehensive economic unit. All sons continued their businesses as a part of the house of Mitsui, obedient to one head, their father. This clan business was evolving into a corporate body, gumi; and would called Mitsui-gumi. As a clan business house of businesses, it was structurally a partnership wherein all Mitsui shops and exchange houses were managed independently but the capital of all was pooled and under centralized authority.

Following upon Hachirobei, his eldest son, Takahiri proved to be a good leader and established a great main headquarters, Omotokata, to guide the Mitsui-gumi. The house-of-houses authority imposed good business practices on all Mitsui businesses, including a double-entry bookkeeping system. A central financial reserve was established to help the house survive periodic vicissitudes and financial crises in the government. This was important because the relatively new money system that was evolving was unstable with frequent recoinages by feudal lords. By pooling the family resources, the pool was big enough for the family to survive such changes.

Hachirobei's sons reflected upon the proper managing of business and wrote business principles as the Chonin Kokenroku, Merchants Observations. At the centenary of Hachirobei's birth, a grandson, Toakahaira redrafted his father's will and prepared a house constitution. In the constitution were business principles, such as

- Thrift is the basis of prosperity, but luxury ruins a man.
- Be diligent and watchful, or your business will be taken away by others.
- Farsightedness is essential; do not miss great opportunities by pursuing trivial ones close at hand.
- Avoid speculation of all kinds, and do not touch upon unfamiliar lines of business.

Meanwhile as Japan was being firmly ruled by the Tokugawa shoguns, it was being isolated from the rest of the world. The third shogun, Iemitsu (1623–1651) closed the country to all foreign trade and forbade Japanese to leave the country (any entering foreign missionaries were killed). He required all daimyo to spend several months of each year in Edo, leaving their families there as hostages. This was expensive and increased the demand for money-lending. During these times, Hachirobei had expanded the family businesses further into money lending.

Mitsui businesses prospered during the Tokugawa peace until the 1860s, when violent change was to impact the stable, peaceful, isolated kingdom. European navies were beginning to dominate Asia, forcing uneven trade and European colonialism on the region. Japan's isolation was coming to a forced end, and the event that signaled the end was the arrival into Edo Bay of U. S. warships under Commodore Perry in 1853. He encountered a feudal society, with samurai wearing armor made of silk, leather, and thin plates of metal. Swords and lances and the bow and arrow were still the main weapons. While guns also were being used, the guns were antiquated flintlocks and muskets which had seen service in past European wars.

The feudal lords of Japan, daimyo, were divided. If they refused contact, the West would force it upon them by conquest with superior military arms. If they engaged in trade, their stable world would end. On March 31, 1854, the government signed the first diplomatic treaty with a Western nation, the Treaty of Kanagawa, giving American ships access to two ports and the reception of a United States consul. The door was opened, and the daimyo began visiting Western countries to see the new world. A U.S. steamship with side wheel transported eighty samurai to San Francisco, where they were shown western science, technology, government, and military weapons.

Still in Japan, national feelings were mostly for continued isolation and hatred of the foreigners. Patriots blamed the shogun for admitting the foreigners; and political turmoil continued into the 1860s, finally resulting in major governmental change. Two major families (han) of the Satsuma and Choshu reached agreement that the ruling house of the Tokugawa must be overthrown. This alliance was made between four leaders of these houses, Okubo, Saigo, Komatus, and Kido Koin. They were the persons who were to establish the new government which would be called the Meiji Restoration. (But it could more accurately be called the Meiji "revolution"—a revolution of Japanese society that was to be imposed from the top by the Sat-Cho oligarchy). In July of 1867, the leaders of this Sat-Cho alliance met and signed a pact to carry out a coup de'etat. They then proceeded to overthrow the old Tokugawa shogunate and establish the ancient imperial family as symbolic rulers of Japan, with the Sat-Cho group running this new imperial government.

After the successful coup and a new imperial government was established. Its first need was for money. Since the Mitsui had long standing business relationships with the Satsuma and Choshu clans, the house of Mitsui was immediately called upon for contributions. A samurai messenger from the coup was sent to Mitsui and told them of the need for finance. Late into the night, Mitsui men counted out money, filling chests with a treasure of two thousand ryo. In the morning the samurai soldiers carried this off to the new Imperial government.

The national problem facing the Meiji reformers was to jump the social conditions of Japan from a feudal structure directly into a modern industrial structure. Their competitors in Europe needed six centuries (from the 1300s to the 1900s) to create this transition, moving from feudalism through mercantilism into laissez-faire enterprise to industrial capitalism (while continuing socially to evolve as the twentieth century began with new basic technologies and political struggles between democracy and various authoritarianisms, such as fascism and communism).

The political agenda of the Meiji government was to modernize Japan, rapidly catching up with the Western nations in military and economic might. The new government understood that they could not redress the unfair treaties, that had been forced upon them by militarily superior foreign governments, until Japan caught up to the new industrial "civilization." This meant to conform to European and American standards; and the Meiji government began the enormous task of "westernizing" the country, economically, politically, and culturally.

The first social reforms of the Meiji government were to establish universal education for a literate society, to abolish the caste structure of the samurai status, to create modern governmental structures and modern military organization, and to foster international trade and industrialization through the import and improvement of foreign technologies.

At the center of power of the new imperial government was this Sat-Cho oligarchy, which included Inoue Kaoru from the Choshu han. Inoue headed the new Ministry of Finance and took a strong interest in Mitsui. He appointed Mitsui as agents for the government mint, to exchange new coins for old money. Thus Mitsui's old money-exchange business now positioned them as agents in the new government's financial structure.

Meanwhile, Mitsui needed to organizationally restructure. Fortunately at that time (over two hundred years after the founding of Mitsui-gumi by Sokubein, Shuho, and their son Hachirobei), Mitsui had a competent leader, Minomura. Minomura had not been born a Mitsui but had been recognized by the clan for his exceptional merit and had been promoted to run Mitsui-gumi. When Minomura took control of Mitsui, the first thing Minomura did was to separate the textile branches from the money exchanges. Minomura intended to make Mitsui-gumi into a great banking firm. He anticipated a new law (the National Bank Act), and formed one of the first banks in Japan, as a partnership between Mitsui and another house, Ono. This new bank was called the Dai-

Ichi Kokuritsu Ginko (First National Bank). The financial exchange traditions of the feudal house of Mitsui began to evolve into a new form of financial services—a modern bank.

Meanwhile, Inoue Kaoru as an important official in the Meiji government also was a friend of the House of Mitusi. He took an active interest in Minomura's transformation of Mitsui. Inoue also saw Mitusi as a loose conglomeration of semi-independent operations. Inoue and Minomura, together revised the Mitsui's charter toward a form that would evolve toward a modern corporate form.

Inoue's interest in Mitsui was a part of the general pattern of the Meiji government policies which encouraged the development of Japan's trade and industry—to provide the economic basis of a modern state with a modern military. Inoue encouraged Mitsui to expand its trading capabilities. The government organized a Tokyo Commerce and Trade Company to facilitate foreign trade as a joint enterprise between the government and Mitsui. Mitsui aggressively expanded its retail shops and banking operations. Minomura consolidated trading activities in a new company of Mitsui, called Mitsui-gumi Kokusan-kata (National Products Company).

At first, Mitusi's National Product Company supplied silk and grain to foreign traders and imported blankets for the army. As trade grew, it was reorganized in 1876 as the Mitsui Bussan Kaisha (known abroad as Mitsui and Company). Then its main exports were coal from a state-owned colliery at Miike in Kyushu and surplus rice, and soon the government sold the colliery to Bussan. In this way, Mitsui's Bussan began expanding into production as well as trade. Thus as the twentieth century approached, the House of Mitsui was being organized into the structure of a modern corporation, with operations of banking, retail, trade, and production. It was set to become the foremost zaibatsu in Japan.

After Minomura's retirement from heading Mitsui, Masuda was appointed the new head of Mitsui. He continued the transformation of Mitsui by next acquiring substantial basic production capabilities in mining. The Meiji government had inherited the mineral deposits from the shoganate and the mines continued to be state owned until the 1880s, when the Meiji government turned to a policy of privatization of industry. They sold mining properties at nominal prices to financial houses. The political connections between government officials and financial houses influenced the sales.

In a tight political game, the houses of Mitsubishi and Mitsui bid for the Meiike mines, and Mitsui won. A young manager in Mitsui, Dan Takuma, improved production techniques in the Miike mines to make them immensely profitable. Mitsui next acquired the Kamioda mines (which were literally mountains of lead and zinc ore laced with silver, cadmium, and copper and showing even an occasional glint of gold). Other coal mines and the best iron deposits in the country came to be owned by Mitsui. Minerals production provided the revenue base for the subsequent growth of the powerful Mitsui zaibatsu (combine). Such resources provided a strong cash flow in times of peace and in times of war.

In summary, the partnership between Mitsui and the government, particularly through Innoue's interest, helped Mitsui grow and prosper. At the time the growth of all financial houses in the modernizing Japan were helped with political connections. The Meiji government saw the financial houses as tools of national policy to build the new industrial Japan. Later heads of Mitsui (after Minomura's retirement) continued reforms in the organization and management of Mitsui, such as improving salaries to off set the inflating cost of living, introducing a promotional system based upon merit, beginning twice-yearly bonuses as an incentive, and establishing a pension fund.

SOCIETAL STRUCTURES

We pause in this case to review the theoretical concept of societal structures. What we are seeing illustrated is the origin of a major commercial firm from a merchant house in a pre-industrial, feudal society. The transformation required strategic insight and managerial competence in successive managements of the business to meet the challenges of the times and to find profitable enterprises in their changing situations. Change in enterprises over time is always necessary because the environments of enterprises change over time.

Also we see an illustration of the total environment of a firm, its societal structures, economic structures, governmental structures, etc. The purpose of scenario planning in strategic management is to attend to these environmental structures to identify trends and changes that will impact the conditions of business in the future. And in this case, the structural changes were enormous. First a society of feuding warlords in the early 1600s was stabilized into a long-lasting and peaceful feudal shogunate until the mid-1800s.

Under the Tokugawa shogunate, the government, bakufu, created the conditions for peace; and in peace, economic prosperity often grows. However, the isolation of the country from the world also caused the country to fall seriously behind in the advance in technology in the Western nations from the 1600s through the 1800s. This left Japan in a militarily weak state, subject to the dictates of militarily superior countries. The reforms set in place by the new government under the former clan houses of the Sho and the Chu began the transformation of Japan into a modern industrial and military power. And in that transformation, the financial house of Mitsui (as a governmental favored private concern) became a modern industrial conglomerate giant, Mitsui Gumi.

We can use this illustration to see the complexities required to describe societal change and its implications for a businesses future. For the purposes of building a planning scenario that captures this kind of complexity and completeness of

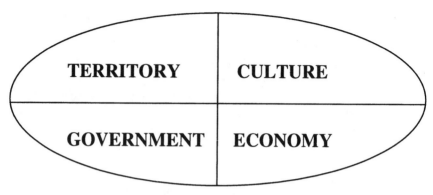

FIGURE 4.1   SOCIETAL MODEL

possible future change, we need to have a classification of all the societal environments of a business. The major changes in the environments of a business are changes in structures of the society in which the business operates. Before we continue our case history, let us now classify the kinds of societal structures within which a business exists within a modern industrial society.

In all human societies, traditional or modern, there have been four classes of social patterns that create societal structures: territory, culture, economy, and government, as illustrated in Figure 4.1.

Territory

All societies have human populations organized upon the basis of territory. The control of a geographical territory by a human group provides the physical basis for of people's lives in terms of residence space and natural resources. Populations consist of young, middle-aged, and old people who compose the society, with varying degrees of wealth dependent upon environment, technology, and military capabilities. These populations unite and divide under different cultures within a territory.

Human warfare has focused on territorial control since time immemorial. Territorial control still operates as the major source of warfare in modern societies, as for example in the major wars of the twentieth century (e.g., World War I, World War II, the Vietnam War, Operation Desert Storm, etc).

In the case of the House of Mitsui, it was the struggle over territorial control that motivated the civil wars among the feudal war lords; and the final feudal victory by Tokugawa Ieyseu that established the long-lived *shogunate* from 1616 to 1863.

Government

Government is another universal social pattern in all societies, for the internal control of the population in a territory. For administration of the controlled territory, a formal organization of government is necessary. Is the case of Mitsui, the Tokugawa shogunate established a feudal government called the *bakufu*, which ruled Japan. Only after the *bakufu*'s failure to defend the country from dominance by foreign governments was a clique enabled to perform a *coup d'etat*, replacing the government with a new form at the Meiji restoration. The form and conduct of the government provided major influences and opportunities in business for the house of Mitsui, during the *bakufu* and after the restoration.

Culture

Culture is another major source of social patterns found in all societies. Culture consists of the ways people identify with each other into bonding groups. The three primary sources of cultural bonding in preindustrial societies were kinship, language, and religion.

Economy

All societies have some form of economy by means of which the population in a territory supports itself and upon which a government taxes in order to pay for the government of peace and war. Ancient and primitive forms of economy were hunting and gathering tribes. Later, agriculture and herding became the primary basis of preindustrial economies. Production of material artifacts in pre-industrial societies is predominantly local, except when needed materials for the artifacts is not available in a territory. Trade therein is mainly barter for uniquely located materials and artifacts produced from such materials. As territories expanded under military conquest in ancient empires, trade also expanded, added by a form of money (gold, silver, or bronze) minted by a government of a dominant territorial empire. Production of food and artifacts and trade in materials and goods form the basis of economies in all societies. The Arab tribes, which emerged as conquering armies of Islam, traditionally had desert economies in the Arabian peninsula agriculture at an oasis or in a watered valley, with herding and hunting in desert and mountainous areas. In the case of Mitsui, preindustrial Japan had a typical feudal economy of an agricultural base cultivated by a peasantry caste, with taxes in the form of rice product collected by the ruling samurai caste. Peasants also grew silk worms and wove silk for garments, and mined and refined iron for weaponry. The peasants who migrated to the artisans and merchant, *chonin*, castes, mainly supplied products to the aristocratic caste.

Societal Models

In summary, the history of a territory can be described in terms of the interactions between societal patterns of government, economy, and culture within the region and the changes in these patterns. In the case of Mitsui, it was a major change in the political environment, the dominance of the Tokugawa shogunate, that stimulated the Mitsui Sokubei to choose to become chonin. As we also saw, it was the military arrival of Perry's American warships that forced the governing bakufu into a crises, which eventually created the Meiji reformation coup. In both these major changes in political structures, the business of Mitsui was founded and profoundly altered.

CASE STUDY: House of Mitsui, Continued

We continue the case of Mitsui. The principle focus of the new Japanese government of the Meiji restoration was to regain military potency (as summarized by the Foreign Minster Okuma Shigenobu in 1897):

> If we enquire what points are practically most important in the foreign policy of the Meiji Era, we find that to attain an equal footing with other Powers, as declared in the Imperial Edict at the Restoration, has been the impulse underlying all the national changes that have taken place.
>
> —(Pittau, 1967, p. 39)

By 1900, progress in Japan's new military powers became evident to the international community at the time of the Boxer Rebellion in China. The rulers of Japan's giant neighbor, China, had chosen a strategy of isolation and had fallen prey to the commercial exploitation of China by Western countries—partitioning China in spheres of influence. The British opened China militarily to enforce a trade of opium grown in the British colony of India for tea grown in China and sold by British traders at a high profit in England. The French, Germans, Russians, and Americans joined in the open China policy. Then a populist rebellion in China, called the Boxer Rebellion, tried to eject the Western invaders; and Western powers sent in troops to suppress the rebellion. A modernizing Japanese army joined in with other international forces and its discipline and performance impressed other nations.

The Meiji restoration government pursued the build up of military power vigorously and the same pattern of colonialism of the Western nations as an economic basis upon which to build industrialization. Then the global scenario of the world was economic colonialism. During the Boxer rebellion, the Russians moved forces into Manchuria; and they completed the Trans-Siberian Railway, seized Port Arthur, and were strengthening their position in Korea. A secret society in Japan, the Black Ocean Society, murdered the Korean queen; and the Korean king took refuge in the Russian ligation. The Japanese government sought an alliance with the British government, against Russian expansion. This alliance was signed in January 1902, and Britain acknowledged Japan's special interests in Korea, while asserting Britain's interest in China. This alliance was the first time in modern history that an Asian country was accepted as a full military ally by a European power. From the Japanese perspective, it should have legitimated in European eyes Japan's ambition to stake its claim to parts of the Asian continent.

This was important because Southeast Asia had been subjugated by the British, French, and Dutch; and now among them, they were contending (along with Germany and Russia) for control over China. Russia wanted control of Manchuria, and Japan and Russia next tried negotiating for control of both Manchuria and Korea. But the negotiations failed, and on February 6, 1904, they broke diplomatic relations. Russian troops crossed the Yalo River from China into Korea. At Port Arthur near the norther border of Korea, Japanese boats launched a torpedo attack at Russian ships in the harbor. On February 10th, Russia and Japan declared war.

For five months, the Japanese army besieged the Russian garrison at Port Arthur, and the Russians withdrew. Loses were great, with the Japanese losing 60,000 soldiers. Fighting in Manchuria was also intense. At the Battle of Mukden, a total of 750,000 soldiers engaged, with the Japanese losing 40,000 soldiers, but again being victorious. Mitsui's Bussan was the principal supplier for the Japanese army and was also developing the shipping capability to supply the army over the seas.

The final battle between Russia and Japan occurred at sea. Russia's Baltic Fleet had been sent out from the North Sea, through the Atlantic, around Africa into the Indian Ocean into the South China Sea, appearing off the coast of Indochina in May 1905. Observed by intelligence officers of the Japanese Navy, information was sent to the Japanese fleet that the Russian fleet was likely to pass the narrow passage between the island of Tsushima and the Korean coast. There the Japanese ships lay in wait. The Russian admiral ignored warnings of a possible trap, did not even make any reconnaissance, and sailed the Russian fleet into the ambush. Early on the morning of May 27, 1905 the battleships of the new Japanese navy, commanded by Admiral Togo, appeared on the horizon and immediately opened fire upon the Russian Baltic Fleet. For two days, the Japanese warships devastated the Russian fleet, destroying many ships and 18,000 Russian sailors. The Japanese fleet lost only three torpedo boats and 116 sailors killed.

Russia had lost the war. Japan now began to get the military and political respect of the world and became firmly committed to a military path of national development. On September 5, 1905, the Portsmouth Treaty between Russia and Japan (1) recognized Japan's "paramount interest" in Korea, (2) yielded control of the southern section of the Manchurian Railway, and (3) provided a lease on the strategic Liotung Peninsula. This opened the way to the Japanese colonialization of Korea, its penetration into Manchuria, and it placed Japan on the edge of China. Japanese military expansionism plans for the twentieth century were thus laid out by the treaty. Although Russia had yielded some of its power in Asia, other Western nations (e.g., England, France, Germany, and the United States) had not yielded power. In this way, the first step was taken down the path that would lead Japan into the second world war of the twentieth century.

Major resources in Manchuria were coal and iron. Near Mukden was one of the largest open-cut coal mine in the world. The Japanese constructed the Yawata Iron Works in Lyushu, using an indemnity forced from China. Coal and iron fueled the industrial revolution of the nineteenth century and was essential to any industrial nation in the twentieth century. Mitsui's trade firm, Bussan, brought in the foreign machinery for the railway and mining activities. Bussan set up offices in Mukden and other Manchurian cities and controlled the export-import trade between Japan and the Manchurian colony.

This territorial conquest stimulated an economic boom in Japan. For example, ship building to carry goods across the sea expanded rapidly. Commercial ship building also developed military production capabilities. In 1910, the world's largest warship at the time was finished in a shipyard in Yokosuka. Thus Japanese industrialization was creating the military power for external aggressive expansion, and the addition of new territories was providing the economic opportunities for continued economic growth.

As a historical societal pattern, this industrialization of Japan was following exactly the same pattern as the industrialization of Europe during the previous century. Japan developed aggressive new military technology to acquire new colonial territories for economic expansion.

During this time, the firm of Mitsui was evolving into a mighty commercial and industrial combine, a zaibatsu. A zaibatsu was a financial clique of companies, controlled by a Japanese family, and Mitsui was the prototype and prominent example of this. Mitsui represented a form of societal organization transitional between feudal merchant clan and a modern corporation. The zaibatsu were family-dominated trading empires, run as interconnected groups of modern business corporations.

In 1909, Mitsui formally transformed itself into a holding company, a zaibatsu of the name Mitsui Gomei Kaisha. It was capitalized at fifty million yen and contained fifteen companies in banking, mining, and commerce. Moreover, as Japan still remained behind Europe in the progress of technology, zaibatsu, such as Mitsui, made strategic alliances with foreign firms to acquire and use new technology. The relationship of zaibatsu to the government was strong. For example, Inoue Kaoru of the Meiji government continued to see Mitsui as not only a private family holding company but also as an instrument of national policy. Kaoru asked Mitsui to enter the munitions industry to make weapons for the Japanese military. Mitsui established Nippon Seiko, Japan Steel Works, in a joint venture with Britain's two largest makers of arms, Vickers and Armstrong.

One can see just how strongly were the societal patterns between territory, government, economy, and culture interconnected in Japan at this time.

The first world war in Europe began. Germany, Austria-Hungary and Turkey were allied against England, France, and Russia. The conflict began generally in colonial ambitions between Germany and England, particularly over the continent of Africa. After that war, Europe was in turmoil, particularly with a communist revolution in Russia and a destablizing inflation in Germany. Meanwhile, Japan had further progressed to new strength and prosperity. By the late 1920s, Mitsui had become the dominant zaibatsu controlling capital of over 500 million yen and 130 companies.

In November 1921, there next occurred a key political event for the Japanese military. This was an international armament conference about battleships. Agreements between the United States, Britain, and France set naval ship tonnage ratios for the respective navies, which left Japanese navy restricted to half the total tonnage of the U.S. and British navies. This infuriated the Japanese military and government. Then adding racial insult to injury, the United States government passed a immigration law in 1924 that excluded further Japanese immigration. These incidents strengthened political positions of the extreme military cliques for further military expansionist policy.

Mitsui personnel supported expansionist views. For example, in 1926 Yamamoto Jotaro, obanto (head) of the Mitsui Bussan empire in Manchuria, was given a high post in the Tanaka cabinet in Japan. In 1927, Yamamoto was appointed president of the South Manchurian Railway and began to build five new railway lines. In 1928, he was invited to the Imperial Palace to present his "new economic plan," which included the construction of improved railway and harbor works in Korea, a major plant for making fuel oil from coal in Manchuria, and expansion of agriculture and forestry and industry in Manchuria. Japanese national prosperity was deeply tied to the colonial expansionism in Manchuria and Korea, and Mitusi was prospering in the expansionism.

The 1930s depression was a world wide phenomena. One impact of the earlier inflation and subsequent depression in Germany was to bring the Nazis into power in Germany, which laid the path to World War II. Also the world depression severely hit the Japanese economy, which depended upon exports to finance the needed imports of new technology products and know-how. The depression further concentrated economic control in Japan in the few zaibatsu families. In 1931 to fight the effect of the depression, the government passed the Important Industries Control Law, which organized large producers into cartels. By the mid 1930s, the eight zaibatsu groups would control 50% of Japan's financial capital. Government policies became focused upon national "preparedness," which meant getting ready for war.

At the same time, the Japanese government was also fostering cultural focus on an extreme form of nationalism, which finally eliminated any vestiges of a possible democracy in the pre-World War II period, and the zaibatsu supported this direction. They cooperated with the government is suppressing activities that restricted their economic power, such as suppressing labor organization movements. Overall, the government established a kind of police state, similar then to the fascist governments of Hitler in Germany and Mussolini in Italy. Security and thought-control police, the Kepeitia, expanded in scope and activities. School children were indoctrinated in "moral education," which consisted of Shinto mythology (the state religion), worship of the emperor, and racist superiority. Newspapers were controlled and enlisted to promote militarism. Dissenters were warned; and if they persisted dissenting, they were arrested and tortured and, without trial, imprisoned and even executed. About 60,000 people were arrested for "dangerous thoughts" between 1928 and 1937.

Thus in the patterns and accidents of history, the reformation of Japanese society into Western industrialization that began in the 1860s grew into a *military-oriented industrial state* by the 1930s (wherein a modernized military ruled Japan, replacing the earlier feudal miliary rule—but still military rule). Control over the government by the military was finalized in 1936–37, beginning with the appointment of Hirota Koki as prime minister, who did the Army's bidding. Government policy aimed toward a "total defense" economy with a domination of Asia under Japanese hegonomy (as a Great East Asia Association). Thus Japanese government policy become one of international expansionism, after the earlier model of the Western nations and in a similar pattern to the fascist governments of Germany and Italy of the 1930s.

It was in 1937, when these two general patterns of industrial and military expansionism in both Europe and Asia paved the way to World War II. In Europe, the German Nazi government under Hitler began a series of aggressive moves into the European Rhineland, Czechoslovakia, and then Poland. The invasion of Poland resulted in a declaration of war of Great Britain on Germany, and the second world war began a little over thirty years after had the first world war. Also in 1937, the Japanese Army began a series of aggressive moves in Asia (first invading China) that would lead to World War II in the Pacific.

The immediate political interests for the beginning of the Japanese/Chinese war in 1937 arose from economic competition between two cliques of military/economic groups—the Manchurian clique involving a new zaibatsu of Nissan and the older zaibatsu (including Mitsui) with economic interests in China. The new Nissan zaibatsu was officially named Nippon Sangyo and consisted of Hitachi Ltd., Nippon Mining, Nippon Marine Products, Nissan Motors, Nissan Chemical Industries, and hundreds of subsidiaries. The Japanese Army in Manchuria was headed by General Tojo Hideki and formed an alliance with Nissan, whose managers controlled the Manchurian Heavy Industries. The economic power of the new Nissan zaibatsu in Manchuria led the older zaibatsu to cooperate in exploiting the trade and resources in China. In turn, this extension of the old zaibatzu into China alarmed the new Nissan zaibatsu-centered military clique, and they decided to invade China.

The military clique fabricated a shooting incident between Japanese and Chinese troops at the Marco Polo Bridge on the outskirts of Peking. On this pretext, Japan declared war upon China. The Japanese army quickly overran China along the coast and into the interior. From 1937 to 1940, the war continued in China at low gear and benefitted the Japanese economy. The cost of Japan's occupation was borne by the subjugated Chinese. Japan obtained an inflow of raw materials from China at a low cost that were exchanged for a relatively higher-priced Japanese manufactured goods.

This kind of exchange of the raw materials and agricultural products from an occupied nonindustrialized country for manufactured products of an occupying industrialized country was the heart of colonialism. The European nations had used this form of colonial exchange to drive the industrialization of Europe, and it had led to the European World War I. Firms in industrializing countries benefitted from victorious wars, producing military products for their governments. This was also true for the firm of Mitsui, who prospered in producing munitions and weapons for the Japanese military.

However, Japan's resource needs could not be completely met by its colonial occupation of Manchuria, Korea, and northern China. It still lacked sufficient petroleum, with only 8 percent of the nation's needs produced domestically. This need led the military-dominated government to next launch an invasion of northern Indochina (then a French colony) in September 1940. The United States responded with an embargo on the sale of scrap iron and aviation fuel to Japan. In response, the war minister of Japan, General Tojo, then proceeded to plan a total war, and in the summer of 1941, Japanese army expanded their occupation into southern Indochina. This area had supplied raw materials to western nations such as rubber, tungsten, tin, copra, silk, jute, and shellac.

The United States continued its embargo upon strategic fuels and raw materials, ending its trade with Japan, and Japan then had to give up its new colonies or go to war with Britain and the United States. Japan, the former land of the samurai, chose war. Its military forces bombed the U.S. Naval base in Hawaii and, in a series of successful invasions, conquered the British colonies of Singapore and Hong Kong and the American dependency of the Philippines. Thus began the great war of the twentieth century in the Pacific, between Japan (allied with Germany) and the United States (allied with Great Britain).

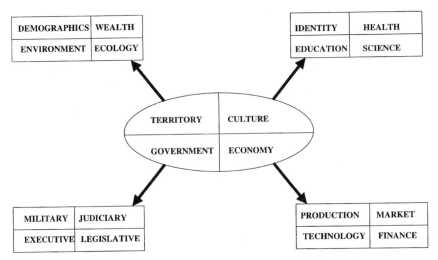

FIGURE 4.2 MODEL OF SOCIETAL STRUCTURES

MODEL OF AN INDUSTRIALIZED SOCIETY

In this historical scenario of a society's history, we continue to see the many and entangled interconnections between its governmental policies, its economic development, its cultural development, and the conflict over territory with other national powers. A historical review of the history of any society shows similarly entwined patterns. Now we pause again in this case to provide further theory about societal models. We detail the four classes of societal structures (territory, government, economy, culture) into more detailed substructures. In a modern industrialized society, the societal structures of governance, economy, territory, and culture are further specialized into substructures, as sketched in Figure 4.2.

CULTURAL SUBSTRUCTURES

The cultural structures of a modern industrialized society become refined into patterns of social identity, health, education, and science substructures.

In social identity, the sociological patterns of kinship, language, religion, and patriotism provide foci for organizing groups and for individual identification with cultural groupings.

Kinship is the first socio-biological basis of organizing groups in society. In preindustrial societies, the family is the basic, which is then extended to larger family units of relationship, such as clans. Tribes are clans organized by linguistic groupings. In feudal societies a militarily superior tribe subdues other tribes. A tribal society is a living group of humans sharing a linguistic form and regional basis. The earliest forms of human society are tribal, and many tribes still exiting in the early twentieth century were studied by anthropologists, who found that some had patriarchal structures and some matriarchal. A feudal society exists when one group of tribes militarily dominates another group of tribes and subjects the subjugated tribes into a lower caste.

Feudal structures were the way empires were constructed in classical times, such as the Egyptian, Babylonian, Assyrian, Persian, Alexandrian, Roman, Chinese, Mayan, Aztec, Byzantium, and Ottoman empires and the European feudal states in the Middle Ages. All feudal societies have had caste structures that reflect the decedents of the dominant ruling tribe as currently an aristocratic caste lording over the decedents of a previously subdued tribe as lower castes. In the case of the house of Mitsui, the founder Sokubei was a descendent of a ruling caste in Japan, the samurai caste. When he gave up his upper caste status to become a lower caste of shop keeper and merchant, a *chonin*, it was a major step down the social ladder. However, it was a worthwhile sacrifice for Sokubei (and an understandable choice since his wife, Sokubu, was already of the *chonin* caste) for in the ensuing peace of the Tokugawa regimes, a low-ranking samurai had fewer opportunities for wealth than a prosperous *chonin*.

Language is the basic tool of human organization. Primitive tribes are cultural groups who members share a common linguistic form, language. Modern nations were organized around shared languages. In fact, in modern countries encompassing more than one linguistic group, this factor is the most basic source of politics in the nation. In the case of Mitsui, the territory of Japan was determined by the shared language of people in that territory speaking Japanese. By the time of the Tokugawa *shoganate*, the Western countries of modern Europe had emerged in the Middle Ages formed around territories occupied by English-speaking, French-speaking, Spanish-speaking, Italian-speaking, Polish-speaking, Russian-speaking peoples.

Culture in preindustrial societies is also provided by religions, which can bond together different linguistic groups. In the case of Mitsui two religions were common in Japan, Shintoism and Buddhism. For the Western natures, the common religions were Roman Catholicism in Western Europe, Greek Orthodox Christianity in Eastern Europe, and Islam in the southern and eastern Mediterranean territories. In the fifth century after the fall of the Roman empire, there arose under the new religion of Islam a conquering tribal nation of Arab tribes. It created conquering armies that subdued all the middle eastern cities, eastern Eurasian cities, and cities in northern India. The Islamic armies established new feudal empires, with the conquering Arab families as the new ruling aristocracy (e.g., in India, they established the Mogul empire). All traditional empires were based upon conquered agricultural areas (e.g., the great river valleys of the Nile, the Euphrates, the Indus rivers, etc.) ruled with an aristocracy, many of whom came from conquering herding tribes (e.g., from the Eurasian plains).

For example, in the case of Mitsui, we saw how the older feudal structure consisted of the castes of (1) samurai, the ruling military aristocracy, (2) *chonin*, merchants and artisans, (3) peasants, farmers, and laborers, and (4) foreigners, whom the Tokugawa shogunate excluded from the country. The founder of the house, Sokubei, was of low samurai ranking, who had married a merchant's daughter of the *chonin* caste. He gave up the samurai status to become a *chonin*, when he saw that he lacked proper feudal connections to the new ruing feudal house of the Tokugawa. Under the Meiji reforms, the castes of samurai, *chonin*, and peasants were abolished and universal education for all citizens was required through the eight grade.

An education and science infrastructure is necessary for universal literacy and for advanced skills that a population needs to survive in an industrialized culture. In the Meiji reforms, universal education through the eighth grade was required of all citizens and national patriotism was reinforced through "moral" education. Universities were also created in the Western model, and scientific literacy was cultivated in education and industry. Public health and medical infrastructures are also built up in a modern industrialized culture for a healthy population, using scientific knowledge in medicine and health.

Economic Substructures

In a modern industrial state, the economic structure, its economy, becomes specialized into economically functional areas of production, trade, technology, and finance.

In the case of the policies of industrializing Japan, we saw how the economic stimulus toward colonial expansion and aggression in Korea, Manchuria, China, and Indochina was based on the economic need for resources of coal, iron, oil, and on the need for expanded agriculture and trade. We also saw how the industrialization and military growth was based on acquiring and using Western technologies in weaponry and industrial production, artifacts, and services.

Also in this case, we saw how Mitsui strengths in trade and finance were the commercial foundations of the house, which later grew stronger in production as the industrialization of the country began. Mitsui added its Bussan organization to provide international trade capabilities and then added production capabilities under Bussan in mining. Later Mitsui add light industry, as government sponsored heavy industry, which later was transferred to the *zaibatsu*. Mitsui extended its finance capacities in acting as a bank for the modern government finance and money policies and also expanded to overseas banking connections for international trade and finance.

And we saw how the economic interests of individuals and business cartels, commercial leaders and military leaders, became involved in government policies, so that the actual war with China was triggered by internal rivalries between army leaders and the old and new *zaibatsu*.

Government Substructures

In a democratic industrialized society, the governance structures are divided into military, judiciary, executive, and legislative functions and organizations. In the United States, the legislative branch of the federal government consists of the House of Congress and the Senate. The executive branch is headed by the President of the United States, and all officials in the administrative units of the federal government ultimately report to the President. The judicial branch of the federal government consists of federal courts to the highest level of the Supreme Court. The miliary branch of the federal government consists of the Army, Navy, Air Force, and Marines, controlled by a Joint Chief of Staffs, all also reporting to the President as the Commander-in-Chief.

Territorial Substructures

In a modern industrial society, social patterns of a geographic territory can be decomposed into demographic patterns, patterns of wealth distribution, environmental systems, and ecological systems. Demographic patterns trace the changes in the size of a population and its age distribution. For example, throughout the twentieth century the population sizes of territories all over the world increased enormously, and in industrialized societies, the average life span increased, with an increasingly large percentage of populations becoming old people.

Patterns of wealth distribution describe how wealth is distributed by caste, ethnic, or class groupings. In traditional feudal societies, about 10 percent of the population belongs to the ruling caste, 80 percent to the peasant class, and about 10 percent to the artisan/merchant caste. In industrialized societies, the population divides generally not into caste or ethnic groupings but into the socio-economic classes divided by wealth and income. How the population divides by wealth has varied in the time and place of industrialized nations.

Environmental ecological systems of a territory describe the physical and biological properties of the territory, such as patterns of climate, rainfall, ecological systems, and so on. The patterns of effects of industrialization of the world over the last two centuries on biology has lead to a vast extermination of natural species, replaced by narrow and rather homogeneous agricultural-based ecologies and by cities.

SUMMARY OF SOCIETAL MODELS WITH SUBSTRUCTURES

For the purpose of writing scenarios, the model of Figure 4.2 provides a way of classifying the societal environments of any business or firm, in order to examine trends and changes that may impact future activities. Now the taxonomy of structures in the economy sector of the society provides the immediate contexts for strategic planning in businesses and in firms. Attention should be paid to the sectors of governance, population, culture in scenario planning.

CASE STUDY: House of Mitsui, Continued

Returning to the historical scenario of the case of Mitusi, we pick up the narrative after the Second World War. After the initial Japanese military successes of 1941, the fortunes of war turned upon the Japanese military. By the end of 1942 and beginning of 1943, the Japanese forces were on defense. The United States forces were beginning to move inexorably, step by step, across the Pacific toward Japan.

By 1945, the United States had captured islands close enough to Japan to extensively bomb Japanese cities. For example, on May 25, 1946, 500 American B-29 long distance bombers dropped incendiary bombs, destroying most of central Tokyo and adjacent residential areas. Finally on August 6, 1946, the United States bombers dropped a terrible new weapon, the atomic bomb upon Hiroshima. This 20-megaton explosion of light and air-pressure waves instantly obliterated a central mile circle of the city, killing more than a hundred thousand civilians and making thousands more sick with nuclear radiation, a new deadly peril in the world.

The news of this final turn of the terrible war spread shockingly and in puzzling confusion through the Japanese nation. In Mitsui, communications between Tokyo headquarters and branches in Hiroshima were suddenly cut off on August 6, and headquarters was then aware that some kind of catastrophe had happened. Two days later news came from Manchuria that the Soviet Army had invaded. On August 8, there was news of another great cataclysmic explosion at Nagasaki.

The Emperor and government also were struggling with this news, and the Emperor decided upon a surrender. Still, some army personnel were committed to national suicide, and after an aborted army coup on August 15, the emperor was able to broadcast to his people that the war was ended. The Emperor announced that he had accepted the Allied Joint declaration for unconditional surrender. He explained to his people:

> We declared war on America and Britain out of our sincere desire to ensure Japan's self-preservation and the stabilization of Southeast Asia . . . But now the war has lasted nearly four years. Despite the best that has been done by everyone—the gallant fighting of military and naval forces . . . The war situation has developed not necessarily to Japan's advantage.
>
> —(Roberts, 1989, p. 364)

After the surrender, the United States President Truman ordered American forces under General MacArthur to occupy the country, begin democratizing the Japanese economy and to dissolve the zaibatsu:

To this end, it shall be the policy of the Supreme Commander . . . To favor a program for the dissolution of the large industrial and banking combinations which have exercised control of a great part of Japan's trade and industry.
—(President's Directive of September 6, 1945).

This news shocked the commercial leaders of Japan, as it would completely alter the country's economic structure. The officials in the zaibatsu had simply assumed that they would lead the reconstruction of the nation's war-ravaged economy. But General MacArthur immediately issued orders to the four largest zaibatsu—Mitsui, Mitsubishi, Sumitomo, and Yasua—to plan the dismantling of their holding companies.

In 1945, the ten largest zaibatsu held 35% of the nation's paid-up capital, 55% of bank assets, 71% of loans and 67% of trust bank deposits. Mitsui then was probably the world's largest private business organization, with the eleven branches of the controlling Mitsui family having a wealth of about six hundred million yen and owning about 336 companies.

The American Supreme Commander for the Allied Powers (SCAP) moved into action. On October 8, 1946, U.S. Army trucks with U.S. Military police arrived at Mitsui headquarters to seize Mitsui financial valuables. They loaded forty-two wooden cases that contained Mitsui owned stocks and bonds of a value of 1.2 billion yen. At other Mitsui locations, they also seized another 260 million yen worth of certificates. These impounded certificates made their owners powerless in the board rooms of their own companies. SCAP ordered a purge of the leaders from government and industry in Japan. More than 220,000 leaders of the military, bureaucratic, political, and economic cliques who had been running Japan were banished from their positions. SCAP drafted a new constitution for Japan, vesting state power in the "will of the people" and providing a bill of rights making all equal under the law and enfranchising women. The new Diet in 1946 created a steeply graduated tax on personal assets. The Mitsui clan was hard hit by these changes.

SCAP began carving up zaibatsu groups. The Mitsui Bussan was divided into 170 companies and the Mitsubishi equivalent Shoji into 120 companies. But disagreement within Washington politics over the concentration of economic power in Japan quickly altered policies, and the zaibatsu bands were not made subject to deconcentration. Eventually, forty-two holding companies were dissolved. Yet the largest banks were still intact, such as Fuji, Mitsubishi, Sumitomo, Snawa and Dai-Ichi. These provided nuclei to replace the old holding companies, the zaibatzu, to reconstruct financial and industrial groupings in the post war economy. But these new economic groupings of companies around the banks turned out to have similar membership and size as the prewar zaibatsu.

In effect, the institutional framework and financial system, centered on the zaibatzu, survived the American occupation. Moreover, this old/new structure was even encouraged by the Americans, after the Korean War began with the North Korean communist army invasion of the south. The United States army intervened with United Nations allies to save South Korea. It used Japan as a base for pursuing this war, which helped revive the Japanese industry and economy.

SCAP was then told to reorganize Japan for strength rather than "peace and democracy." In 1951, more than 2500 former imperial army and navy officers were depurged and moved into positions of leadership. Also former high-echelon zaibatsu managers were depurged and let back into running companies. Despite the disaster of the war, Japan's infrastructure was back in place. Economic progress in postwar Japan had begun to restore the nation to one of the mighty industrial nations of the world. When the Allied occupation of Japan ended in 1952, economic opportunity was booming. Older companies regrew, and new companies were started. Entrepreneurship and opportunity in Japan flourished again in a new generation.

In the case of Mitsui, the organizations that had been controlled by Mitsui Honsha drew back together to promote a new Mitsui Group. Promoting this cohesiveness were former Mitsui financial institutions and mining and chemical production companies and real estate. For example, one special Mitsui manager was Tashiro Shigeki, who headed Toyo Rayon. He knew that Japan needed to produce the new synthetic fabrics invented in the United States, such as Nylon. He traveled to DuPont and negotiated for a license to produce DuPont's Nylon. Tashiro also bought licenses to polyester fiber innovations from Britain's Imperial Chemical Industries. He began to build Tory into a company that would become the world's third-largest producer of synthetic fibers.

This example illustrates the pattern of economic reconstruction of Japan. Entrepreneurial managers reached out to the global world for advanced technology, brought it home, implemented it and improved it and became world-class manufacturing and financial institutions. Even the name of Mitsui Bussan was eventually restored. On August 5, 1958, members of a fifteen Mitsui companies met and witnessed a business agreement among the companies that restored Mitsui Bussan Kaisha (Mitsui & Company, Ltd.). Once again Mitsui was Japan's largest trading firm. For example, in 1985, the Mitsui Group comprised sixty-nine companies, with 400,000 thousand employees and transactions of forty-two trillion yen. The House of Mitsui from the 1600s had spawned a long lasting firm.

All this historical scenario had provided the changing contexts for the House of Mitsui: the Tokugawa shogunate, the Meiji restoration, the industrialization and militarization of Japan, the postwar recovery, and the growth of Japan into a mighty, modern industrial giant.

If one could go back to 1722 knowing all that was to unfold, one could have witnessed a significant celebration of the House of Mitsui on the first hundredth anniversary of Hachirobei Takatoshi's birth. Remember that Hach-irobei was the youngest son of Sokubei and Shuho and had begun the building of house of Mitsui. It was Hachirobei who was sent by his mother Shuho to Edo to assist his elder brother in the first Mitsui draper's shop in Edo. Then Hachirobei opened a second shop in Edo and went on to lay the foundation of the clan's business empire. So it was that a hundred years later, the House of Mitsui celebrated his birth by publishing his will as a code of conduct for Mitsui managers. It was called *Code of Regulations* for the House of Mitsui, and in part read:

> The (following) articles are the instructions which I leave as my will. These instructions are to be obeyed strictly and without fail. In the seventh year of Kyoho, the Year of Water and the Tiger, the eleventh month, first day.

- The members of the House shall promote the common welfare with one accord.
- Unless a merchant is diligent and attentive, his business will be taken over by others.
- Farsightedness is essential to the career of a merchant.
- All kinds of speculation . . . shall be strictly forbidden.
- Do not forget you are a merchant. You must regard dealings with the government always a sideline of your business.
- The essential role of the managers is to guard the business of the House.
- In order to select worthy managers, keep an eye on the young . . . and train promising candidates. . . .

Case Analysis

We can see in this case of the very long evolution of the House of Mitsui how a business needed to change to adapt from the old world of a feudal society to the new world of industrialized society. The clan business of the House of Mitsui evolved into the corporate firm of a holding combine under the name of Mitsui Gumi.

One can see in this dramatic case the theoretical point of strategic change itself—in the great transition from a feudal society to a modern industrialized society all the strategic changes that were needed in a business to survive over time. Mitusi began as a family (clan) oriented house of business in a time of stable political organization and peace. To prosper, it diversified into both retail of value-added goods (silk cloth) and finance. When outside political forces with superior military technology drove a whole transition of the society in the Meiji restoration, Mitsui needed to become a modern corporate organization. In becoming a holding company, one of the zaibatsu, Mitsui Gumi diversified into banking, retail, trade, and production. Always the political forms and ideology of the changing society involved the leadership of Mitsui in participation in the political and commercial life of the country. As the Japanese government aggressively expanded territorial conquests, Mitusi exploited commercial advantages to the firm of the colonial policies and military adventures of the government.

After the end of World War II and occupation by foreign forces, Japan's commercial and governmental structures of the country were rebuilt, with forced guidance by U. S. government agencies. As an occupation force, the United States can be credited with contributing to the democratization of Japanese government institutions and the stimulus of Japanese industrial redevelopment, rather than merely punishing a former enemy.

In the rebuilding of the country, some parts of the former Mitsui companies were resurrected, transformed, and recombined to a reconstituted Mitsui. The postwar model of reconstruction was again the active seeking out by Japanese firms of new technology from global sources and implementing and improving these into new businesses and selling goods abroad. This model of acquiring and innovating new technology-based products and aggressively competitive trade and marketing produced a rapid economic recovery and created in the second half of the twentieth century an industrial giant of Japan among the countries of the world.

Although the change in Japan from feudal to industrial society was dramatic, all human societies on Earth have had to transform from tribal and feudal societal forms to industrialized forms from the 1700s through the 2000s.

We also can see illustrated in this case the scenario issue of how, over time, the societal contexts of all businesses change and all business firms must adapt to new societal environments or perish. Over the long term, all modern societies change, and all industrial firms must adapt to changes. This is the importance of scenarios in strategic planning.

We can also see in this case how major worldwide trends of industrialization and colonialization swept all nations into the world scenario. From about 1765 to about 1865, the principal industrialization occurred in the European nations of England, France, and Germany. From 1865 to about 1965 (the second hundred years) other European nations began industrializing; but the principal industrialization shifted to North America. By the middle of the twentieth century in the 1940s, the U.S. industrial capacity was alone so large and innovative as to be a determining factor in the conclusion of the second world war of that century.

For the second half of the twentieth century, U.S. industrial prowess continued to grow, and European nations rebuilt their industrial capabilities that had been destroyed by the war. But for the pattern of industrializations, significant events occurred in Asia. From 1950 until the end of the twentieth century, several Asian countries began emerging as globally competitive industrial nations: Japan, Taiwan, South Korea, Singapore. Other Asian countries, including Philippines, India, China, Indonesia. were also moving toward globally competitive capabilities.

Asian industrialization actually began in Japan in 1865 with the Meiji restoration. But it was diverted principally to a military-dominated society and production. It was only after the World War II in the reindustrialization of Japan that democracy, free-markets, and world-class industrialization actually occurred. So it was that the major growth of world economies and industrialization for the second half of the twentieth century and into the twenty-first was principally focused in Asia.

SCENARIO RELEVANCY MATRIX

Planning scenarios are stories told in the present about the future, while the case of Mitsui is a story of the past told in the present. This is a big difference—the difference of a planning perspective on the future and the historical perspective on the past. Stories of the past and the future are both kinds of stories, scenario stories, but one is of a creating the times (the future) and one is of re-creating the times (the past).

So the case of Mitsui has been told as a historical story—a scenario story of the past—with characters, plot, action, and drama. But it was informed by narratives looking backward, from a historian's perspective. What did the participants of the time see of their history? It was certainly not this neat historical study, seeing the past backwards from the future. The participants saw only the present, a time of confusion and uncertainty. How then could they have told the story of Mitsui as it unfolded? We have clues to this perspective of the participants from the stories they told of their time.

For example, we recall that the samurai, Sokubei, who changed the fortunes of Mitsui by giving up caste status for the then lower caste of *chonin*, had told a kind of scenario story at the time to his family: "A great peace is at hand. The shogun rules firmly and with justice at Edo. No more shall we have to live by the sword. I have seen that great profit can be made honorably. I shall brew sake and soy sauce, and we shall prosper." Sokubei told a story of the future as one of challenge and opportunity.

As another example, we recall that his descendent, Hachirobei, left a scenario story for his Mitsui descendants in the form of a will with admonitions of how to behave in the future (e.g., the members of the House shall promote the common welfare with one accord; unless a merchant is diligent and attentive, his business will be taken over by others, etc.). These admonitions were warnings that the future always brings challenge to a merchant.

We also saw that the participants in the Meiji restoration that overthrew the Bakufu and established a new reforming government also saw the future as a challenge, one of great threats to Japanese freedom from foreign governments. We also saw that the leadership of the new firm of Mitsui during the reforming period continued to see the future as one of challenge and opportunity. For example, we recall that Minomura separated the textile branches from the money exchanges and purposely intended to make Mitsui-gumi into a great banking firm. Also Inoue pushed for even a greater reorganization of Mitsui creating policies to make the "House of Mitsui . . . a model of progressive business organization," foreshadowing the modern concept of corporate management.

So we can see from this historical case of Mitsui that the participants of the time, living in times of great societal change, saw the present and future as times of trends, challenges, and opportunities. All they could be certain of in their time was that the times were changing.

But this lesson is true of all times of change. All the participants in any historical period can see in change are the directions of change (trends), the needs to change (challenges), or the freedom to change (opportunities). Uncertainty about the future is the nature of living in a time of change.

The basic perceptual forms of seeing change in a time of change are perceptions of the future as trends, challenges, and opportunities.

Now how can one use the model of societal change and the perceptual forms of the future systematically in scenario planning? The strategic technique is to first construct a scenario relevancy matrix of the structures of society that show trends of changing and within which will be found the challenges and opportunities of the future.

We saw in the case of Mitsui, examples of the theoretical interactions between changes in the structures of government, economy, culture, and territory. In Figure 4.1A, the arrows indicate that any of these features can interact in societal change. One can use connecting arrows to visibly indicate the major kinds of changes that can occur in a society.

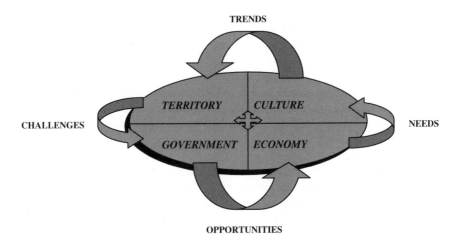

FIGURE 4.1A   SOCIETAL MODEL INTERACTIONS

For example on Figure 4.1A, we have indicated the case of Mitsui several of the important connections as curved connected arrows between areas. At the time of the Meiji restoration one can indicate with a curved arrow connecting Territory to Government how the challenges of foreign governmental control by Western nations over the territory of Japan stimulated the coup of Meiji restoration which changed the government structure. One can also indicate with a curved arrow connecting *Government* to *Economy* how the government policies of the Meiji oligarchy were aimed at industrializing the economy of Japan in order to gain military parity with the Western nations. These policies created the great economic opportunities for the businesses in Japan. With a connecting arrow between *Economy* and *Culture* one can indicate how the new economic activities of the reforming society created a need for universal literacy and new institutional structures for education, science, and health. The changing economy needs also led the Meiji government to abolish the samurai caste, changing the identity groups of the country. Finally, another curved arrow connecting *Culture* to *Territory* can indicate the enormous thematic trends that took place in the population, the wealth, environment, and ecology of the Japanese islands under the cultural impacts of the Meiji restoration.

As another illustration, consider a portion of these interactions as sketched on Figure 4.2A. Therein, new technology provides a means for production of hard goods that the military needs, such as weapons and supplies. Improved production from new technology also provides international markets for Japanese industry that brings in international money as Finance, which in turn provides the resources to acquire new Technology from the international community. It was this dynam-

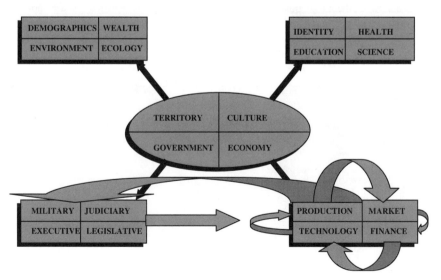

FIGURE 4.2A MODEL OF SOCIETAL STRUCTURES

ics of societal change—of the military and economy, through technology, production, international markets, and finance through acquiring new technology, which provided the formula for the successful growth of industry and military power in Japan after the Meiji restoration.

These kinds of symbolic pictures as in Figures 4.1A and 4.2A can provide a basis for a scenario descriptions of historical pasts or anticipated futures of the impacts of societal change upon business Moreover, one can see that the way to describe such societal changes indicated in such scenario pictures is in terms of stories—as the story of changes in the house of Mitsui.

> *This illustrates and emphases the earlier point that telling stories in scenario planning is a much more powerful format for strategic planning than merely using bullet lists of planning points.*

As a guide to what kinds of scenario stories should be told in a given strategic planning context, one can use the taxonomy of model of societal structures to construct a matrix of relevant interactions for scenario planning, as shown Figure 4.3. In such a matrix, one can indicate for scenario planning, the kinds of societal interactions that one anticipates to change and their impact upon economic activities of production, market, technology, and finance.

For example, in early 2000, the largest software house in the world, Microsoft, lost an antitrust case prosecuted by the U.S. federal government, and the penalty was then being debated in the federal agency:

> In the government's proposal to break up Microsoft . . . the beneficiaries are supposed to include other software companies . . . but so far, reaction . . . seems muted,

FIGURE 4.3 SCENARIO PLANNING RELEVANCY MATRIX FOR ECONOMY

		ECONOMY			
		PRODUCTION	MARKET	TECHNOLOGY	FINANCE
TERRITORY	DEMOGRAPHICS WEALTH ENVIRONMENT ECOLOGY				
GOVERNMENT	MILITARY JUDICIARY EXECUTIVE LEGISLATIVE	Anti-trust challenge	Mp3.com Copyright Case		OPEC impact on inflation
ECONOMY	PRODUCTION MARKET TECHNOLOGY FINANCE				Merger of Exchanges
CULTURE	IDENTITY HEALTH EDUCATION SCIENCE				

FIGURE 4.3 SCENARIO PLANNING RELEVANCY MATRIX FOR ECONOMY

or outright skeptical. . . . Executives at some other small companies . . . said they
did not see how a Microsoft breakup would create new possibilities for them.

—(Markoff, 2000, p. C1)

Microsoft had not anticipated it would lose its antitrust case and face a possible
breakup of the company. This is an example of an interaction that could have
been indicated in a strategic planning relevancy matrix in the box labeled *Exec-*
utive/Production.

As another example of strategic change in infrastructure that was happening
in 2000 was the merger of two stock exchanges, London and Frankfurt, across
national borders:

> In a stunning development, Deutsche Borse, the parent company of the Frankfurt
> stock market, confirmed that it would merge with the London Stock Exchange and
> that Mr. Seifert would be chief executive of the new company. It is the most ad-
> vanced step yet taken toward creating a pan-European securities market, uniting the
> Continent's two biggest exchanges into a single corporation. . . . In also announcing
> an alliance with Nasdaq, the new exchange could become the European anchor for
> a global trading system available 24 hours a day."
>
> —(Andrews, 2000, p. C1)

This merger was one example of a long-term trend of globalization of industry
and commerce that accelerated toward the end of the twentieth century. In terms,
of the scenario planning relevancy matrix, this type of event and issue can be
indicated in the box labeled *Finance/Demographics.*

Another example of major structural change that occurred 2000 was the sudden
reinjection of inflationary energy prices into the world economy:

> For weeks an accelerating inflation rate was confined to fuel and energy, but in
> March [2000] it spread to other areas of the American economy. . . . Rising energy
> prices made the biggest contribution to the rise in the overall rate. . . . What caught
> the market's attention, however, was the so-called core inflation rate, which mea-
> sures all price increases except those for energy and food, which are considered too
> volatile to provide a clear picture. This core rate rose 0.4 percent in March, double
> the monthly increase over most of the last two years.
>
> —(Uchitelle, 2000b)

This news of inflation spreading early in 2000 came as a result of the ree-
mergence of the political organization of OPEC, in a position to once again control
production of oil, since its first major success in the late 1970s. In terms of the
relevancy matrix, this kind of event can be noted for examination as to its consequences on businesses by an entry in the box labeled *Finance/Executive.*

Another example of societal change impact upon business in 2000 was the court cases over copyrights between traditional music companies and e-commerce music distribution services:

> The record industry's legal campaign for control of the Internet music business got a boost yesterday when a federal judge ruled against MP3.com Inc in a closely watched copyright infringement suit. . . . U.S. District Judge Jed Rakoff ruled that the San Diego–based Web site was infringing on the rights of musicians and record labels with a service that allows users to listen to copies of albums stored in online "locker rooms" . . . Labels and artists have been waging a fight against a handful of music sites, accusing them of infringing on copyrights by fostering the trade of unlicensed MP3s, a software format that allows computer users to send and receive songs online.
>
> —(Segal, 2000, p. E1)

This kind of example of changes in the legal rules of business can be marked for examination in the relevancy matrix in the box labeled *Market and Judiciary.*

These kinds of examples show that the use of a strategic scenario relevancy matrix is to systematically examine all kinds of impacts that societal change can make upon a business and the proceed in the scenario planning to tell stories of what kinds of change can likely occur, what trends are happening and the challenges or opportunities such changes will present to the future of the business.

SUMMARY: USING THE STRATEGY TECHNIQUE OF PLANNING SCENARIOS

The topology of societal models of Figure 4.2 can be used to systematically explore trends and anticipated changes of the business environments to construct planning scenarios.

1. Form a planning scenario team
    A planning scenario team for the company should be formed and consist of planning staffs and selected managers from both the firm level and strategic-business-units levels (company divisions or company businesses).
    • The size and complexity of the scenarios team depends upon the business diversity of the firm.
2. Divide the scenario team in societal scenario teams
    Form the scenario teams into groups (territory, culture, government, and economy) focused upon identifying and forecasting changes in these different societal sectors.

3. Use appropriate external experts to assist societal groups

> The groups should identify and use as consultants appropriate external experts to help identify trends and describe structures underlying trends.

4. Prepare appropirate forecasts

> Trends that can be partially expressed in quantitative measures should have extrapolated forecasts prepared by the groups, with narratives of underlying structural features upon which the extrapolations depend.

5. Present summary of planning scenario to company executives

> A summary presentation of planning scenarios should be made and presented to executives of both the firm and divisional units for review and adjustment.

6. Modify planning scenario by strategic modeling teams

> Strategic corporate modeling teams for the company and divisional units should next review and possibly modify the planning scenario, particularly identifying the parts of the scenario relevant to constructing a strategic company model.

7. Extract strategic issues from planning scenario

> Each strategic modeling team should extract and summarize from the planning scenario, the strategic issues particularly relevant to constructing a strategic company model:
>
> - Markets and innovation
> - Competition and economy
> - Operations and control
> - Information and knowledge

8. Use the planning scenario to construct appropriate strategic models

> Using the strategic modeling techniques of Chapter 3, the modeling teams should construct appropriate strategic company models for the businesses of the company.

For Reflection

Read a history of the industrial revolution in the nineteenth century. When and where did it begin? What were the first industries to be industrialized? How did the industrial revolution spread through Europe? How did economic competition and military competition interact that century? What were the roots of World War I?

CHAPTER 5

STRATEGY THEORY

STRATEGY PRINCIPLE

Modern strategy theory facilitates the ability to think strategically and plan effectively.

STRATEGY TECHNIQUE

Use strategy theory to create a strategic management capability as a core competence:
- Establish a strategic process
- Train personnel in strategic thinking
- Fit strategic planning into annual budget planning
- Implement strategic change

CASE STUDIES

Steve Jobs' First Exercise in Strategy: Apple Computer
Steve Jobs' Second Exercise in Strategy: PIXAR
Planning in Henkel KGAA

INTRODUCTION

We have presented modern strategy formulation and implementation processes in which information strategy plays a key role. We have also examined how to construct strategic business models and planning scenarios. But before we proceed further, we should pause for a thorough review of the traditional business literature on strategy to see if and how our depiction of the strategy processes does capture the basic and important universal lessons about strategy theory and practice.

Normally, the presentation of theory proceeds with the exposition of techniques of the theory. However, in the case of strategy theory, its development in the business literature has proceeded piecemeal with different schools of thought describing different pieces of strategy theory. The reason for this appears to have been (1) the complex nature of the whole idea of strategy as (2) the narrow disciplinary perspectives of academics—which together has resulted in disciplinary explanations of pieces of the whole of strategy.

Henry Mintzberg (one of the theorists of strategy) along with his colleague, Joseph Lampel, has nicely expressed this problem about strategy theory in business literature:

> We are the blind people and strategy formation is our elephant. Each of us, in trying to cope with the mysteries of the beast, grabs hold of some part or other. . . . Consultants have been like big game hunters embarking on their safaris for tusks and trophies, while academics have preferred photo safaris—keeping a safe distance from the animals they pretend to observe. Managers take one narrow perspective or another—the glories of planning or the wonders of learning, the demands of external competitive analyses or the imperatives of an internal "resource-based" view. Much of this writing and advising has been decidedly dysfunctional, simply because managers have no choice but to cope with the entire beast.
>
> —(Mintzberg and Lampel, 1999, p. 21)

Therefore, what we must do is to carefully survey all the pieces of strategy that has been described in the different schools of strategy theory. We need to draw from these pieces the important lessons about strategy theory and practice and then incorporate them into a modern whole theory of strategy. Just because the "elephant" of strategy theory has been difficult to describe in its whole gestalt, it is still necessary for practical managers to proceed with a complete set of strategy techniques to cope with the entire beast. And it is a compete set of strategy techniques for which we have aimed in this book.

Theory and Practice

Theory is important in any practice—management, medicine, engineering, and so on. In management practice, theory is sometimes called "management principles" and provides the critical base for ensuring that the appropriate lessons of successful practice can be transmitted from one commercial context to another. For example, David Besanko, David Dranove, and Mark Shanley nicely expressed the importance of theoretical principles in management:

> There is a keen interest among serious observers of business to understand the reason for profitability and market success. . . . However, observers of business often uncritically leap to the conclusion that the keys to success can be identified by watching and imitating the behaviors of successful firms. (And this is often called "benchmarking" or "best practices.") . . . However, uncritically using currently successful firms as a standard for action *assumes* that the successful outcomes are associated with identifiable key success factors, and by *imitating* these factors, other firms can achieve similar successful results.
>
> —(Besanko et al., 2000, p. 4)

The important idea here is the *critical analysis* of just what are the key success factors in any strategy. Uncritical imitation of a prior success by a different company may not prove successful in a new situation. In any action, no two situations nor actors are ever absolutely identical. Action is always a particular set of factors and activities, all of which together explain a particular success (or failure). Teasing out of a benchmarking case what is really general and transferable is what critical analysis is intended to accomplish.

Theoretical principles are the result of critical analysis over a range of particular and unique benchmarked cases in order to identify, abstract, and understand the generalizable *key success factors of practice.*

Now we turn to summarizing the theoretical principles of the cognitive activity of strategy that are generalizable to all cases of practice.

CASE STUDY: Steve Jobs' First Exercise in Strategy: Apple Computer

To begin understanding the theory and practice of strategy, we will look at a case of a particular industrial leader's experiences with strategy in the information industry, Steve Jobs. Since strategy is a leadership activity, one needs to look at cases of the experiences of business leaders or of governmental leaders in their successful or unsuccessful strategies. In the case of Jobs' strategy experiences, one can clearly see many of the ideas of strategy.

Steve Jobs was one of the many successful business leaders in the evolution of new information technologies in the 1980s and 1990s. He played three important commercial roles, first in the early growth of the personal computer industry by founding Apple, second in the continuing information technology progress in the movie industry through Pixar, and third in the attempt to rescue

Apple as the personal computer industry began maturing at the beginning of the twenty-first century.

Jobs displayed strategic vision very early. As a young man, he and a friend, Steve Wozniak, started a historically important company, Apple Computer. This was one of the new personal computer firms started at the end of the 1970s. In 1976, Wozniak had visited WESCON, an industrial trade show for consumer electronics, and purchased one of the first computers-on-a-chip, the MOS 6502 microprocessor.

As background, Chuck Peddle, an electrical engineer at MOS Technology, had designed the 6502. Along with another early chip, the Zilog Z80, it powered the first personal computers. Later in the 1980s Intel took over the CPU market with its series of chips beginning with the 8088. Strategically speaking, Intel was never a technology-leader but a technology-follower—but a technology-follower strategy has often been commercially successful.

Wozniak was working as a technician at Hewlett-Packard in Palo Alto, California, when he used the 6502 to build a personal computer, which he called the Apple I. His bosses at Hewlett-Packard had no interest in a potentially new market of personal computers, so in spring 1976 Wozniak showed it the Homebrew Computer Club, a local amateur computer club. It was only a partly complete computer—no keyboard, no case, no power supply, no external memory, no printer, no software. Yet two friends in the club, Steve Jobs and Paul Terrell, were impressed. Jobs formed a company with Wozniak to produce the computer, and Terrell ordered the first 50 units to sell in his Byte Shop (Ahl, 1984)). This is where Jobs got his first important strategic business vision.

The concept of strategic vision denotes the ability to imagine where a future could go from only partial and meager evidence in the present.

To grow their new business venture in 1977, Jobs and Wozniak obtained $40,000 of venture capital from A. C. Mike Markula. Markula was an electrical engineer who had become wealthy on another early start-up in the electronics industry of the 1960s and 1970s, Intel. Markula also offered management assistance (and actually dominated the Apple Board until the late 1990s).

At first, Jobs' new company Apple did well, with an open architecture policy that allowed other companies to write software or market peripheral equipment for the Apple. *This "open architecture" policy is an example of a technology strategy.* Apple gained a 27 percent share of the very new personal computer (PC) market in 1981. But in 1982, IBM entered the market and immediately gained a 27 percent share of the PC market, matching Apple's share. IBM had entered with a technically superior product using a 16-bit word-length microprocessor, allowing memory address to 640 K (compared to Apple's 8-bit microprocessor, which limited memory address to 64 K). But as the IBM PC was priced higher, Apples continued to sell. This give the company time to respond to the IBM's technical challenge.

Case Analysis

This case illustrates an important theoretical point that strategic vision results from experience. A vision and experiential excitement about personal computers as a new artifact was shared by Steve Jobs, Steve Wozniak, and other amateur computer enthusiasts of the time. As a group, these enthusiasts had experience with computation with the mainframe computer and minicomputer. Their vision was to have their own personal computers to use and play with. This required a new kind of computer of low cost, made possible by advances in applied knowledge in the form of the microprocessor chip.

With this vision, many of the amateurs became entrepreneurs and started new companies that were to form the new personal computer industry. Jobs and Wozniak started Apple, helped financially and managerially by Mike Markula. Since the personal computer was of little value to the market without software applications, Apple's strategy to open their operating architecture to software developers facilitated the growth of the software suppliers for Apple. Finally, we see that the entry into the new market of a major computer manufacturer, IBM, caught Apple strategically unprepared to meet the competitive challenge.

Any theory of strategy must include ideas about vision, and the experiential basis of vision and of the process of creating strategy from experiential-based vision.

SCHOOLS OF STRATEGY

What is strategy? This is the basic scholarly question of strategy theory. As was illustrated in this case, strategy is really an idea about the future. It is an idea for a future direction, requiring a vision of the future as opportunities and challenges. The ideas of a strategy succeed and fail in the face of competition in that future. So why is the concept of strategy a difficult concept? Why is it a term so frequently used yet so unfrequently understood? The reason is the complexity of meanings in the use of the term *strategy*.

For example, one of the important scholars on strategy, Lowell Steele, nicely summarized the two common different meanings of the term *strategy* as the ideational content of strategy, and strategic planning as the process of formulating strategy:

> *Strategy* is the array of options and priorities with which one elects to compete (offer superior value to the customer) and to survive (sustaining a level of financial

performance that will continue to attract capital and to retain the autonomy of a business) . . .

Strategic planning addresses the continued viability of strategy; it probes the need for change . . .

—(Steele, 1989, p. 181)

Ideas within a strategy, strategic issues, provide the content of a strategic plan—what factors of change should be anticipated and how addressed? Strategic planning is the way these content ideas are formulated. Strategic processes are the procedures for formulating plans within an organization.

Yet even within this distinction, the many observers of strategy have found additional complexity in the meaning of strategy, so much so as to have divided into many different schools. As we noted earlier, Henry Mintzberg and Joseph Lampel have been concerned with trying to capture the whole of strategy theory. And to do so, they have used the approach of classifying all the writers in the business literature on strategy into ten "schools of strategy" as follows:

School 1: **Design School**
> This school focused upon the formulation of strategy—as matching external conditions to internal opportunities of the organization—achieving clear, simple strategies that can be implemented by all in the organization. (This school dates from Selznick through Chandler and Andrews.)

School 2: **Planning School**
> This school emphasized strategy formulation as formal and decomposable into steps, characterized by checklists and supported by formal techniques. (This school dates from Ansoff's writings in 1965.)

School 3: **Positioning School**
> This school emphasized strategy as general positions selected from analyzes of industrial situations. (This school dates from Porter's writings in 1980.) The role of analysis in specifying the industrial situations uses techniques such as value chain analysis, game theoretical structuring, and so on.

School 4: **Entrepreneurial School**
> This school focused primarily upon the role of the chief executive in strategy and saw strategy formulation primarily depend upon the cognitive function of intuition in the executive. This school shifts the focus of strategy theory from planning to vision.

School 5: **Cognitive School**
> This school focused upon the cognitive base of strategy, adding to the analytical concepts of the planning school an emphasis on intuition. It emphasized the role of information and knowledge structures in formulating

strategy and included a constructivist view of the strategy process that sees strategies as creative constructs of what reality could become.

School 6: **Learning School**

This school viewed strategy as a kind of learning process in which formulation and implementation interact for the organization to learn from past planning and experience. (This school dates from the writings of Lindblom, Quinn, Bower, and Burgelman.)

School 7: **Power School**

This school focused on the power relationships in the situations in which strategy is formulated. They saw strategy formulation as involving processes of bargaining, persuasion, and confrontation among the actors in an organization. Also externally, an organization can use strategy as one of its tools of power to negotiate strategic partnerships.

School 8: **Cultural School**

This school emphasized the role of culture (as opposed to power) in the formulation and implementation of strategy. (This school dates from writings by Rhenman and Normann and from Hedberg and Jonsson.)

School 9: **Environmental School**

This school focused upon the environments of organizations, seeing organizations as principally reacting to and responding to threats and opportunities in their environment It includes approaches such as contingency theory that classifies responses expected of organizations facing particular environmental conditions.

School 10: **Configuration School**

This school focused upon the nature of organizational structure as influential upon strategy. For example, it saw formal planning as prevailing in organizations with machine-type structure in conditions of relative stability, and it sees entrepreneurship as prevailing in organization in situations of start-up or turnaround. It emphasized that the conditions of stasis or transformation impacted the forms of strategy processes within the organization.

Scanning these brief summaries, we can see that all these schools do still group within Steele's distinction between emphasis upon the common component ideas in any strategy and in the process of strategy formulation:

1. *Component Ideas in Any Strategy*
   • Herein lie the schools which emphasized the ideas and cognitive activities in formulating strategy, particularly the Design, Planning, Positioning, Entrepreneurial and Cognitive schools.

2. *Strategic Processes of Strategy Formulation*
   - Herein lie the schools which emphasized the processes for strategy formulation, particularly the Learning, Power, Cultural, Environmental, and Configuration schools.

From this scan of the literature by Mintzberg and Lampel, we can conclude that a complete strategy theory does need to cover both the component ideas in a strategy and the process of strategy formulation.

Furthermore in addition to the distinction between strategy content and process, Steele also emphasized the importance to see the implementation of a plan as another part of strategy practice:

> "*Strategic management* is the implementation of modifications in the fundamentals of how one competes and survives . . . (controlling) actions and behavior required to implement change."
>
> —(Steele, 1989, p. 181)

Accordingly to characterize the whole of strategy theory, we have used the terms of the strategic plan, strategic process, and strategic implementation as the three aspects of strategy. The strategy process of the first and second chapters do cover these aspects:

1. Strategy component ideas. The ideas of any strategic plan are built upon *strategic issues explored in the planning scenario* (about changes in economy, territory, government, and culture) and *upon issues explored in the strategic business model* (about changes in markets, competition, operations, information).
2. Strategy process. The strategy process emphasizes *vertical interactions of top-down and bottom-up perspectives* and *horizontal interactions of forming a strategic vision and then a strategic plan.*
3. Strategy implementation. After a strategic plan is formulated from a strategic vision, then it needs to be *implemented as operational plans of strategic pathways.*

A complete theory of strategy needs to emphasize strategy content issues, strategy formulation process, and strategy implementation process.

CASE STUDY: Steve Jobs' First Exercise in Strategy: Apple Computer, Continued

Let us look now at the component ideas of strategy, strategic issues. To do so, we will continue the case of strategy in the Apple Computer company to examine how Jobs' strategic ideas saved Apple in its first competition.

In the face of the new competitive challenge of the IBM PC, Apple needed to introduce a next-generation personal computer, based on a 16-bit word-length central processor, as IBM had used. Wozniak took up the task of upgrading the Apple II to keep it marketable, while Scott began the development of a business product model, the Apple III with a 16-bit microprocessor to compete with IBM's PC. But Scott's product development leadership was poor. The Apple III was slow to market, full of bugs, and without any superiority to the IBM PC.

Next, Jobs asserted strategic leadership, knowing that Apple needed new technology. But since Apple did not do research, Jobs had to look outside. Jobs heard of outstanding research being done in computer science by Xerox at its Palo Alto Research Center (PARC). Jobs visited it and saw a technological vision of the future PC: PARC's Altos distributed computer system. PARC's Altos research project had developed the world's next-generation PC system, with a graphical user interface, mouse, local area network, computer connection to a laser printer, object-oriented operating system—nearly everything a PC was to become in the 1990s was invented back in the late 1970s.

Lawrence Tesler was a researcher at PARC who helped develop the world's first object-oriented programming language, Smalltalk, which was used to program Altos's operating system. He later commented on the Apple team's visit to PARC; "Their [Apple's personnel, including Jobs] eyes bugged out." (Uttal, 1983, p. 98)

To move Xerox's research strategy into a product strategy, seven months later Jobs hired Tesler and began the product development effort that was to result in Apple's Macintosh:

> The Apple III office computer was a bug-infested flop. And Lisa, precursor to the Mac, was an expensive dud. Job's masterpiece, however, the 1984 Mac, was a stunner. . . . It was also severely underpowered and limited in expandability. The market balked, and in May 1985, Jobs was pushed out of daily operations."
> —(Rebello et al., 1996, p. 36)

Jobs' visionary ability continued as brilliant information technology strategy—he saw the PC future from the experimental model of Xerox's Altos system. Back at Apple, this vision was implemented by Jobs' in the Macintosh computer. At first, the Macintosh had slow sales, but a new software vendor introduced a new functionality to use the Mac's features and launched desktop publishing. This was a business vision Jobs had not foreseen, but he got lucky. This application saved the Mac by bringing in corporate customers. In the information industry, a commercially successful new business application is often called a "killer application."

Killer applications are new, commercially successful uses of information technology.

In the information technology of the 1980s, computer applications in graphics were the cutting edge of strategic vision (complementing the earlier progress in computer applications in computation and text). With its desktop publishing application, Mac was finally succeeding. In 1987, Apple was saved and had a market niche in corporate computing. It was not a market Steve Jobs had envisioned. Apple was using what is called a "technology-leader" business strategy, and Jobs' strategic strength was in this kind of strategy. However, any business operated in a technology-leader strategy must stay ahead of its competition or its competitive strength will erode. And this is what eventually happened to Apple.

As Apple's Macintosh computer was pioneering the market niche of desktop publishing, most of the rest of the personal computer industry was following an IBM-PC brand strategy. In this brand strategy, firms were using a technology-follower strategy, but it was working because of the major cost of software in information technology. It was turning out that continuity and availability of software, as opposed to the hardware part of the personal computer system, was really behind customer demand. And in the brand-name technology-follower competitive strategy, two companies were reaping the lion's share of IBM's brand-name recognition: Microsoft and Intel.

Here is an important lesson about business strategy and information-technology strategy. In 1987, the IBM-PC world was technologically far behind the Mac. Yet because of the marketing clout of IBM and its clones, it continued to dominate the PC market. Bill Gates, CEO of Microsoft, was successfully executing a brilliant business strategy. He was using IBM's market reputation to build the MS-DOS world, with Gate's operating system dominating, even with a technology-follower strategy. It would not be until 1995 when Microsoft released its Windows 95 operating system that MS-DOS information technology would finally begin catching up with the Xerox PARC/Apple Macintosh information technology.

Yet when Microsoft caught up, Apple was failing, declining in market share, and losing money. How did Apple come to lose its competitive edge as a technology leader? Why did it simply throw away a ten-year lead in information technology? The answer was leadership vision,or, in this case, failure of leadership vision. In contrast to Jobs' early visionary leadership in Apple, all the subsequent Apple CEOs failed in vision about information technology strategy.

In 1987, after the Macintosh was beginning to succeed in the desktop application market niche, Jobs became chairman of the board of Apple and looked for a new president to succeed him. He picked John Sculley, who had many years of marketing experience at Pepsi Co. Yet within a year, there were policy clashes in Apple's team. Marcula backed Sculley, and Jobs resigned from Apple.

Scully's leadership of Apple provided the company with both a poor information strategy and a poor business strategy:

> But Apple [under Scully] entered the 1990s with an overpriced product line and a bloated, over-perked executive staff. Microsoft Windows was gaining ground and Apple's rate of innovation was slowing. . . . Determined to catch the next technology wave, Sculley put himself in charge of research and development— and came up with the Newton personal digital assistant, a marketing and technical fiasco.
>
> —(Rebello et al., 1996, p. 36)

We note that Scully's product idea did succeed later in another product called the Palm Pilot. In June 1993, the Apple's board fired Sculley. Marcula next chose Michael Spindler, who headed the successful European division of Apple to become CEO: "Michael Spindler started off with a 2,500-employee layoff, the first move toward a new, low-margin business model." (Rebello et al., 1996, p. 37). He produced inexpensive Macs for the home market and introduced a new higher-power Mac line with a new chip, the PowerPC chip, developed jointly with IBM. But in 1995, Apple stumbled dramatically when Spindler's large inventory of lower-priced and lower-powered Mac's were ignored for their PowerPC line, and not enough were produced for the Christmas season sales: "The [1995] Christmas quarter was a disaster. . . . January 1996 brought news of a last quarter loss of $69 million. Laying off 1,300 workers is just the first step in an overhaul that could include Spindler's ouster and/or even a sale of Apple." (Rebello et al., 1996, p. 37)

Markula called an emergency meeting for January 31, 1996 at the St. Regis Hotel in New York City. Spindler was surprised by the request for his resignation and argued for more time:

> The board was firm: Spindler had contributed much over his 16 years at Apple, but directors had been surprised by plunging gross margins, throwing into question management's credibility.
>
> —(Armstrong and Elstrom, 1996, p. 29)

The Apple Board next selected Gil Amelio from National Semiconductor to replace Spindler. Amelio had transformed National Semiconductor from its worst loss in 1991 of $151 million to a best year profit in 1995 of $262 million. Amelio finally began shopping for a new operating software system:

> Apple was known to be casting about for a new operating program, the software that serves as the computer's master-control panel. Apple's in-house develop-

ment effort, code-named Copland, had collapsed. For Apple, shopping for an operating system was a humiliation akin to General Motor's having to buy engines from another company."

—(Lohr, 1997, p. 16)

Earlier when Steve Jobs had been ousted from Apple, Jobs had set up a company called, Next to develop a next-generation PC. The hardware Next produced had not been successful, but the operating system was advanced. In fact, it was an example of an object-oriented programmed operating system that Jobs had seen back in 1980 at Xerox's PARC but had failed to innovate in the Macintosh. Unfortunately, Next was not commercially successful. Jobs was still showing excellent vision in information technology strategy but not in business strategy. Next developed good software but sold few hardware platforms. Jobs then scaled back Next's strategy to focus on building Internet sites.

When Jobs heard that Apple was shopping for a new operating system, he met with executives of Apple on December 2, 1996 in Cupertino, California, and explained that adopting Next's operating system would be Apple's best choice of a new system:

On Dec. 20, 1996, Apple's C.E.O. and chairman, Gilbert F. Amelio, announced that the company would buy Next Software Inc. for $400 million. For that price, Apple [gets an advanced operation system software and] also gets Steven P. Jobs . . . So Jobs becomes the computer era's prodigal son: his return to Apple after more than a decade in exile is an extraordinary act of corporate reconciliation, a move laden with triumph, vindication, and opportunity."

—(Lohr, 1997, p. 15)

Apple's problems continued. Under Sculley and his successors, Apple had failed to establish a visionary corporate research laboratory (as Xerox had done a decade earlier in establishing PARC). Although Apple's strategy was as a technology leader, its successive leaders did not establish a scientific technology capability in the firm to continue that strategy. Apple coasted too long on borrowed innovative technology, which Jobs had found in 1980 at Xerox PARC. By the mid-1990s, Apple was still not a technology creator, only a technology borrower, and one cannot continue for long on a technology-leader strategy without becoming a technology creator.

On July 10, 1997 after only 18 months on the job as Apple's CEO, Gilbert Amelio resigned after a "confrontation with the company's Board of Directors over the company's faltering performance" (Corcoran, 1997, p. E1)

Apple's board of directors asked Steve Jobs back to return to the board, which he had left eleven years earlier after losing control of Apple to John Sculley. (Later in an interview, Jobs was asked to comment on Sculley. Jobs responded, "What can I say? I hired the wrong guy.") Back in control at Apple, Jobs began developing a new product model for Apple computers. He cut Apple's product lines down to four: a laptop and desktop for consumers and a laptop and desktop for professional users. He ordered Apple's design team to redesign the case to look exciting. He also replaced about three quarters of Apple's management team.

The new iMac, eventually added a digital video publishing application and included video editing software. Still after the company turned profitable in 1999 and Jobs became permanent CEO again of Apple, the market share of Apple in the PC market remained small: "Will it be enough? Apple's 12 percent home-computer market share is a big improvement over 6 percent, but it still leaves the Max on the margins. . . . (Krantz, 1999, p. 68)

So it happened that the long struggle of the late 1980s and early 1990s between Microsoft and Apple for dominance of the PC market's operating systems had fallen to the different visions of CEOs of the two companies. In 1997, even Jobs admitted "the era of setting [(PC operating systems] up as a competition between Apple and Microsoft is over as far as I'm concerned." (Chandrasekaran and Shannon, 1997, p. A1)

Case Analysis

Microsoft had won the competition. Apple, after all, had given Microsoft more than ten years to catch up on the brilliant Xerox research that Jobs had innovated in the Macintosh computer.

Give a rival a long enough time to catch up with your information technology strategy, and they will.

Information technology strategy, however initially powerful, is always only a temporary advantage. It is business strategy that succeeds in the long run.

What counts in the long run for competitive strength is the integration of business strategy and information strategy.

Jobs had envisioned the correct information strategy in personal computers and implemented it in the Apple and Macintosh PCs. But Jobs never did get Apple's business strategy correct. Moreover, the CEOs that followed Jobs never got the information strategy *or* the business strategy correct. They allowed Microsoft to catch up in information technology and never built a significant market share for Apple. Jobs counterpart at Microsoft, Bill Gates, finally got the information technology correct, but the big difference from the start was that Gates kept getting the business strategy correct. In the 1990s, Microsoft had become one of the giants in the world's industrial firms, while Apple struggled for its survival with smaller shares of the personal computer market.

This case nicely illustrates the theoretical point that the content of strategy, its strategic ideas, are indeed very critical to long-term competition and corporate survival. When Steve Jobs cofounded Apple as a new firm producing personal computers in a new industry for a new market, his strategy was delineated as a set of open-ended boundaries:

1. Personal computers were technically possible.
2. Personal computers were useful.
3. Customers would buy personal computers.
4. An industry infrastructure would develop to support the product.

But this is usually true in a business strategy that uses a technology-leader strategy. In innovating new markets through new technology, people are seldom able to fully envision the eventual form of the whole market. One sees in any historical case study about innovation that strategic issues of the time were always open-ended. In the 1980s at the beginning of the personal computer industry, no one how these assumptions would play out. This is why strategy is a direction and not an end point

Personal computers were technically possible but severely limited in memory. It wasn't until IBM's PC that memory capacity began to allow sophisticated programs. Personal computers were useful for business but not until spreadsheets and integrated database/spreadsheet programs became available. Customers would buy personal computers first for hobbies, next for games, and then for business uses. An industry infrastructure would develop in which Microsoft and Intel would become giants of American industry.

So strategy never needs to know all the answers before time, but it does need to anticipate what is possible and likely to develop. Jobs' strategy projected experience with computers moving down in price: mainframes ($1 million) to minicomputers ($100,000) to personal computers ($2,000).

What was solid in this strategy (and not merely silly or wishful thinking) were the two contemporary experiences, first, that microprocessor chips were available and, second, that a large segment of the population already had technical experience with larger and more expensive computers. These two facts—available technology and potential customer base—were the facts upon which then Jobs' vision was projectible. And the visionary projection was correct. When Apple went public in the early 1980s, both Jobs and Wozniak became millionaires, and Marculla substantially increased his millionaire status.

The next step in Jobs' strategy occurred after Apple's technology failed to meet IBM's competitive challenge. (When the IBM PC entered the market in 1982, it took 37 percent of the market, the same percentage Apple had built—and after that Apple's share declined steadily.) Jobs took action after Apple failed to meet the competitive challenge of giant IBM by developing a technologically look-alike product. Job understood even then that a small competitor cannot beat a market giant with look-alike products, and as a visionary leader Jobs went out to look for new much-higher-quality product through new information technology. He saw his new strategy at Xerox PARC and implemented their technology strategy in the Apple Macintosh. Jobs' vision and leadership saved Apple from IBM and its clones.

Then Jobs made his first business mistake. He thought he needed a manager with marketing experience to be president of Apple. He hired Scully, who had been a very successful marketing executive at Pepsi Cola. But Scully's experiential base in marketing had been in a mature-technology industry, and his market vision was not projectible into the different market conditions of a rapidly progressing high-technology industry. Accordingly, Scully missed the correct long-term business strategy, which should have been to never let IBM and Microsoft catch up with Apple's technological lead. But Scully neglected this—as did the succeeding CEOs—until Apple was on the financial ropes by the mid-1990s. In 1995, Microsoft's Windows 95 caught up with the Mac. By then Apple's share of the PC market was down to 8 percent and dropping. (A market-share less than 20 percent of a market is never a happy strategic place to be in for long-term survival.)

The quality of strategy is dependent upon the correctness of the projectibility of management's vision based upon prior experience.

STRATEGY CONTENT SCHOOLS

Now we will review the key ideas in strategy, that have been emphasized by the schools that emphasized the content in any strategy:

design, planning, Positioning, Entrepreneurial, and cognitive. These had all focused on the concept of strategy in terms of its ideational components—the key ideas within a strategic conception.

One important contributor of these schools was H. Igor Ansoff, who wrote that strategy is "the concept of the firm's business" (Ansoff, 1988, p. 10). Another important theoretician, James Brian Quinn, wrote of strategy as "the pattern or plan that integrates an organization's major goals, policies, and action sequences into a cohesive whole" (Quinn, 1988, p. 3). In fact, Henry Mintzberg himself had divided strategy into five parts: plan, ploy, pattern, position, and perspective (Mintzberg, 1988, p. 13). By this Mintzberg meant that strategies contain plans, ploys (maneuvers intended to outwit a competitor), positions as tactics, and patterns as consistency of behavior.

In these ideas of strategy, we can see that they include:

1. An idea about the totality of the firm ("concept of . . . the business", "cohesive whole")
2. An idea about the vision of the firm ("pattern . . . sequences", "pattern, position, perspective")

If we look at other writers of these schools, which emphasized the component ideas in any strategy, we see also an emphasis on the idea of vision as important in strategy. For example, C.K. Prahalad and Gary Hamel, emphasized that effective management needs to have a strategic vision for the total organization, and they noted a historical change as to what has been regarded as proper leadership vision:

> During the 1980s, the top executives were judged on their ability to restructure, declutter, and delayer their corporations. In the 1990s, they'll be judged on their ability to identify, cultivate, and exploit the core competencies that make growth possible—indeed, they'll have to rethink the concept of the corporation itself.
> —(Prahalad and Hamil, 1990)

Prahalad and Hamel saw that strategic vision should establish a capability in a company of a core competency and argued that a vision of the core competencies of an organization provides a cross-cutting vision for the whole firm.

Another important writer on the content of strategy was Michael Porter who urged viewing the business as the idea of a "value-adding transformation" and of viewing industrial structure as a "value-adding chain." The importance of these ideas was to help a company become competitive by appropriate positioning with transformations and within the industrial chain:

> Competition is at the core of the success or failure of firms. . . . Competitive strategy is the search for a favorable competitive position in an industry . . ."
> —(Porter, 1985, p. 1)

Thus two strategic ideas that Porter is emphasizing is an idea of the totality of the business (value-adding transformation) and the totality of its competitive context (value-adding industrial chain). Moreover, implicit in Porter's idea of a search for a favorable position is that this is an idea about strategic control—control within the competitive environment of business.

Porter asserted that two central strategic issues underlie this search for competitive control. First there is the condition of the common profitability of the industrial sector in which the business lies. Second there are the factors for successful competition in the industry. The first issue is not controllable by any particular business in the industry (except that it be a monopoly), while the second issue is controllable.

In the second issue, control through competitive factors, Porter identified three kinds of generic strategies: cost leadership, product differentiation, and market focus. Thus Porter suggested that the purpose of strategy In business is the strategy component of control, focusing upon the generic strategies (cost leadership, product differentiation, market niche) that can be controlled to "search for a favorable competitive position in an industry".

Thus we see that Porter's writing on strategy has emphasized the strategic ideas of totality and control.

A later strategic writer, Lowell Steele was strongly influenced by Porter, also using his idea of the totality of a business context in an industrial value-adding chain. However, Steele added to Porter's views by emphasizing the need to strategically distinguish between the ideas of control and change:

> The fundamental question that must be addressed in strategic management is whether and how the enterprise must be changed in order to survive and achieve its potential."
>
> —(Steele, 1989, p. 263).

In summary, the strategy writers who focused upon the content ideas of strategy have listed several important types of strategic ideas:

Totality
Vision
Environment
Control
Change

We will see that these types of strategic ideas appear in the strategic content of any case of a successful organizational strategy, business or governmental. Together they provide a systematic way to analyze any strategy as to whether the strategy is complete or has missing components.

Logical completeness is a necessary (but insufficient) conceptual test of a good strategy.

CASE STUDY: Steve Jobs' Second Exercise in Strategy: Pixar

These content ideas provide a way to ask the basic kinds of strategic questions that one should ask when thinking strategically about a company. To see how these kinds of ideas do provide a systematically way to think basically about a company's future, we will continue with the case history of Steve Jobs's experiences in business strategy, but this time focus on his experience with another new company, Pixar. The scene of this case shifts from the personal computer industry to the entertainment industry. But the rapidly changing information technology of the time ties the two scenes together.

Early in the twentieth century, the movie industry was begun when cinema cameras were invented (a kind of information technology). Hand-drawn animation was innovated, and Walt Disney pioneered in the early cartoon films. Next, sound was added in the early 1930s. In the 1970s, video was added to cinematic information technologies. In the 1980s, computer-aided graphics were added to the cinematic tools.

In the late 1970s, the film director George Lucas made a commercial success in his science fiction film for youth, *Star Wars*. He used extensive animation in the film. Meanwhile, computer-aided graphics information technologies were being developed by government agencies in the military and space agencies, using minicomputer platforms. By the mid-1980s, researchers in information technology advances had begun applying parallel processing computers (small computers that used not one central processing unit but several within the same computer) to greatly speed up picture-processing. Lucas then began to develop and use this new kind of computer in his studio, Industrial Light and Magic. It developed the first parallel processing computer devoted to graphics and animation, which was called Pixar. Lucas' company of the same name, focused on contract services for producing animation sequences using and developing the new information technology of computerized animation. However, Pixar lost money, and Lucas sold it to Steve Jobs.

We recall that after the bitter struggle for power at Apple in 1986 between Jobs and his chosen successor, Scully, Jobs resigned from Apple and sold his remaining Apple stock. Jobs looked for new business opportunities, and he still depended upon his vision based primarily on advanced information-technology strategy. Jobs started a new personal computer business, Next, to compete directly with Apple, and he bought Pixar from Lucas for $10 million. But Pixar continued to lose money. Over the next five years, Jobs invested an additional $50 million in Pixar (which at the time was about 25 percent of his total wealth).

But the information technology for animation was still rapidly progressing at Pixar. Jobs had purchased a good information technology team in Pixar. For example, a key engineer, Catmull, loved animated films as a kid but had little drawing talent. He did have technical skills, and he studied computers. In 1975, he was employed at a vocational school, New York Institute of Technology in Old Westbury, New York. There he teamed with some artists to try to develop computer-assisted animation, but the computers drew very slowly then. In the 1980s, this team left New York and went to work for Lucas at Industrial Light and Magic in San Rafael, California. When Jobs bought Pixar, he acquired Catmull and his team's expertise.

The business break for Pixar finally came in 1991, when Disney gave Pixar a three-film contract. This was Pixar's first venture from contract animation support to producing full-length animated films. The first film it was to produce was called *Toy Story*. As it neared completion in early 1995, Jobs' strategy for Pixar became clear to him; and he decided to take a bold strategic step. It was clear to him by then that Pixar couldn't prosper by selling its information technology to others, so Jobs decided to alter the business vision of Pixar. Rather than remain an animation contractor, he decided Pixar should become a major movie studio. For this new business vision, Jobs next raised additional capital for Pixar from investors such as Seagram's CEO Edgar M. Bronfman, investment banker Herbert A. Allen, Disney CFO Robert Moore, and movie agent Michael Ovitz.

But the key step Jobs took was to remake the strategic alliance with Disney. *Toy Story*, which Pixar was producing and Disney was to distribute, would give Pixar only a small percentage of the profits. The film became a big children's hit in 1995. To carry out Jobs' new strategy, he then took Pixar public at the height of *Toy Story*'s success. The initial price of $22 quickly went to $33. It was reported that Jobs called one of his best friends, Lawrence J. Ellison, CEO of the high-tech database company Oracle, to tell him he had company in the billionaires' club (Burrows and Grover, 1998).

In 1998 with *Toy Story*'s success, Jobs was able to cut a better deal with Disney. For the next five years, Pixar would get an equal share of the profits with Disney (after a 12.5 percent distribution fee), and Pixar had the assistance of Disney's powerful marketing and distribution capabilities. Profits in the children's movie industry comes not only from film sales but also greatly from merchandising deals. In 1999, Pixar had made about $53.8 million from *Toy Story* but was anticipating more than $200 million in merchandising royalties, video sales, and box-office receipts from a successful new movie called *A Bug's Life*. For that film, many companies had cut merchandising deals with Disney (e.g., Mattel Inc. and McDonald's Corp).

To produce full-length feature children's films such as *Toy Story* and *A Bug's Life*, Pixar developed the production capabilities of a full studio but focused upon the information technology of computerized animation. First a story line and script were developed and approved by Pixar and by Disney. Pixar's landscape artists painted lush backgrounds for the scenes in the film, and character sculptors created 3-D computer models of characters for the film, and animators drew the characters. For example, a cartoon character, Flik, from *A Bug's* Life was first sketched by the animater, then computerized as a wireframe model, upon which the computer can next develop a sold-surface representation (using computerized polygon mathematical equations), and finally rendered as a character with texture and appropriate lighting in each scene. Within the computerized graphic information technology, the characters are placed in landscapes and animated through the motions required by the script. Finally, human voice-overs are added.

The information technology of computer-animation provides much cheaper and faster film production with fewer animators and flexibility to alter and hone the story. Such films cost at least one-third less than traditional animated films, using one-third the staff. Moreover, since everything is stored digitally, it is easy to alter. For example, *A Bug's Life* was completely changed after more than a year's work on it was expended. The story was originally about a troupe of circus bugs that tries to rescue a colony of ants from grasshoppers:

> But because of a flaw in the story—why would the circus bugs risk their lives to save strange ants?—codirector Andrew Stanton recast the story to be about Flik, the heroic ant who recruits Flea's troupe to fight the grasshoppers. "You have to rework and rework it," says [the film's director John] Lasseter. Indeed, one scene was rewritten thirty times."
> —(Burrows and Grover, 1998, p. 146)

One of Pixar's business strengths (and its potential value to the giant Disney) was the ability to continue to develop information technology. For example, in one scene in *A Bug's Life*, the director Lasseter was dissatisfied about the fact that the crowds of ants in the movie's scenes all had look-alike faces. Pixar' engineer, William Reeves, developed new software that randomly applied different physical and emotional characteristics to each ant in a scene. As another example, the writers of the script brought a model that had been created of one of the butterflies (called Gypsy) to show the Pixar's researchers and asked them to write software to make the animated butterfly show hairs pressing down and popping back up when the butterfly rubs her antennas. They did just that.

It was the original vision of Pixar, which Lucas founded and Jobs developed, that the technology strategy and the production strategy would continue to make possible cheaper animation with higher quality—attention to the visual details of animated films. And this integrated technology and business strategy is what has made Pixar a valuable studio that was ahead of other competing animation studios in the 1990s:

> (Pixar) has turned out ever more lifelike short films, including 1998's Oscar-winning Geri's Game, which used a technology called subdivision surfaces. This makes realistic simulation of human skin and clothing possible. "They're absolute geniuses," gushes Jules Roman, co-founder and CEO of rival Tippett Studio. "They're the people who created computer animation really."
> —(Burrows and Grover, 1998, p. 146)

Part of the business strategy of Pixar was to continue to develop its creative and talented staff. Each new employee spent ten weeks in training at "Pixar

University," which included courses in live improvisation, drawing, and cinematography.

But movie production is art and storytelling. The person in charge of storytelling is John Lasseter. He was born in Whittier, California, and loved cartoons since childhood. He decided to become an animator as a freshman in high school after reading a book on the making of Disney's famous children's film *Snow White*. After graduating from the California Institute of the Arts, Lasseter was first employed by Disney. In 1984, he joined Pixar. Lasseter was credited with helping Pixar make the transition from short-subject films to full-length films. In the partnership between Pixar and Disney, Disney insisted that Pixar sign Lasseter to a seven-year contract and paid half his salary.

In 1999, Jobs' vision for Pixar was that it should grow into a major movie studio, rivalling Disney:

> "I think Pixar has the opportunity to be the next Disney—not replace Disney— but be the next Disney," [Jobs] says. . . . So how will Jobs achieve his dream? Not surprisingly, he's tapping into his Silicon Valley roots and using computers to forge a unique style of movie making.
> —(Burrows and Grover, 1998, p. 142)

Yet competitors always acquire new technology for their own survival. For example, Disney in 1999 was improving its own technical staff and information technology capability. About one hundred employees were working what would be Disney's first completely computer-animated film, *Dinosaurs*. Thus Pixar would find itself in direct competition with its strategic partner, Disney. This is the nature of business.

One can see in this part of the case that Jobs did grow in ability for leadership vision beyond information technology strategy to business strategy. In Pixar, he effectively integrated information and business strategy (something that was never well done at Apple). This integration required him to reformulate Pixar's business mission and to create a strategic business alliance with Disney to implement it. When Jobs had purchased Pixar he still was operating with a strategy predominantly on information technology. Pixar was at the cutting edge of computational graphics (an advanced area then of information technology,) but not making money. Lucas had been losing money at Pixar. After Jobs paid $10 million for it, Pixar continued to lose money, requiring Jobs to invest another $50 million. It was this experience of continuing to lose money (and yet having an exciting information-technology strategy) that stimulated Jobs to come up with a new business vision for Pixar to become a major movie studio.

So it happened that as the twenty-first century began, Jobs was CEO of two information technology firms, Pixar and Apple:

It's 3:00 pm in Richmond, California, and Steve Jobs is micro-managing. He's sitting around a conference table at his Pixar Animation Studios with a gaggle of Pixar producers and Disney marketing types, poring over the color-coded, small-print, stunningly elaborate "synergy timeline" for the upcoming *Toy Story 2*, which Pixar made and Disney will distribute. Ah, the endless promotional arcana of a $100 million aspiring blockbuster: trailers, press junkets,. . . . At 44, the Pixar chairman and Apple Computer interim-CEO-for-life finds himself a leading force in not one but two iconic late-90s American Industries. . . . Jobs, after years spent pacing the sidelines, was suddenly at the top of both his games."
—(Krantz, 1999, p. 64)

Case Analysis

This case illustrates the theoretical point that the components of a strategy provide the perspective of the basics of a business, the perspective upon which a business model rests and is constructed.

Jobs' strategy for Pixar was formulated concretely through a business strategy that specified that:

1. The strategic totality of the business enterprize of Pixar should expand from a contract animation house to a full movie studio.
2. The strategic vision of Pixar's new products of full movies should focus on the animated children's film market.
3. The strategic environment of Pixar focused on the market dominance of Disney.
4. Strategic changes in the practices of Pixar would require the scripting capability for a full movie, budget control, and distribution capability.
5. Strategic control over production and distribution of full-length feature films and merchandising would be accomplished in a business partnership with Disney.
6. Information strategy required Pixar to continue as an information technology leader for a competitive edge.

We can see in this case that these ideas that in Jobs business strategy for Pixar (of totality, vision, environment, change, control, information) are examples of the six key components of any complete strategy.

The principles of totality and environment are ideas about the strategic domain over which strategic thinking is focused. The principles of vision and change are principles of strategic direction upon which strategic thinking is focused. The principles of information and control are principles of strategic decision capability upon which strategic thinking is focused.

STRATEGIC IDEAS AS THE BASICS OF A BUSINESS

The ideational components of a strategy can provide a systematic way to pose the strategic questions about basics of a business, thereby establishing the basis for the strategic business model. These ideas together can provide an exploration of the complete set of business assumptions upon which a business model rests and is constructed. To accomplish effective change, strategic management requires focusing on and reexamining the basics of the organization. Asking strategic questions about the key components of strategy can provide a systematic list of the kinds of questions to ask about business basics, such as:

1. What is the totality of the enterprise?

   What businesses are and should be those of the firm?
2. What is the vision of a business?

   Who are its customers? How should a business profitably add value in its products/services to its customers?
3. What is the competitive environment of the business?

   How should the business compete against competitors for the customers sales?
4. What changes will or should occur, which can affect the current businesses of the enterprise?

   What changes need to be made to create a desirable future for the enterprise and ensure the firm's survival?
5. What about the business will continue to be properly controlled, and what control will likely fail in the future?

   What kind of controls are needed to operate a business successfully in the future?

And as we see clearly in the case of Pixar, to this last category of control, we now need to specifically look at the kind of control progress in information technology can provide. So to the list of component ideas of strategy, we now need to add a six component idea in strategy, information:

6. What should the information strategies be within the enterprise?

   How can new information technologies improve competitive control in the future?

All strategic thinking is based on a basic vision, a concept of the enterprise, but that vision may not always be clear or well articulated. It is the completeness of considering all the basic strategic issues that can assist in the construction of an effective and successful business model.

ANALYSIS AND SYNTHESIS IN THE PLANNING PROCESS

Now let us turn to the cognitive school of strategy, which emphasizes the important role of intuition in formulating strategy. This is an important idea, needed to explicate the key concept of vision in the strategy process.

As we saw illustrated in the case of Jobs' business ventures, strategy does not create vision but is derived from vision. Experience is the basis upon which strategic vision is projected. So to speak, vision is a kind of "intuition thing." This basic fact about vision as an intuitive outcome of the mind has also contributed to complexity in strategy theory. This is because to occidental writers on strategy, it has appeared easier to describe analytical techniques than intuitive techniques. Consequently, the complexity of the theory of strategy is partly because of the elusiveness of the idea of intuition as a formalized organizational process—intuition in strategic planning. Thus although vision is a key ideational component of creating strategy, more written has been on planning than on vision in the business literature.

For example, Henry Mintzberg emphasized that the formulation of strategic vision is intrinsically an intuitive activity (Mintzberg, 1990). For this reason, Mintzberg categorized the kinds of strategic theorists who have primarily emphasized the role of analysis in the strategic process as the design school of strategy:

> The Design School represents an extreme end of a spectrum, where strategy is derived from deliberate thought. Grass-root emergent strategy is intuitive. It is just as extreme, but it is at the other end of the spectrum. To understand strategy, you have to accept both ends of the spectrum."
>
> —(Campbell, 1991, p. 109)

Accordingly, Mintzberg emphasized that both analysis and intuition play critical roles in strategy:

> I am not saying that analysis is bad or unnecessary. There is a danger, though, that you can preclude synthesis with too much analysis. The reason that strategic planning failed as a technique was because it was based on the assumption that you could get synthesis from analysis."
>
> —(Campbell, 1991, p. 109)

Analysis and synthesis are complementary cognitive functions in human thought. Analysis is the cognitive processes of taking into parts a complex concept as simpler component ideas. Synthesis is the complementary cognitive process of putting together different ideas into a unifying idea, of which the assembled sub-ideas now become components of the unifying idea. Analysis divides, and synthesis unites.

Analysis and intuition are complementary tools of cognition—of thinking and of recognition.

The ideational components of strategy that are its strategic totality and strategic vision require a management intuition that both synthesizes what is the boundary (totality) of the firm and the purpose (vision) of the firm.

In some cultures, it is easy for organizational planning processes to emphasize analysis, while in other cultures it is easy to emphasize synthesis. Accordingly, planning processes must take into account the cultural setting of the organizations.

After synthetic formulation of a planning strategy is presented in analytical form as a plan, a strategic plan. The topics of the presentation express the analytical components of a strategy. Accordingly, the formal logic in the *analysis* of a strategic plan is called the planning logic. As we shall see in Chapter 6, the deductive logical format of a plan will include such topics as

- Mission and stakeholders
- Objectives and metrics
- Scenarios and knowledge
- Strategy and goals
- Tactics and competition
- Organization and resources
- Budgets

But this is the format for expressing a plan, not for formulating a plan. We will see in Chapter 7 that the intuitive act of formulating a plan requires a different kind of inductive logic, containing

- Perception
- Commitment
- Preparation
- Policy

The analytic logic facilitates the expression and communication of a plan in a systematic manner:

1. One should state the mission, or general purposes, of the activity.
2. One should state the strategy, or long-term direction, in the activity.

3. One should state the goals, or desirable immediate outcomes for the activity.
4. One should state the tactics, or immediate steps as means of accomplishing the outcomes.
5. One should state the resources (such as budget, personnel, equipment, logistics, etc.) needed and available with which to implement the activity.

The synthetic logic facilitates the formulation of a plan in first suggesting that the perception of the future opportunity and challenge is the beginning of a strategic vision and then a commitment to some kind of existence within that perceived future. Then one must prepare and obtain the resources for initiating the action to which one has strategically committed.

There are two kinds of cognitive logics in strategic planning: an analytic logic for expressing a plan and a synthetic logic for formulating a plan.

Now referring back to Mintzberg's classification of the schools of strategy, we can see that the following schools (in addition to placing different emphases of the components of strategy or the processes of strategy) also place different emphasis upon the importance of analysis or synthesis in strategy:
Emphasis on the planning logic of analysis:

Design School
Planning School
Positioning School
Environmental School
Configuration School

Emphasis on the planning logic of synthesis:

Entrepreneurial School
Cognitive School
Learning School
Power School
Cultural School

We conclude that strategy is first intuitive in envisioning and, second, analytical in planning.

CASE STUDY: Formal Planning Process at Henkel KGAA

Next we will look at the nature of strategy as a planning process, as emphasized by some of the schools on strategy. We look at planning at Henkel, which provides a typical case of a planning process as practiced toward the end of the twentieth century.

Henkel produced chemical products and was headquartered in Frankfurt Germany, and in 1989 had 169,000 employees worldwide with $24 billion sales (Starr, 1992). Its capital expenditures then were $1.7 billion, and its R&D budget was $1.4 billion. In 1980, 73 percent of its R&D was performed in Germany, but by 1989, 50 percent was performed outside of Germany. In 1981, Henkel was organized by product areas, geographical areas, and functional areas and with close attention to the market: "Thinking and activity within the company is . . . strongly oriented to the careful observation of market activities. . . ." (Grunewald and Vellmann, 1981, p. 20)

Annually, Henkel's planning process began a new cycle with their management board posing a new set of overall planning targets and by reviewing corporate purpose. The targets used information from a forecast (prepared by a planning staff) that extrapolated past performance into the future. The targets expressed desired levels of cash flow, return on investment, and levels of investment. The corporate purpose as a planning element was summarized in four interrelated areas of fields of activity for the company with overall statements on the kinds of products, technologies to be used, consumer groups, and (4) geographic orientation.

The second step in Henkel's planning process was for corporate staff to take the targets and express them in a corporate-level strategic plan with an accompanying environmental analysis. This was provided to the firm's divisional and functional units. We see that Henkel's process was primarily "top-down."

The environmental analysis consisted of an economic forecast of the condition of the national economy (and other relevant international economies), including possible changes in government regulations. Also, baseline market forecasts that estimated trends in the sales of product lines by application and customer were provided. In addition, technological change was discussed. The planning environment linked three kinds of forecasts: economic, market, and technology. Henkel's environmental analysis emphasized the underlying factors affecting the company's businesses. We have called this a "planning scenario."

The Henkel's profit centers reviewed the corporate targets and planning environment in order to formulate profit-center-level goals and objectives. With these goals and objectives, Henkel's divisions then formulated tactics to reach the goals and budgets to fund the tactical activities. The individual plans of the profit-center-level activities were assembled into a total corporate plan and reviewed by Henkel's management board. We see that because the process was only top-down, Henkel's businesses did not necessarily re-examine their business models.

Case Analysis

This case illustrates a formal planning procedure in a large commercial organization, typical of 1970s American and European practice. This kind of top-down planning process has been common in corporate practice and lacks the important ability to periodically change strategic business models.

STRATEGY PROCESS SCHOOLS

Let us briefly review the key ideas about the strategy as a process that can be gleaned from the schools on strategy that emphasized the process of strategy formulation. By the 1980s and 1990s, these schools had many of their views incorporated in textbooks on strategic management.

For example, one textbook said that strategic management "consists of (1) SWOT analysis, (2) mission formulation, (3) objectives formulation, (5) determination and (6) implementation and (7) control of organizational strategy" (Boseman and Phatak, 1989, p. 6). We see this is a procedural prescription to articulate what Ansoff, Quinn, and Mintzberg required of the strategic view of the firm as a totality and direction. This text emphasized a process to articulate strategy in the organization.

Other management texts also emphasized a procedural, or methodological, approach to strategic management. For example, another textbook of the time emphasized strategy as a methodological procedural framework in which (1) strategic planning, (2) resource requirements, (3) organizational structure, and (4) strategic control are all integrated into (5) strategic management (Rowe et al., 1989, p. 1).

The emphasis in strategic management textbooks upon the process, procedures, and methodology in formulating an organization's strategy continued throughout the 1990s.

For example, a popular text by Fred R. David defined *strategic management* as: "the art and science of formulating, implementing, and evaluating cross-functional decisions that enable an organization to achieve its objectives" (David, 1998, p. 5). Another strategy text defined *strategic management* as a procedure setting the strategic mission of business, measurable objectives and targets, formulating and implementing strategy, and evaluating performance (Thompson and Strictland, 1995, p. 3).

FORMAL PLANNING PROCEDURES

However, in any strategy process in a large organization, there still remains an issue of how complicated the formalism of the planning procedures should be. The kind and degree of formalization of the formal planning process is important to the quality of strategy capability in an organization—correctness in foresight.

For example, a study of the planning practices of 120 companies published in 1980 found a spectrum of formalism in the planning processes in these companies (Gluck et al., 1980):

1. Basic financial planning
2. Forecast-based planning
3. Externally oriented planning
4. Strategic management

Basic Financial Planning

At the first level of planning formality, planning procedures in some organizations only formulate an annual budget. Since basic financial control in any organization requires an annual budget formulation, all organizations plan at least at this ground level. The limitation of this kind of planning—budget formulation—is that within a budget format, many elements of a plan may be hidden or neglected, such as explicit strategy, strategic assumptions underlying the budget, strategic benchmarking with competitors, unanticipated changes in the economic and commercial environments, and so on. A budget presents only the allocation of resources for some plan, explicitly formulated or only implicit in the budget. A budget is never itself a complete and articulate plan.

Primitive planning procedures that focus only on the budget level of basic financial planning mistake only part of the outcome of a planning process (e.g., the budget) for the whole outcome, a full plan (which includes a budget).

Forecast-Based Planning

At the second level of formalism in planning procedures, a plan is formulated in addition to the budget, and the plan emphasizes forecasts of sales and targets for profits. Forecast-based planning is often begun because of the need to estimate future capital requirements, in a addition to an annual budget. This level requires managers to confront longer-range issues than the next year, presented in annual budget plans. Financial forecasts are an attempt to anticipate the longer-term financial future of the corporation.

Forecasts are the essential heart of operational planning and budgeting.

Externally Oriented Planning

The weaknesses in forecasts, of course, are the assumptions underlying the forecasts, and a forecast-based plan does not always adequately and explicitly identify these assumptions nor does it explore the viability of the assumptions. Accordingly, a third level of planning, externally oriented planning, is often implemented next. It explicitly addresses the external environments that determine the basis of forecasts. Externally oriented planning tries to understand the basic marketplace phenomenon that is creating changes, such as changes in natural resources in availability and pricing, changes in the market (due to changes in life styles, demographics, socioeconomic statuses), changes in the regulatory, financial, or international structures, changes in technology as in information technologies, etc. The value of externally oriented formal planning is the focusing of strategic thought upon chan ges in markets, competition, environments, technology.

However, the experience of assembling and reviewing all these external environments often proves burdensome in the volume of plans. A fourth level of formalism, strategic management, is begun when strategic planning is distinguished from operational planning. Operational planning can be projected from continuing current operations; and forecasts of this continuation in trends and activities can be made in the detail needed for annual budgets. This alone can create a volume of data in planning that needs to be distilled in understandable charts and budgets—but focused upon short-term operational plans. Strategy that can focus upon a few important long-term changes. The planning process at Henkel illustrated this level of planning.

Externally oriented planning that intertwines operations and strategy usually becomes burdensome in detail and obscuring to strategic thinking.

Strategic Management

This is addressed at the next level of strategic management planning procedures, which explicitly distinguishes between operational forecasts and strategic change. This often occurs in a firm when executive management realizes that all the formality in the externally oriented planning has begun to outweigh benefits. Annual planning meetings are consuming too much time, producing too many details, requiring too much review, fragmenting the view of the corporation into a bunch of unconnected profit centers, and confusing decision-making responsibility.

The first step in strategic management is to modify the formal planning procedures to distinguish between operational planning and strategic planning, while maintaining their appropriate linkages.

Strategic management is a formal planning procedure for identifying and focusing on the major changes to a corporation that affect more than one profit center. Strategic management should attend specifically to changes in directions—not to changes in goals. Goals are part of operational planning. *Direction* is the heart of strategy. Operational planning consists of the assumptions and projects of continuing operations as they are—assuming projected budgets, sales, operations. But what are the major structural assumptions, that is, the ones that if changed all projections will be in error? This is the heart of the formal procedures in strategic management. The techniques for performing this separation of continuation and change lies in distinguishing forecasts from scenarios.

Planning scenarios are essential to strategic planning, while forecasts are essential to operational planning.

How can strategic planning and operational planning be effectively distinguished and yet properly linked? This is where modern information technology becomes an invaluable tool in strategic management.

RECENT HISTORY OF STRATEGY PRACTICE

Theory and practice are not always the same thing. Let us briefly review the recent history of how management practices in strategy have changed in the corporate world. We review two historical views of this by Lowell Steele and by Bernard Taylor.

Lowell Steele

Looking back at the last quarter of the twentieth century, Lowell Steele noted that the idea of strategic planning became very popular in management thinking in the 1970s:

> The decade of the seventies will no doubt be regarded as the era of strategic planning. The concept appeared, flowered, and by the end of the decade was generating a backlash in the business press. It was accused of overpromising, of being simplistic, and of failing to achieve implementation."
>
> —(Steele, 1989, p. 182)

Steele judged that whatever the limitations of the form of strategic planning in the 1970s, specific attention to planning issues and procedures did have the value of helping to systematize planning activities and provide tools of increased planing rigor. He thought the business literature on planning in that period failed to make evident the dynamism inherent in planning processes. This dynamism arises from the way in which planning processes evolve over time and are altered to accommodate different managerial styles and the need to change focus and priorities over time.

Steele saw that three important strategic concepts were emphasized in the 1970s:

1. Defining the business segment for which it made sense to talk about strategy (e.g., the strategic business unit)
2. The strategic value of high market share
3. The matching of business strengths with market opportunity

While these strategic ideas were important, Steele also observed that

> In retrospect, these concepts would better have been regarded as helpful diagnostic tools to initiate strategic planning, because they did help bring order out of a chaos of information. They did not provide much guidance for what to do.
>
> —(Steele, 1989, p. 182)

Accordingly, what Steele thought attention then needed to turn to was the *process* of planning as it assisted corporate learning:

> Strategic planning takes on a life of its own. With the passage of strategic planning issues are addressed that were never even conceived of at the start. In part, this is because strategic planning is a powerful tool for management learning. As initial issues are resolved, new ones appear or attention can be turned to problems less easily addressed. In part, this results from the fact that strategic planning generates a new kind of data base that reveals features of the business not known before. These new insights call attention to new issues."
>
> —(Steele, 1989, p. 182)

Bernard Taylor

A second chronicler on the changes of practice in strategy, was Bernard Taylor, writing in 1997:

> In the past 30 years Strategic Planning has appeared in many guises. We can trace its development as it has moved from Long Range Planning to Strategic Planning in the 1960s, from Strategic Planning to Strategic Management in the 1980s and from Strategic Management to Strategic Leadership in the 1990s.
>
> —(Taylor, 1997, p. 334)

Talyor viewed the practice of strategy as changing from adding different concerns:

- Long range planning focused on projected budgeting and extrapolation forecasting.
- Strategic planning focused on a two-phase process of both strategy and operational plans and budgets.

- Strategic management focused on implementing strategy through transformations of organizational structure, culture, process.
- Strategic partnerships and alliances focused on forming stronger partnerships with suppliers and distributors and upon alliances with competitors to access new markets and technologies.

Taylor observed that before the 1990s, corporate planning was highly formalized:

> Before the downsizing of the early 1990s, as part of the strategic planning process in a large divisionalized company, the Corporate Planning department would prepare planning guidelines—including economic and market forecasts, corporate objectives and policies, and management priorities for investment. The Business Planning team in the division would review their market sector, benchmark the company's products and services against the competition and analyze relevant political, technological and economic trends. The Finance department would prepare risk and sensitivity analyses, and the Personnel department would produce forecasts of wage levels and manpower requirements and assess the implications of new legislation."
> —(Taylor, 1997 p. 340)

Taylor noted that as part of the downsizing to increase profitability, corporate planning teams were often removed and staff groups were reduced in the centralized business functions of marketing, manufacturing, information technology, and human resources. To replace these functions in the planning process, Taylor saw that the planning activity was reassigned to business unit teams in many instances. Three important planning process requirements were

- A need to coordinate strategy between corporate-level and business unit levels
- A need to establish formal strategic agreements between firm-level management and business-level management
- A need for operating managers to use a practical, team-based approach to strategy and planning

To coordinate firm-level and business-level strategy processes, the firm's board and top management needed to understand the nature of the businesses in each of their operating units and share common standards and priorities about product innovation and quality, customer service, and employee participation. Taylor saw the formal strategy agreements between firm and business-level managements as a kind of strategic contract between them.

Taylor saw this change in planning practices occurring for several reasons: (Taylor, 1997, p. 342)

1. Large central planning teams were found to be costly and ineffective.
2. The annual planning round proved to be too cumbersome and inflexible to cope with fast-moving high technology markets.
3. The strategic planners, along with other central staff groups, were putting a break on the line managers.
4. When the plans were finally approved, the business unit teams did not own them, consequently the plans were not implemented and stayed in the file.
5. Planning as a bureaucratic process did not produce innovative thinking, and (in general) the plans were simply an extension of business as usual.

From this history, Taylor concluded that the new lessons for the practice of strategic management included the following (Taylor, 1997, p. 341):

1. Strategy should not regarded as an annual event but as a *continuous dialogue* that takes place throughout the year.
2. Strategy discussions should not focus on operational plans but around a few strategic issues.
3. Large strategic planning teams have been replaced by directors of corporate development who work on corporate projects (e.g., strategic alliances and joint ventures).
4. The (corporation should be a) profit-accountable organization (in which) the top executive team is increasingly working toward targets that have been agreed upon the board of directors, and the divisional or business unit teams are working toward targets that have been agreed with the group chief executive.
5. Strategy consultants earn their fees by working with the top management team to benchmark the company's performance, collect evidence of major external trends, do some "blue-sky thinking" about the future, and so on.
6. Management should face the task of aligning the organization behind and implementing strategy through
   - Management training in the (optimizing) shareholder value
   - Developing learning companies that focus on the development of their human assets
   - Adopting participative management and employee empowerment
   - encouraging honest, upward feedback from the employees to the management.

Summary of Historical Practices

These histories illustrate how the process of strategy became modified from experience in practice:

- It is important to recognize that strategy at the firm level and at the business level have very different concerns and perspectives and that they need to be coupled in some strategic contractual framework.
- It is important to make the strategic planning process a kind of learning process for the organization, emphasizing and capturing experiences with stasis and change.
- It is important to distinguish operations planning from strategic planning, so that strategy focuses upon a few, selective issues of long-term change.
- It is important not only to formulate strategy but to ensure it is implemented with honest kinds of feedback from those who implement it.

STRATEGY PROCESS

Now after this review of the literature on the theory and practice of strategy, we can look back at the strategy process outlined in Chapter 1 to see how this scheme encapsulates the many important lessons and of theory and practice.

Let us look again at Figure 1.3

First, we can see the emphasis in the process of the top-down and bottom-up strategic perspectives in strategy does capture Taylor's summary of practical requirements of the strategy process:

- A need to coordinate strategy between corporate-level and business unit levels
- A need to establish formal agreements between firm-level management and business-level management about what resources and services the firm will provide and about what the business units will deliver
- Operating managers will need to use a practical, team-based approach to strategy and planning

Second, we can see that by having the techniques of a planning scenario and a strategic model as conceptual results of an interactive top-bottom cognitive process we have used Mintzberg's classification of schools of strategy as ideas and processes. For example,

1. Some have emphasized the ideas and cognitive activities in formulating strategy, particularly the Design, Planning, Positioning, Entrepreneurial, and Cognitive schools.
2. Others have emphasized the processes for strategy formulation, particularly the Learning, Power, Cultural, Environmental, and Configuration schools.

Third, by identifying the strategic plan as a result of strategic vision, we have also used Mintzberg's insistence that one distinguish between and use both the cognitive functions of analysis and intuition in the planning activity. For example,

I am not saying that analysis is bad or unnecessary. There is a danger, though, that you can preclude synthesis with too much analysis. The reason that strategic planning failed as a technique was because it was based on the assumption that you could get synthesis from analysis.

—(Campbell, 1991, p. 109)

Fourth, by having a strategic plan as an outcome of vision and implemented in operational plans, we are also using Steele's important definitions about strategy, strategic planning, and strategic management:

Strategy is the array of options and priorities with which one elects to compete

Strategic planning addresses the continued viability of strategy; it probes the need for change. . . .

Strategic management is the implementation of modifications in the fundamentals of how one competes and survives.

Fifth, the strategy process does use the theoretical summary of the key ideas for strategy of environment, totality, control, vision, information, and change. The ideas of environment and totality are explicitly expressed in the planning scenario and strategic business models used by the process. The idea of vision is used in the process as an intuitive synthesis that unites the environmental and business totalities. The idea of change is the focus of both the planning scenario and of the strategic models, with changes in markets and innovation, competition and structure, and operations and control providing the strategic issues for constructing the strategic business model. And also the idea of information is made an explicit strategic issue in the strategic models.

Thus we can see that this technique of the strategy process was constructed specifically to incorporate the important theoretical ideas of the business literature on strategy and the lessons of the evolution of strategic business practices. It is a useful way to capture the whole elephant of strategy in its size and complexity and to systematically guide it through the fundamental management activity of strategic thinking within an organization.

SUMMARY: USING STRATEGY THEORY IN ESTABLISHING STRATEGIC MANAGEMENT IN A COMPANY

Strategy theory can be practically used as follows:

1. Establish a strategic process
   • Devise the formal procedures, schedules, and team structure for putting

into place a business process of strategic thinking, integrating top-down and bottom-up perspectives on change.

- Go beyond basic financial planning, forecast-based planning, externally oriented planning to strategic management and alliances.

2. Train personnel in strategic thinking

- Provide training for personnel in the key ideas and processes of strategic thinking.
- Develop a common use of strategic terms, such as strategy, strategy process, strategic management, planning scenarios, strategic models, environments, totality, control, vision, information, change, analysis, and synthesis.

3. Fit strategic planning into annual budget planning

- Differentiate short-term operational planning processes and techniques from long-term strategic planning processes and techniques.
- Focus annual budget processes on operational planning as short-term projections from current operations.
- Modify the annual operational plans, according to needed next-year incremental changes required for implementing a long-term strategy.

4. Implement strategic change

- Establish an implementation team to monitor the progress of strategic change.
- Measure the progress of the strategic plan against performance in the annual budget operations planning procedures.

For Reflection

Read a study of Microsoft's origin and growth under Bill Gates. Read also a study of the origin and development of the personal computer industry. What were Gates's successful strategies that guided Microsoft to dominance in the PC software industry? Do his strategies illustrate or contradict tenets of strategy theory? Also read about the U.S. Government's antitrust suit against Microsoft in the 1990s. What is your opinion about the merits of the suit?

CHAPTER 6

STRATEGIC PLAN

PRINCIPLE

Planning is the analysis of future action into means and ends and is only as good as it guides successful action.

STRATEGIC TECHNIQUE

1. Choose the appropriate plan format
2. Keep the strategic plan as brief as possible
3. Use the plan to monitor progress
4. When devising a plan, keep in mind why plans fail
5. Use the planning process to know when and how to revise action

CASE STUDIES

Planning D-Day
GE Capital Expands in the 1990s
Rise and Fall of Osborne Computer in the 1980s

INTRODUCTION

We are using the term strategy to mean change in direction, strategic vision to mean direction, and strategic plan to mean steps-in-a-direction. We need next to look carefully both at the concept of vision and at the concept of plan. A strategic plan implements a strategy based upon a strategic vision.

But before we examine strategic vision, we will look at the logic of planning. Planning and vision are related as the analytic and synthetic aspects of strategy. Analysis is often an easer first concept to grasp than is synthesis, because of the heavy emphasis in Western cultural education upon analytical techniques. However, in practice, neither analysis nor synthesis come first in conceptual priority. Both analysis and synthesis are cognitively interactive in experience—going round and round in actual experience and learning—from analysis to synthesis to analysis to synthesis, and so on. But we will first explore the analytical logic of strategic planning before we explore the synthetic logic of strategic vision— just because, for many of us, analysis is a more familiar concept than synthesis.

Going round and round in strategic thinking in business practice, strategic vision and strategic planning interchange as time goes on. A strategic vision analyzed as a strategic plan begins an organization (e.g. A new business venture plan). Then as the organization grows and competes, strategic direction must change and a new strategic vision needs to be formulated—upon which a change in plan, a new strategic plan, needs to be formulated.

Strategy (change in long-term direction) is implemented as a strategic plans (steps in a long-term direction) based upon a strategic vision (a long-term direction).

INTELLIGENT ACTION

The basic reason for formulating strategy is to conduct long-term intelligent action; and all intelligent action is planned action. Thus all practical strategic thinking is concerned with planning future action. For example, we recall (from Chapter 1) that the importance of strategy resulting in action was nicely expressed by the CEO of Solectron, Koichi Nishimura, who commented upon his experience of being a CEO:

> Four things, I think, are important. First, communicate a vision of where the company is and what you are doing. The second is that when you communicate, you want to be able to motivate people. Third, you want feedback. And fourth, you want to take action.

In the strategy process, after constructing planning scenarios and strategic corporate models and after synthesizing an intuitive strategic vision for change, a strategic plan should be created. Now we will look in more detail at the formal activity of planning.

Furthermore, we recall the idea of strategic planning has always been a core idea in traditional strategic management theory. We recall both the design schools and planning schools of strategy emphasized the role of formal planning in strategy:

Design School. This school focused on the formulation of strategy as matching external conditions to internal opportunities of the organization—achieving clear, simple strategies that can be implemented by all in the organization.

Planning School. This school emphasized strategy formulation as formalizable and decomposable into steps, characterizable by checklists and supported by formal techniques.

Also we noted that there are two kinds of planning, strategic planning and operational planning:

• Strategic planning is a concern for and laying out of the directions for the long-term future;
• Operational planing is a concern for and laying out of the directions for the short-term future.

Also, there are different organizational levels for planning: firm level and business level. The single-business company finds its principle competition in the marketplace—face-to-face with customers and competitors also directly in contact with customers. In contrast, the multiple-business company is primarily a financial holder of businesses, so that its performance is not in the customer market but in the financial market.

And also we noted the different organizational levels of management attention to process:

• Operations, which control of activities
• Procedures, which control of operations
• Policies, which control of procedures.
• Strategies, which control of policies

Accordingly, we will look carefully at planning (operational and strategic planning) and at how strategic plans provide the lead for operational plans (strategic planning setting policies and operational planning setting operations).

Strategic plans and operational plans are related to one another as the integration of long-term and short-term planning—setting policies and setting operations.

CASE STUDY: Planning D-Day

We will look at a military case of strategy and planning in the Second World War to see the importance of good strategy and planning in decisive action in situations of extreme conflict. Of course, the cases in this book have been cases in business activities, since the focus is upon business strategy. However, we will temporarily deviate from this style to use a specific case of military strategy—the planning of the British and North American Allied Forces invasion of France in 1945.

In military action, one starkly can see both the importance and risks of planning.

We saw in Chapter 4 how business and economic activities always occur within societal contexts that include government. And in modern history, one can see how important good governments are to long-term economic growth. Democratic and honest governments are, in fact, essential to efficient and effective industrial economies. In the early 1990s, the terrible, corrupt, inhuman, tyrannical communist empire of the Union of Soviet Socialist Republics simply fell apart when its leadership stopped ruling with terror. Earlier, the communist regime in China began moving under Deng toward a free economy after the terrible excesses of Mao. The major political lessons of the twentieth century were about how good democracy is as a government form for industrialized societies over the long term. But in the 1930s (as we saw earlier in the history of the industrialization of Japan) this issue had not yet been decided.

Still, in retrospect from the perspective of the twenty-first century, one can see that it was no accident of history that modern industrial capitalism began, grew, and thrived in democratic forms of government. The world politics of the twentieth century had two major themes: the need for industrialization of a country to avoid colonial exploitation by other industrialized nations and the military clash between countries with democratic forms of governments and countries with totalitarian forms of government (fascist or communist).

World War I arose from conflicts in the first theme—the struggle of the industrializing nations of England, Germany, France, and Japan for the control of colonies all over the world. World War II arose from conflicts in the second theme—the struggle between totalitarian governments with democracies. Thus in planning scenarios during the twentieth century, both colonialism and totalitarianism were major strategic themes.

Also turning to the strategy and planning, one can find important lessons in both wars. For example, earlier we used an example of warfare in the First World War to illustrate the importance of the idea of the experiential base of strategy. We noted how all the generals in that war—German, English, French, or Russian—failed repeatedly to understand the real tactical importance of machine-gun-defended trench warfare. This was all due to their lack of an experiential base in this battle mode, when they were young lieutenants they had fought in the front lines of older battles—before the machine gun had been innovated.

The lessons about planning in the military context stand out clearly because of the terrible consequences of action in war—triumph or defeat, life or death, survival or extermination. Although not so starkly portrayed in business planning, still good and poor planning there have serious consequences—profits or loses, wealth or bankruptcy. But it is in war that lives as well as property and wealth are at risk.

In World War II, much was at risk. Could the democratic governments survive the military forces of tyranny? Could enslaved people be freed from a new kind of terrible government—police states of efficient bureaucracies ruled by mad, evil dictators? It was over these great issues that the great armies of the world battled.

As you read this case, keep in mind how very general is the fundamental concept of planning. Planning is essential to all actions in life, commercial or military, in peace or in war.

Now we look at the case of military planning for the invasion of Europe by Allied forces in World War II. At the time of this invasion on June 6, 1944, the Allied governments of the United States and Great Britain called the day "D-Day." The planning for this day had been code named "Overlord." It was the final goal of a war strategy devised by Winston Churchill, Franklin Roosevelt, and Dwight Eisenhower—respectively, Prime Minister of the United Kingdom of Great Britain, President of the United States of America, and Supreme Commander of the combined European armies of Britain and the United States. Operation Overlord was the largest planned, logistical war operation in military history.

Winston Churchill, on that day, reported to the British parliament about the action:

> Our long months of preparation and planning for the greatest amphibious operation in history ended on D-Day, June 6, 1994. During the preceding night the great armadas of convoys and their escorts (about 4000 ships) sailed, unknown to the enemy along the swept channels from the Isle of Wright to the Normandy coasts. . . . There was no doubt that we had achieved a tactical surprise. . . . Reports are coming in rapid succession. So far the commanders who are engaged report that everything is proceeding according to plan.
> —(Churchill, 1953, p. 4–5)

The military operation was the largest ever undertaken, transporting a modern mechanized army across the English Channel and landing them under enemy fire upon the coast of Normandy in France. As Churchill reported:

And what a plan! This vast operation is undoubtedly the most complicated and difficult that has ever taken place. It involves tides, winds, waves, visibility, both from the air and the sea standpoint, and the combined employment of land, air and sea forces in the highest degree of intimacy and in contact with conditions which could not and cannot be fully foreseen. The battle that has begun will grow constantly in scale and intensity for many weeks to come"

—(Churchill, 1953, p. 4–5)

With the D-Day invasion of France from the west by U.S. and British armies, World War II was rushing to a climax. On the eastern front, an equally great Russian army was already defeating the major part of the German armies. Together, these allies—England, United States and Russia (and an unlikely combination of two democracies and one dictatorship, thrown together by Hitler's rapacious invasion of Russia)—would destroy the fascist dictatorship in Germany. Under the truly evil Adolf Hitler, Germany was under one of the most terrible governments in the history of the world. (And yet the butchery of innocent peoples by Hitler was only second place in evil to the communist government in Russia under Joseph Stalin, as measured by the number of people these tyrants murdered.) From the perspectives of the governments of the United States and Britain, the Second World War was really about a mission for the two democracies to overthrow an aggressive and evil tyrant. This was the view of both Churchill and Roosevelt.

But a long trail of actions lay ahead from the formation of the Allied alliance in January 1942 to the successful D-Day invasion in June 1944. For until 1942, the world's events since 1937 had made a very bleak picture of the possibility for democracy to survive in the twentieth century:

- From 1937 to 1942, Hitler had overrun and conquered all of Europe (except England) and invaded Russia.
- In the Atlantic, German submarines were attacking and sinking merchant ships (carrying critical supplies from America to England) faster than they could be replaced.
- German armies had aided the Italian army in North Africa and threatened Britain's control of the Suez Canal and its access to resources of India.
- Japan had colonized Korea, Manchuria, and conquered the coastal area of China.
- Japan had seized Indochina, Malaya, and Thailand, threatening both Australia and India.
- Japan had defeated the British army and naval forces in the Far East.
- Japan had attacked and sank a substantial number of U. S. battleships in the Pacific, leaving the West Coast of the U. S. relatively unprotected.

Events then were looking so bleak that on January 27, 1942, British Prime Minister Winston Churchill asked for a vote of confidence in his government. The bad news in Europe was that the British army had barely escaped from capture with the defeat of France. The bad news from the Far East had been the fall of Hong Kong and Singapore and the sinking of the two British battleships in the South China Sea. The bad news in North Africa was that Rommel's German army was winning. The bad news in the Atlantic was that German submarines was sinking many, many British ships trying to gain supplies from the United States. As Churchill summarized the bleak picture on January 1942: "In the two and half years of fighting we have only just managed to keep our heads above water." (Churchill, 1950, p. 69).

Yet Churchill urged the British Parliament to press on to ultimate victory, and his government survived the parliamentary vote. Churchill continued to lead the British war effort. Then the basis of his confidence for ultimate victory was that the United States had entered the war both in the Pacific and in Europe only a month earlier in December of 1941, after the Japanese attack upon Pearl Harbor in Hawaii. Also Russia had survived the German invasion of the summer of 1941. Yet when 1942 began, times were very perilous for the allies— Britain, Russia, and the United States.

PLANNING AND ACTION

Planning, either strategic or operational planning, is thinking about future action. Planning makes concrete what actions one needs to perform in the present to bring about a future state that one desires. Action is the essential nature of business—actions as in production, marketing, sales, product development, new business venture, and so on. Action is the basis of all productive organizations, whether commercial, governmental, or military. We will briefly describe action theory and use it to define the logic of a plan (for both strategic and operational plans).

Underlying all strategy and planning is the theory of action.

The idea of action is fundamental. It is also so familiar that, like all very familiar ideas, the depth of its implications may not be fully appreciated. Familiarity may not always breed contempt, but it frequently breeds inattention. We need to take time to stop and carefully analyze what is so familiar about action. The idea of action basically distinguishes the animate from the inanimate.

We recall that in science all matter can be classified as living or nonliving. Living matter alters its environment through willful and purposeful activity, whereas inanimate matter makes up environment in a purely physical, unintentional way. Thus, biologically speaking, action is an animate organism's willful engagement of its environment.

For example, the action of an amoeba is to move its psuedopods out to encompass organic matter to obtain its food. The action of a plant is to grow leaves and orient them toward sunlight to absorb the energy of light, while also growing roots into the soil to absorb the materials of water and nitrogen for growth. The action of a herbivorous gazelle is to graze on the green grasses of the plains; and the action of a carnivorous lion pack is to seize a gazelle for breakfast, lunch, and dinner. Thus the actions of the laborer going to work and punching in the computerized time clock, or that of the white collar worker signing the computerized biweekly time sheet, or that of the executive exercising stock options are all like that—of the noble lion, or the graceful gazelle, or the peaceful plant, or the lowly amoeba, seeking their breakfast, lunch, and dinner. Action is the basic unit of animate life, necessary for survival and prosperity.

Life is organized into different levels of complexity in terms of action, but the essential purposes of action are similar.

For example, even bacteria (a step of organizational complexity above the amoeba) execute elaborate operations to gather matter and energy from their environments and to construct proteins and lipids and sugars for its own material being and biological processes. As organisms evolved in complexity, they developed capabilities of sensing and motion in their environments, to actively seek out materials and energy.

The seeking out of materials and energy for living is the ground of animate purpose.

The capacity of organisms to display purpose has traditionally been called "will." Most of the will displayed by biological organisms is programmed in their DNA, and we call such displays of DNA-programmed will "instinct."

Instinct provides the primitive hard-wired biological instructions for attaining the basic purposes of animate beings.

Instinct provides the basic motivations for action. Yet for more complex organizations of life, learning and planning add to their instruction set for action. In the complex cognitive and social capabilities of humans, their purposes are not all purely instinctual but are also learned and deliberate. Instinct combines with learning and reason in the human, so that human will and purpose is a mixture and synthesis of instinct and reason. This mixture conceives of action as means and ends.

The means of an action is the way the action is carried out by an actor and the ends of an action is the purpose of the action to the actor.

This is where planning impacts human action. Humans can reason about possible futures as means and ends:

- In order to satisfy both instinctive and learned needs as *ends*
- To plan action as *means* to bring about such futures

The reasoning about future action is called planning. The means-ends description for future action from such reasoning is called a "plan."

Reasoning about action and futures is facilitated by a logical dichotomy of means-ends applied in a time line from the present into the future; this is illustrated in Figure 6.1.

The organism doing the planning reasoning and performing the action of the plan can be called the "actor" (or, more commonly in management science, the "decision-maker"). For a firm, this actor/decision-maker is the manager of the business organization.

In a business plan, the description of the stakeholders of the organization are the participants in the plan.

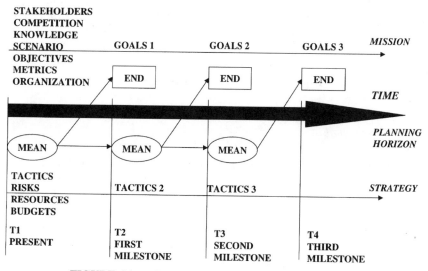

FIGURE 6.1   ACTION PLANNED AS MEANS & ENDS

Means are the activities performed in the present to create a desired future outcome.

Ends are the purposes of the desired future outcomes planned to occur from conduct of present activities.

Means and ends are conceived from the present knowledge the actor/decision-maker has about its wants and needs and capabilities.

Knowledge is the awareness and understanding by the actor/decision-maker of its desires and needs and capabilities.

In Figure 6.1, means are labeled as the activities performed at a given period of time (shown as T1 to T4) to bring about desired future outcomes, ends. In a plan, the future ends are usually described as "goals" of the successive milestone periods, and means are described as "tactics."

Goals are the descriptions in the plan of the desired outcomes at the milestone times of the plan.

Tactics are the descriptions in the plan of the actions planned at the milestone times.

Ends are the purposes of the goals and so are realized at the time of the achievement of the goals—Goals2 attained at time T1, Goals3 at time T2, Goals4 at time T4, and so on. To attain these goals, means of action, tactics, need to be taken at the appropriate time. Thus Tactics1 are performed at time T1 to attain the Goals2 at time T2. Tactics2 are performed at time T2 to attain the Goals3 at time T3. Prior tactics produce subsequent goals over a time period.

The time period over which a plan is formulated is called the "time horizon" of the plan.

This is the key idea of planning: tactical means must be performed *before* goals of ends are attained. This is the essence of the theory of action—goals must be envisioned before being experienced and tactics must be performed in the present to attain goals. Planning is thinking and acting in the present to attain desired futures as successive goals over a time horizon.

Now since active life consists of a *series* of actions at times T1, T2, T3, T4, and so on, the concepts that provide continuity in this sequence and tie these actions together into a pattern over time are the two ideas of mission and strategy.

"Mission" is the description in a plan of an organization of its long-time ends—of the commonality of purpose among its successive goals.

In a business, a mission statement is a statement of what kind of business is the company engaged.

"Strategy" is the description in a plan of an organization of its long-time means—of the commonality in the successive actions of tactics.

Strategy is the direction of the successive tactics of action that ties these actions together in a coherent direction to achieve the mission of the organization.

In the real world of the organism and nature, action will always be opposed by competitors. Mission and strategy need to be formulated to overcome the expected opposition of competitors.

"Competition" is a description in a plan of the competitors and their strategy—those who will oppose and compete against the actor/decision-maker's mission and strategy.

Since in an organization, the mission statement of a plan is always very general in order to encompass the different purposes of the stakeholders, it is usually necessary to make the mission more concrete by expressing the commonality of successive goals as objectives and metrics of the plan.

"Objectives" are general aspects of future outcomes, goals, which realize the mission of the plan. "Metrics" are measures of the degrees of attainment of the objectives.

To prepare for planning, the actor/decision-maker needs to know something about the environment in which the planned actions are to occur. This description is usually called the "planning environments," and such environments are often usefully expressed as "planning scenarios."

A "planning scenario" is a dynamic description (describing change) of the environments in which organizational action can occur and may be opposed by competitors.

Finally, in any plan, the organization of responsibilities in carrying out the plan and the resources required need to be thought out and described.

Organization in a plan assigns authority and responsibilities in carrying out a plan. Resources estimates budgets and facilities required for performance of a plan.

This view of action involves the important idea that all action, although planned, will occur in a present time of the action (the milestone times of T1, T2, etc.) This implies that all preparation for any current action must have preceded it in time. Accordingly, planning is present knowledge applied to future action. An important implication of this is that present knowledge must always be incomplete and not perfectly accurate about the future conditions and environments in which action will occur. From this arises the basic idea of risk in planning. At any time, complete knowledge of any future time is never possible. Consequently, planning will always provide incomplete knowledge for future action, and risk is inherent in any plan. Risk is inherent in any action because planning for that action must have be reasoned about with incomplete knowledge of the future.

"Risk" is a description of the kind of difficulties that may be encountered in a given set of tactics.

For this reason, there is a very important tradition about the experience of planning that all plans are risky.

While planning is essential for intelligent action, nothing ever goes exactly as planned.

This is a universal truth that must be kept in mind about planning. Planning is essential to intelligent and effective action, but by its very nature it is incomplete and risk in action is fundamental and universal.

The quality of planning is measured by its effectiveness in facilitating action to bring about a planned future and reduce the risk in action toward that future, not by everything going according to plan.

Thus in planning action, future outcomes are classified according to the purposes, or ends, of the person/organization performing the action, or the actor; and the activities performed are classified as means to those ends. The means-ends dichotomy is a fundamental way to think about planning a future. One must be clear as to the means possible and available and to the ends desired in choosing and performing any action. This means/ends conceptual dichotomy provides the fundamental logic for planning and for describing any plan:

- Mission and stakeholders
- Objectives and metrics
- Scenarios and knowledge
- Strategy and goals

FIGURE 6.2   LOGIC OF A PLAN

- Organization and resources
- Tactics and competition
- Budgets

Figure 6.2 indicates that the first part of this logic (from mission and stakeholders to strategy and goals) is focused on the longer term view of the organization's actions and so is the heart of the strategic part of a plan. The latter part of the logic (from strategy and goals to budgets) focuses upon the shorter term view of the organization's actions and so constitutes the operational part of a plan. The part of the plan logic with expresses strategy and goals is the section that ties together (1) the long-range strategic plan and (2) the short-range operational plan in (3) the implementation of a strategic vision of change.

We recall that we discussed (in Chapter 1) how a long-term strategic plan is implemented in a near-term operational plan. Therefore, the logic of Figure 6.2 places the planning topics of the longer-term part of the plan first (strategic plan) and the planning topics of the short-term implementation (operational plan) last.

A plan is a prescription for action, oriented from long-term mission and objectives to implementing short-term goals and tactics.

We will next examine how these two strategic and operational parts of a plan interact.

CASE STUDY: Planning D-day, Continued

We will illustrate this interaction by continuing the case of the strategic and operational planning of the Allied invasion of Europe in the Second World War.

Mission and Stakeholder   The second major war of the twentieth century was also called a "world war" by its participants because there were so many nations in the global conflict. The *stakeholders* of the Allied military alliance were the governments of Great Britain, the United States, China, and the Union of Soviet Socialist Republics (USSR). Their *competitors* in the opposition Axis alliance were the governments of Germany, Italy, and Japan. The *customers* of the Allied powers were the citizens of their governments and also the citizens of governments which had been invaded and conquered by the Axis powers— such as the citizens of Ethiopia, Norway, France, Denmark, Belgium, Holland, Greece, Yugoslavia, Indonesia, Vietnam, Korea—and the persecuted minority of the Jewish people in Europe.

The *mission* of the Allied governments was to defeat the military powers of the Axis governments (Germany, Italy, Japan). The Second World War began as a series of military aggressions by the Axis governments. In Asia, war had begun with the Japanese invasion and conquest of Korea in 1936 and invasion of China in 1937. In Europe, war had begun with the Italian invasion and conquest of Ethiopia in 1936, and the German annexation of Austria and Czechoslovakia in 1938 and with the invasion and conquest of Poland in 1940. At the time of the invasion of Poland, Great Britain declared war upon Germany.

In 1940, Germany invaded and conquered Norway, Denmark, Holland, Belgium, and France. In 1941, Germany invaded and conquered Greece and the Balkan countries and invaded Russia (then under the communist government of the USSR). Also in 1941, Japan invaded and conquered Singapore, Hong Kong, Malaysia, Vietnam, and the Philippines. Japan also attacked the Hawaiian islands, at which time the United States declared war on the Axis powers.

The Allied governments of United States and Great Britain had democratic forms of government and saw the mission of the Allies to defeat the totalitarian governments of the Axis powers. (However, the allied government of the USSR was also a ruthless dictatorship under the communists and Joseph Stalin, and so the Allies's alliance was also one of practical convenience against the German fascist dictatorship. Thus war can make strange alliances.)

World War II is one of those rare cases of warfare in which a clear moral position can be seen, even in historical perspective. The aggressions of the German and Japanese military campaigns of conquest were clearly initiated out of deliberate policies of conquest of adjacent nations and peoples. Japan had attacked and conquered Korea and attacked China and conquered Singapore and Malaysia to establish an empire of a "Greater Asian Co-Prosperity Sphere." Germany had attacked and conquered Poland, Norway, Belgium, Holland, France, and Russia to establish a German empire of the Third Reich. Earlier, the fascist government of Italy under the Benito Mussolini had attacked and conquered Ethiopia. Military conquest and empires are standard events in human history, but what made the German aggression particularly revolting in the annals of human civilization was a deliberate government policy of genocide of European Jews—the Holocaust. Hitler can be classified as one of the most evil villains in history in the sheer madness and scale of his policies.

So the mission of the military alliance between the United States and Great Britain of World War II was very clear. Their business was democracy; and their long-term mission goal was to defeat the totalitarian regimes of Germany, Japan, and Italy.

Objectives and Metrics   The *objective* the Allied governments required an invasion of Europe through France by Anglo/American armies to defeat the German army (while Russian armies fought the German armies in the eastern front). The plan for the invasion of Europe was code named "Operation Overlord"; and that operation would comprise the combined forces of the United States and Britain in a ground and air attack against the German army on the coast of France. After a successful landing, these forces would attack Germany.

Metrics for this objective required a sizable and well-equipped invading army. The size of forces planned for that invasion was to be 16 British divisions and 19 U. S. divisions for a total invading army of 35 divisions. (An army division then was about 40,000 soldiers.) This was intended to be much stronger in numbers and equipment than the 11 German divisions stationed in France. *Additional metrics* important to the effort were dominance and control of the air over France and about 4000 naval vessels necessary to transport the Allied army across the channel from Britain to France. At that time, on the Eastern front, the battles there involved 20 German army divisions against 40 Russian divisions.

Scenarios and Knowledge   The plan for invasion in 1944 occurred in the context of the course and progress of the war from 1940 through 1943. This context and conditions provided the *scenario* for the Allied war effort in 1943, which we briefly review here.

The critical events in the planning environment for British forces had begun back in 1940, when a British Expeditionary Army and the French Army had been overrun and defeated by the invading Germany Army in the first two weeks of June 1940. The British retrieved most of their soldiers from that defeat in an evacuation from Dunkirk but had left their army's equipment on the beach. England was then undefendable and would have quickly fallen to a German invasion, had the Germans controlled the air over England.

Hitler did plan to invade Britain in the summer of 1940, and German planes attacked England. British fighter planes took off in time to meet the daily swarms of incoming German planes from the captured airfields of occupied France and Holland. The British fighter pilots had the advantage of being forewarned with radar (the newly invented and secret weapon of Britain) and could wait for the incoming Germans from a height advantage. Throughout July, August, and September, the world's first major sky battle for control of the air was fought between British Hurricanes and Spitfire fighters and German Messershimdt fighters and Junker bombers. Both British and German losses were high, but Germany failed to gain control of the British sky. In October, Hitler called off his plan to invade England. In historical retrospect, this was Hitler's first major strategic error of the war.

Hitler's second major strategic error occurred in the summer of 1941 when he ordered the German army to invade Russia. Earlier in 1939, another brutal dictator, Joseph Stalin of Russia, had foolishly made a pact with Hitler. Moreover earlier in 1937, the paranoid and evil Stalin had murdered most of his experienced Red Army military generals. Hitler's invasion not only caught Stalin's army by surprise but with inexperienced, incompetent military leadership. The Russian defenses were swiftly overrun and about a million Russian soldiers were killed or captured in the German offensive of summer 1941. Yet because of the great distance into Russia and Hitler's late start that summer, the German armies did not reach or conquer Moscow before the winter.

The bitter Russian winter allowed Stalin to rebuild his armies and defend the front in the spring of 1942, when the German army renewed its offensive. Germany then found itself in a two front war—fighting on an eastern front against Russia and a potential western front against Britain and the United States. The Russians urged their allies to begin that western front immediately.

In that desperate Russian winter of 1941–42, the governments of the United States and Britain shipped Russia military equipment to help rebuild its armies. But if the Russians could not hold out against Germany the critical summer of 1942, there would be no foreseeable way for the U.S. and British armies to invade Europe. Stalin asked the U. S. and Britain to invade Europe that summer in 1942 to ease pressure upon the Russians. The U. S. and Britain declined because then they did not have the military force to launch a successful European invasion and could not even control the Atlantic against the German submarine offense to get a U.S. Army to Europe. Besides, from the U. S. And British perspective, unless the Russians could hold their own, it made no military sense even to try to invade Europe. The U.S. And Britain told Russia that they could not invade France before 1943. The new American army needed time to amass great quantities of weapons and time to gain battle experience to successfully fight an experienced German army.

Historically, the fact that the Russian armies held the German armies in the summer of 1942 was to be the turning point of the war. It was all a series of defeats for Germany from that point on until the final gotterdamerung of the German Third Reich. But to get there was hard. The German armies were tough. Thus in planning for D-Day back in 1942, it was the knowledge of military capabilities and the scenario of past events and present circumstances which together required that first military capability must be built and the

Russian front hold before a strategy for invasion by U.S. And British armies was possible.

Strategy and Goals   Strategy by British armies alone was insufficient. In 1941, England was struggling for survival. Until the United States entered the war, offensive operations against the Germans were not possible. And we recall that it was the attack by Japanese navy upon the U.S. fleet in Hawaii that drew the U.S. into war. Only then could beleaguered Britain see hope: "This new year of the Second World War, 1942, opened upon us in an entirely different shape for Britain. We were no longer alone." (Churchill, 1950, p. 3).

In action, all strategy must first be based on hope. The allies of the U.S. and Russia gave Churchill hope of final victory, so strategy needed to be formulated among the Allies for achieving victory. President Roosevelt had been in continual communication with Prime Minister Churchill, having supported England throughout 1941 with the U.S. lend-lease program providing England with supplies and weapons.

Yet when the U.S. entered the war, it was not strategically prepared for military action. As General Dwight D. Eisenhower, who was to lead the European invasion, wrote: "We (America) were the nation which, from the war's beginning to its end, had achieved the greatest transformation from almost complete military weakness to astounding strength and effectiveness." (Eisenhower, 1949, p. 1)

This transformation was not begun until the U.S. government was committed to war, and this did not happen until the Japanese attack upon Hawaii. After that attack, Americans unanimously committed to war. But still the U.S. military had not been prepared to conduct major wars.

With a declaration of war upon Japan and soon after upon Germany, the U.S. Military began the planning and building up of military forces. But the U.S. would not be fully prepared for war for a time:

> Within the U.S. War Department staff, basic plans for European invasion began slowly to take shape during January and February 1942. . . . As always, time was the critical element in the problem. Yet everywhere delay was imposed upon us! It profited nothing to wail about unpreparedness. It is a characteristic of miliary problems that they yield to nothing but harsh reality.
>
> —(Eisenhower, 1949, p. 28)

While the U.S. could plan for victory, the buildup of forces had to first take place. War ships had to defeat the German submarines in the Atlantic in order to safely transport an army to Europe. An army needed to be recruited, trained, provisioned, and become experienced in battle. Airplanes had to be built, pilots trained, and air superiority achieved over European skies And for the navy, army, and airforce, time was required for American industry to build their weapons and material. All this had to be planned and accomplished. In April 1942 when strategic planning began in the U.S., and Roosevelt sent a message to Churchill, stating that he had just completed a survey of the immediate and long-range problems of the military situations facing the United Nations. He then sent his top aide, Harry Hopkins, and his Chief of Staff of U.S. Military Forces, George Marshall, to Europe to begin planning between the British and U.S. Forces.

The strategic issue they discussed was how to establish a new military front to help pull German military forces from Russia. U.S. planning could not see an invasion of Europe before the following year in April 1943. Russia wanted a western front in Europe in 1942, so they would be disappointed. But in that winter of 1942, it was still not known whether the Russian army would survive. But still the U.S. And Britain had to strike militarily that year. They agreed to have the U.S. army invade North Africa and assist the British army battling the German army. This seemed feasible, because the size of the German army in North Africa was extremely small (but very capable under General Rommel). By superior numbers and supplies, it might be possible for the combination of an inexperienced U.S. army and an experienced British army to overwhelm the highly capable German army in North Africa. In this way, the new U.S. army would gain combat experience, and the vital supply route between Britain and India through the Suez canal would be protected. By July 27, 1942, planning had been completed for the U.S. army invasion of northwest Africa and an expanded U.S. army was being trained and equipped.

Northwest Africa was controlled by French Vichy government, whose headquarters was in a French town, Vichy. This Vichy government had been put together to surrender France to Germany, after the German conquest of France in 1940. The Vichy government was a puppet government, collaborating with the Nazi government of Germany. It was in control of French naval forces and French military forces in the French colonies of North Africa. An important strategic issue was whether the French North African military forces would oppose an Allied invasion of North Africa.

Dwight D. Eisenhower was then staff planner to General Marshall and was given charge of planning the North African invasion, which was code-named "Torch." Marshall appointed Eisenhower to be the Allied commander in chief of the expedition.

In November 1942, the U.S. army landing of North Africa was initially a success. French forces defected from the Vichy government and surrendered. But in Tunisia, the German army stopped the U.S. Army. The crucial North African battle occurred in the fall of 1942, when the British army, under General Montgomery, stopped the German army's advance into Egypt (the Battle of Alamein). After this, both the U.S. and British armies squeezed the German and Italian forces into Tunis. Rommel was removed to Germany, and the North Afrikan Corp of German surrendered in March of 1943.

Meanwhile during 1942, the Russians held on the Eastern Front. In that winter of 1941–42, the Russians had regrouped and attacked the Germans. The army of the Russian General Zhukov surrounded and captured a whole German army commanded by General Pauling. Hitler had refused Pauling permission to retreat from encirclement, and Pauling foolishly followed Hitler's stupid orders, losing 100,000 soldiers to captivity. (This was the sign of what came to be Hitler's creeping madness, under the new experience for him of losing battles.)

Also meanwhile during 1942, the sea battles of the Atlantic between the German submarines and the British and U.S. surface ships also began to turn the tide, using convoys and radar to fight the German submarines. Also very usefully, the British had broken the German military code and began to read German submarine communications, learning when and where the submarines planned to attack.

Thus when 1943 began, the Allies could see the real possibility of defeating Germany. Roosevelt and Churchill decided to meet together in Casablanca to plan what to do in 1943. From the Casablanca Conference came three Anglo/American war decisions: to put off the invasion of France until 1944, to invade Italy in 1943, and to demand an unconditional surrender of all the Axis governments.

In February 1943, the successful Russian defence of Stalingrad would mark the turn of the military tide on the Eastern Front. In May 1943, all the remaining German and Italian forces in the African would be captured. Earlier in the summer of 1942, American naval victories in the Coral Sea and at Midway Island had stopped Japanese expansion in the Pacific. By May 1943, the German submarines had been beaten, and Allied naval forces would control the Atlantic

So the Allies had started from a desperate military position but had held and built their military capabilities and experience to begin strategic conquest. Both in military and business, any strategy first requires having the capability of means before any end can be attempted.

Organization and Resources   After the successful planning of the U.S. invasion of North Africa, the authority and responsibility for organizing the invasion of Europe was also assigned to General Dwight D. Eisenhower. Eisenhower's previous experience and career had helped prepare him for this strategic assignment. With the career officers of his generation, Eisenhower had been a lieutenant in the U.S. Expeditionary Army in World War I. His experience there convinced him and many of his fellow young officers that a lack of a unified command in that war had provided real military weakness. In the 1930s, they watched the next war growing in Europe and in Asia and thought about the problems of command. One of Eisenhower's military mentors and friends was Major General Fox Conner, whose opinion about command was that another great war was coming, and in that second world war, the allies needed to fight with a single command, as they had not done so in the first world war. Conner thought that the 'coordination' concept under which Foch had to command had been a major impediment to military success. Eisenhower agreed with this, since his personal experience as a lieutenant in the trenches of World War I had convinced him also of importance of unified command of a multi-national army.

This widely shared experience and belief had been early used as a command principle in December of 1941 when the British General Sir Archibald Wavell was sent to Java to become the Allied commander-in-chief in Southeast Asia. Eisenhower had participated in the task of writing a charter for a supreme commander. He had to help formulate rights of appeal and scope of authority in operations and service organizations for a unified command.

Eisenhower's personal history up to the U.S. entry into the war was that earlier he had served in staff positions during World War I and afterwards. In 1928, he served in the office of the Assistant Secretary of War as a special assistant, analyzing issues on world-wide military matters and studying issues, such as mobilization and composition of armies, mechanization of war, new roles of air forces and navies, and dependence of military force on industrial capacity.

The new mechanization of military warfare had emerged after the First World War as the key element of military strategy. The experience of that war had led to trench warfare because of the dominance of artillery and machine-gun technologies (which had prevented the successful direct infantry assaults on defensive positions). Young officers from the defeated German army had learned this bitter lesson, and in the rearmament of the new German Army by Hitler's fascist government, army staff had created *new military strategies and tactics* for mobile warfare, highly mechanized with fighters, bombers, dive bombers, and tanks. The German military staff called this new strategy blitz-krieg (lightning-war). Blitzkrieg was the reason for the devastating and swift defeat of the Polish, French, British, and Russian armies in the first two years of the Second World War. The buildup of the American army after 1941 followed that successful German strategy for air superiority through fighter aircraft and ground superiority through tanks. The airplane and the tank were to define the new strategies of war (and at sea, the new aircraft carrier was the dominant naval weapon to launch and recover airplanes at sea).

In his staff positions in the 1920s and 1930s, Eisenhower had devoted his professional time to thinking about the changed future of military strategy. In 1935, he was been sent to the Philippines to serve under General McArthur as military advisors to the Philippine government, which had been promised independence as a nation from the United States by 1946. In this assignment, Eisenhower was to have direct political experience in dealing with officials of a different government and culture (experience he was to use to good advantage when later he became an allied commander). In 1939 in the Philippines he learned of the German invasion of Poland and Britain declaration of war on Germany. He then was certain that the United States would be drawn into war and requested a transfer home to help prepare for war.

From January 1940, he served in a series of positions to build military capability through training and military exercises. First he was assigned to troop duty with the 15th Infantry at Fort Lewis, Washington. There he was involved in planning and executing training exercises involving logistic planning and tactical problems. In November 1940, he became chief of staff for the 3rd Division, and in March 1941 was transferred to the chief of staff of the IX Army Corps. In June of 1941, he was transferred to chief of staff of the Third Army. In the summer of 1941, the Third Army joined with the Second Army for a large exercise of 270,000 soldiers. In this maneuver, U.S. army officers would begin to learn the tactics for a mechanized army.

These staff experiences in 1940 and 1941 for Eisenhower were developing his skill and understanding of planning, recruiting, training, and large logistics maneuvers for huge armies. Still, the battle experience was lacking, since the United States was not as yet in war. Finally in December 1941 after the U.S. had entered the war, Eisenhower received a phone call:

> "Is that you, Ike?"
> "Yes."
> "The Chief says for you to hop a plane and get up here right away. Tell your boss that formal orders will come through later."
> —(Eisenhower, 1949, p. 14)

The call was from Colonel Walter Bedell Smith, and the Chief was General George Marshall, Chief of Staff of the U.S. War Department. Marshall reported to President Franklin Roosevelt and was head of all the war planning and operations of the United States. The reason for the call became evident to Eisenhower when he learned that the Japanese had invaded the Philippines, and because of his assignment there in the 1930s, Marshall wanted his appraisal of the desperate situation. Eisenhower's advice confirmed what Marshall dreaded, that the Philippines would fall to the Japanese forces, with U.S. incapable of reinforcing their soldiers. It was that hour when all allies were in a very grim situation.

In January 1942, General Marshall reorganized the War Department, and appointed Eisenhower as chief of staff of the Operations Division. (It was these series of staff positions that had put Eisenhower in position to eventually become the commander of the European Allied Army.) As Eisenhower than began military planning for European operations in the spring of 1942, he formulated several important strategic assumptions before a European invasion was possible, namely that the Russian army must survive, allied naval forces must gain control of the Atlantic, and allied air forces must gain mastery of the European skies. And as we have reviewed that to get from then in the spring of 1942 to the actual invasion of France in 1944 did require these assumptions to be accomplished. Strategic assumptions must be made real for any plan to be successful.

Next, we recall that it was in April 1942, that General Marshall flew to England to present the U.S. war plans to the British government, and it was agreed that the invasion of Europe was not possible until 1943. We also recall that then the issue was what to do in 1942, and the decision was made for the invasion of North Africa by U.S. Forces. It was then that General Marshall had been given by a draft by Eisenhower of the concept of unified command of European operations as a Directive for the Commanding General, European Theater of Operations; after which Marshall appointed Eisenhower in command of the allied operations in the european theater.

Next we recall that it was in January of 1943, when Roosevelt and Churchill had traveled to Casablanca to plan the next phase of the war. They still felt incapable of invading France, putting it off from 1943 to 1944. They decided next to capture Sicily, as the British/American forces in Tunisia could hop right across to Sicily. And Sicily was successfully invaded and conquered from July 10 to August 16, 1943.

Then on August 17th, 1943 in Quebec, Canada, Roosevelt and Churchill met again. Finally, they thought the allies strong enough to invade France in 1944. They also agreed that the supreme commander for the Overlord operation would be an American officer (at the time, Churchill expected Roosevelt to appoint his Chief of Staff, General Marshall).

In September, 1943, the British-Canadian-American armies invaded Italy from Sicily, crossing the Straits of Messina into the 'toe' of Italy; and on September 8th, a sea-borne assault began on the beaches at Salerno. (This allied decision to invade Italy in the south and move north through the mountainous center of Italy would turn out to be a costly decision. The German army invaded Italy in the north, and war went on in Italy from September 1943 to April of 1945, until the final defeat of Germany.)

Meanwhile on November 23, 1943, while the fighting was going on near Salerno, Italy. Roosevelt and Churchill met once again, this time in Cairo to review the progress of the war and finalize war plans for 1944. On the way back from Cairo, Roosevelt stopped in Tunisia to meet with Eisenhower and inform him that he was to command the invasion of France, Operation Overlord. This was how General Dwight Eisenhower came to plan and conduct the invasion of Europe.

Tactics and Competition Finally by the time of Overlord invasion, allied forces had been assembled in England, consisting of 17 British Empire divisions (including 3 Canadian), 20 American divisions, 1 French division, 1 Polish division. Also there were 5049 fighter aircraft and 8516 bombers. There were 6 battleships and 1068 beachable landing craft for delivering troops, tanks and equipment onto the beaches of Normandy.

On D-Day, June 6, 1944, this massive armada landed troops upon the Normandy coast of France. Five beachheads were established: Utah and Omaha Beaches for American armies and Gold, Juno and Sword beaches for British and Canadian armies. However, the German reinforcements held the Allies to their beachheads. It wasn't until July 25, when a sufficient quantity of soldiers and equipment had been landed to effect a breakout, and this was lead by the U.S. General George Patton. (Also at this time, Hitler did not allow his commanders tactical choice and they were forced to refused to allow his German army to retreat, allowing them to be surrounded and captured.) The German western front collapsed. Anglo-American armies raced for France, Holland, and Germany but outran their supplies.

Then instead of assigning gasoline to Patton (whose tanks had run out at the French/German border), Eisenhower left Patton's army stranded and assigned available supplies to the British General Montgomery. The British forces had wished to be the first to strike into the heart of Germany. Montgomery's plan was to send a tank force across Holland into German, but Montgomery's assault failed because it required the capture of 9 bridges in succession. All were successfully captured, except the last bridge. This ended the Allied military thrust in the fall of 1944, and it would take another year for the allies to win the war against Germany.

Meanwhile on the Eastern Front in the same summer of 1944, Russian armies had major, definitive battle successes. Russian attacks began in the north on June 21, opening the railway from Leningrad to Mumansk (the terminal where earlier U.S. military support had been shipped to Russia back in the critical winter of 1941–42). On June 23, Russian armies also attacked the German front in the center between Vitbsk and Gomel. They captured Minsk on July 6. By the end of July, the Russian armies had advanced 250 miles in five weeks, crushing German resistance. The German losses were enormous, 25 divisions! Also Russian armies launched an offensive in the south between Kovel and Stanislav on July 13. In ten days, the German front here was broken, with the Russians advancing 120 miles. Finally, even further south the Russian armies advanced into Rumania, destroying 16 German divisions. By the end of the summer, the Russian armies had broken the German military might in the east. The German armies were in retreat. The Russians began resupplying for a final drive to Germany in the winter of 1944–1945.

Thus on both the eastern and western fronts in the fall of 1944, the German armies were in retreat. The final defeat of Hitler's Germany was likely in the spring of 1945. Hitler made one last counter-offensive in west in winter, striking across the Ardennes forest on December 19, 1944. This the Allies called the Battle of the Bulge, since German tanks put a temporary bulge in the thin allied lines from Holland though France. Eisenhower again called upon Patton to save the battlefield situation, and by January 16, 1945, Patton's tanks and soldiers had rushed to relieve the besieged American soldiers at Bastogne in

Belgium. This final German offensive was stopped. Hitler had no more military reserves to stave off defeat in the west or in the east.

On January 15, 1945, the Russian armies launched their final attack on German armies, which was to carry them into Berlin. Meanwhile, Eisenhower launched Allied armies against German positions along the Rhine River beginning on February 3, 1945, a broad offensive that was to carry them to the Elbe river where the Anglo-American armies halted. They then linked with the Russian armies along this boundary.

In his underground bunker in Berlin, Hitler committed suicide, the end of an evil and insane man. His last wish was for the total destruction of his own people, the German people, who in his mad view had betrayed him.

Budgets The war efforts in Japan, Germany, Britain, United States, Russia, and China were all funded with the mobilization of their industries and patriotic efforts of their peoples. Germany, Britain, and the United States had underused industrial production capability due to the world wide recession of the 1930s, and their economies produced vast amounts of war materials during the war years.

After the war, the economies of Japan and Germany required total rebuilding. Japan rebuilt effectively to become a global economic power (acquiring and improving technologies and exporting high-tech products). Germany was divided into two territories, West Germany and East Germany. West Germany rebuilt with initial U.S. aid and became a major economic power. East Germany languished under a puppet communist government controlled by the totalitarian communist government of the Russian empire.

Britain had exhausted industries, went through a socialist nationalization of key industries, had high taxes and did not economically begin to grow again until after the 1970s.

Industry in the United States retooled for consumer demand. Returning soldiers had higher education benefits, and higher education began a vast expansion. The cold war began, and the U.S. government funded research and development so that a vast expansion of an innovative economy drove the United States to world industrial dominance in the immediate decades after the war.

Russia continued to develop its industry, but wholly tilted toward military production, so that by the 1980s about three fourths of industrial output went into the military. After Stalin's death, a series of communist dictators maintained totalitarian control with an inefficient economy and corrupt society. A serious attempt at reform was finally undertaken by Gorbachev in the late 1980s, but the whole "evil empire" of the Soviet Union simply felt apart without the continual terrorism of a police state.

Communists took over in China after the war and structured another police state of corrupt officials, run by another brutal, selfish, mad dictator. Finally, after the death of Mao and his last decade of anarchy, China began to slowly liberalize and head toward capitalism and democracy.

Case Analysis

Many aspects of planning were illustrated in this case. First, it is the nature of the action which determines the focus of the plan. Accordingly when planning, the first question to ask is what is the nature of action that needs to be addressed. For example, if the action is to start a new firm, then plan must be about the whole business venture; and accordingly a business plan must address all aspects of the business. In contrast, if the action requiring attention is to sell an product or service, then a marketing and sales strategy becomes the focus of planning. Thus planning is circumscribed and defined by the nature of the action to be planned. Any effective plan is focused upon a clearly needed action.

Second, the case illustrated that a plan needs to be analyzed into topics that cover the means and ends of action. Together these topics provide the logical parts of a plan:

- Mission & Stakeholders
- Objectives & Metrics
- Scenarios & Knowledge
- Strategy & Goals
- Organization & Resources
- Tactics & Competition
- Budgets

As we have seen, the later part of the plan, operational plan, occurs in the context of the earlier part of the strategic plan. Strategic plans provide the longer term part of the plan and the direction and sequence of goals in that direction. Operational plans set the tactical means to achieve the goals as ends. Thus operational plans should follow in sequence within an overall and evolving strategic plan.

We can clearly see this sequence of operational planning within strategic planning in the case of the Allied invasion of Europe. There war planning alternated between strategic and operational planning. The strategic planning of Churchill and Roosevelt set a series of goals (the battles of the war). Then their generals did operational planning for each battle. The series of goals (battles) led to the overall mission of the war effort—the defeat of Hitler. The Allied mission was to defeat the German government of the Nazi dictatorship, expressed in successive goals chosen on the basis of the feasibility of means, so that the following milestones marked the achievement of the goals:

1. First Milestone, Goal 1—Battle of Britain

    British fighters defeated the German aircraft for control of the skies over England, thereby stopping an invasion of England by the German

Army. This was battle was critical to the survival of Britain and to the eventual mission of defeating Germany, for without England as a base to assemble an Allied army, there would have been no platform for an invasion of Europe.

2. Second Milestone, Goal 2—Battle of the Atlantic

   Until the German submarine fleet was defeated in the Atlantic, the United States could not supply Britain or build up an army in England for the invasion of Europe.

3. Third Milestone, Goal 3—Defense of Moscow

   Until the Russian Army withstood the German invasion, the Russian government could not survive as an ally of the British and North Americans.

4. Fourth Milestone, Goal 4—Battle of North Africa

   Until the U.S. and British armies demonstrated their ability to defeat a German Army, there could be no invasion of Europe.

5. Fifth Milestone, Goal 5—Battle of Italy

   Building up the logistics of the Anglo-American army for the invasion of Europe required another year's effort. In the meantime, the Anglo-American army's invasion of Italy to diverted one German army away from the Eastern Front against Russia,

6. Six Milestone, Goal 6—Turning the Battle on the Eastern Front

   Until the Russian army began defeating the German army on the Eastern Front (consisting of two thirds of the German armies and leaving only one third of German armies in France), the Anglo-American chances of a successful invasion of France were greatly reduced.

7. Seventh Milestone, Goal 7—Invasion of France

   The invasion of France succeeded with the combined Unites States and British armies (having first acquired control of the oceans, buildup of a powerful U.S. army in England, air superiority over western Europe by the U.S. airforce and over eastern Europe by the Russian airforce).

8. Eighth Milestone, Goal 8—Defeat of Nazi Germany

   The advancing United States and British armies from the west and Russian armies from the east defeated the German armies under Hitler, ending World War I in Europe.

From this example one can see that strategic and operational planning are related as a succession of short-term tactical goals, achieved under the direction of a long-term mission and strategy.

Operational plans specify the implementation of a series of short-term goals selected and specified by a long-term strategic plan.

We look again at the logic of planning for action, which consists of a set of categories defining:

1. Mission and Stakeholders
2. Objectives and Metrics
3. Scenarios and Knowledge
4. Strategy and Goals
5. Organization and Resources
6. Tactics and Competition
7. Budgets

Now we can see that the first four categories (1–4) are strategically oriented, and the last four (4–7) are operationally oriented. Category 4 (Strategy & Goals) provides the connective overlap between the two planning foci, the strategic plan and the operational plan.

STRATEGIC PLANNING

The logic of a plan provides a way to *analyze* a needed action into its underlying means and ends of the action. A good plan focuses on details for a successful action, the *requirements of its means and ends*.

For example, the case nicely illustrated that military planning was to provide the means of battle to attain the ends of conquest. In the Allied strategy for D-Day, the planning began in January 1942 but could not be undertaken until June of 1944—almost two and a half years later. The reason for this delay were two. First, the Anglo-American allies did not have the military force to successfully conduct an invasion of France. Second, they needed to see if Russia, their other ally, could survive the force of the German armies on the eastern front between Germany and Russia.

The long time required to build strategic capability was first necessary in order to build an adequate size of a military invasion force, not only in terms the numbers of soldiers (more than half a million) but also the numbers and kinds of necessary equipment for an invasion, such as tanks and landing craft. In addition, strategic capability included the necessity of air superiority. The years 1942 and 1943 were also devoted to Allied bombing of Germany (British bombers by night and American bombers by day). Although the bombing never impeded much economic production in Germany, it did eventually deplete the German fighter capability (in addition to German fighter losses over the Russian front), and by 1944, both U.S. and Russian air forces had absolute air supremacy over Europe. Finally, the Russian destruction of German armies on the eastern front in 1943 had reduced the military might of Germany sufficiently to make an Anglo-American invasion for a western front possible.

Strategic planning provides the capability to perform action, even against opposition.

This idea of the length of time for strategic action is also an essential feature of strategic planning. Since present planning is for future action, in strategic management one needs to also be clear about the timeline for planning. Timelines should be established that are essential to the type of required action. Arbitrary timelines in a plan do not make planning clearer and may only obscure the requirements of action.

For example, in the case of D-Day, it was clear to American and British military planning that the time line for the war effort to defeat Germany would be several years. The pressure of their Russian allies, under Stalin, was to begin the invasion of France as soon as possible, to relieve military pressure on the eastern Russian front. Accordingly, the timeline of tactics and goals toward the strategy of invasion and objective of the defeat of Germany required the Anglo-American military plans to set intermediate milestone goals of

1. The invasion of North Africa
2. The invasion of Sicily
3. The invasion of Italy
4. The invasion of France.

We notice that these milestones were opportunistic and set as the strategy of war evolved. What is always concrete in a plan about tactics and goals is the present tactic and next-year goal. All future tactics and goals need to be opportunistic, evolving from the success and failures of current action. Yet, it is the timeline over the clear and steady objectives and strategy which provides the meaning and significance in the intermediate tactics and goals.

In the timeline of sequential action, it is strategy and objectives that provide the coherency over time in planning the future.

As action defines the conditions of strategic planning, so reality measures the quality of strategy in terms of the success or failure. In the case of D-Day, the quality of strategy was measured by the reality of military conquest or defeat. Soldiers were killed or survived, territory was conquered or lost, civilians were enslaved or liberated. Reality was experienced as life or death, victory or defeat. When the assumptions in the military plans were incorrect, reality showed the assumptions as false.

For example, the invasion of North Africa successfully assumed the defection of the Vichy-government French forces to the Allied cause. However, the German military capability in North Africa at first made successes against both the inexperienced American army and poorly equipped British army. It required a year of experience and military buildup and new leadership by Generals Patton and Montgomery to finally create an Allied military success in North Africa. In contrast, the Allied decision to invade Italy in the South in 1943, created a military tactical situation that did not result in a victory over German forces until the final defeat of German in 1945.

Reality determines the quality of planning in terms of the actual consequences of action or inaction.

SWOT Analysis

A particular technique called SWOT analysis became popular in the practices of strategic planning in the United States in the 1990s. SWOT stands for strengths, weaknesses, opportunities, and threats. The technique is of limited use in strategic planning, if properly used within the context of a strategic plan. The SWOT analysis can be properly used to summarize the salient points of an ongoing operation within the **scenarios and knowledge** section of a strategic plan.

Therein the SWOT analysis can summarize succinctly (as a list of bullets) the present capabilities of an operation in light of the planning scenario:

Strengths. A list of the strategic strengths of current operations that should be continued.

Weaknesses. A list of the strategic weaknesses of current operations that need to be corrected.

Opportunities. A list of the business opportunities that need to be added by new operations

Threats. a list of the dangers presented by competitors as they continue of current operations or change to new operations

Within a strategic plan and as a part of the planning scenario, a SWOT analysis is a useful technique for briefly summarizing the implications of a planning scenario on current operations of an ongoing organization.

CASE STUDY: GE Capital Expands in the 1990s

We now turn to how planning is used in a business culture. Formal planning is a kind of management logic that is universal to all action, across countries and across sectors of government, military, or commercial activities. This next case looks at how GE used planning to integrate business practices, as it acquired new firms during its strong growth in financial services in the 1990s.

In 1997, a major economic recession temporarily slowed the previously rapid economic growth occurring in some Asian nations in the second half of the twentieth century—most rapidly in Indonesia, Korea, Thailand, and Japan. One global firm to take strategic advantage of the situation was General Electric, which expanded its GE Capital subsidiary by purchasing Asian financial businesses. As Sheryl WuDunn then reported GE's business strategy in Asia:

> the 1997 Asian financial meltdown and resulting recession turned the area into a vast bargain basement. Here was GE Capital's chance to buy up distressed companies and establish itself in the one part of the world where it lacked a strong presence. "There's no question that financial turmoil has resulted in an environment that facilitates deal creation," Dennis S. Nayde, president of GE Capital, said . . . "we have moved into that opportunity.
>
> —(WuDunn, 1999, p. C1)

GE was exploiting the recession by acquiring distressed firms that could build GE's position as a global power in financial services. In 1998 and 1999, it made eight major investments in four Asian countries, increasing its Asian assets to $20 billion. In Japan, business acquisitions included two consumer-credit businesses, a life insurance company and a leasing company. In Thailand, business acquisitions included a consumer credit business and a portfolio of car loans. In the Philippines, GE Capital bought a life insurance unit. In South Korea, it planned to purchase a major part of a South Korean bank.

This kind of rapid expansion was a rare and strategic opportunity for GE Capital to create a major change in its market position. General Electric had operated in Japan for more than a century, but its subsidiary there had never been strong. Buying established Asian financial businesses at bargain basement prices was, in addition to a strategic opportunity, also a strategic challenge in managing this expansion. For example:

> Six years ago, GE Capital's Japanese presence (in Tokyo) consisted pretty much of Taketo Yamakawa, a one-man show in a small office, scrambling to come up with opportunities in a stifled economy. GE Capital, the biggest and most profitable unit of General Electric, one of America's biggest and most profitable corporations, had just bought a Japanese credit card company that it saw as its gateway to Asia—and had put Mr. Yamakawa in charge.
>
> —(WuDunn, 1999, p. C1)

Yamakawa identified other companies to for GE Capital to acquire and bought the consumer-finance business of Lake Corporation. He had been impressed by Lake's lending business but believed its recent expansion into real estate and stocks had been a big mistake. As Lake's financial problems worsened in late 1997, Mr. Yamakawa had GE Capital acquire Lake. They incorporated Lake's profitable operations into a new GE company named Lake

Company. The other less profitable businesses in real estate, such as golf courses and other properties were left in the original company to sell off.

The new Lake company had 2,800 employees, a computer network, buildings, and its customers. Mr. Yamakawa visited many of Lake's 564 branches. He centralized some of the back-office operations of these branches and set up a call center. This raised productivity and cut costs. Mr Yamakawa also set about integrating Lake's employees into GE's culture. He trained Lake Company's management in the discipline of GE's business planning practices, which included formulating both short-term and long-term strategies and budgeting in detail. Under Yamakawa's leadership, Lake Company's managers planned to become Japan's most efficient and profitable consumer-finance company.

They encouraged the previous company's 1.5 million customers to transfer their accounts from the old Lake to the new Lake, using give-away umbrellas or lottery tickets to coax them. Lake also created new high-value-added services for these customers (even charging 27 percent on its unsecured consumer loans, compared to Japanese bank-deposit rates of around three-tenths of a percent). Lake's customers were also able to use its new automated loan machines. Another added value to the customers was a new loan application procedure using these machines to ease the process of some loan transactions. When unsecured loan applications were approved, the machines dispersed cash immediately. (Borrowers sometimes obtained cash in the morning and repayed in the evening at another machine to avoid lengthy loans at 27 percent.)

Another company GE Capital acquired in 1989 was Toho Mutual Life, then an ailing insurer. GE selected good employees and Toho's most profitable operations and put them into a new company called GE Edison. Here too GE needed to integrate its new employees into its strategic culture. Mr. Miyajima was sent to coach Toho Mutual executives on the GE way. He helped Edison's managers to formulate plans for restructuring the business at milestones of a 90-day plan, a one-year plan, and an 18-month plan. Then if the targets of these plans were not met, GE might turn around and sell that recent acquisition.

These plans also included planning new information strategy for Edison's operations. They redesigned business practices at Edison's headquarters and all 56 of its branches. It was that, in the old awkward operation, a customer seeking a benefit payment at a branch office would fill out a form and staff then entered the claim into its computer, while also sending the form by mail to the head office, where staff would again retype it into its computer. Mr. Miyajima's group had the claims operation transferred to a central location in the city of Yokohama, outside Tokyo to which branch offices transmit all information to the new customer care center. This has produced a 30 percent increase in productivity and reduced overhead.

Thus important to GE's integrating new acquisitions into its culture was encouraging the use of GE's planning formalism along with improvement of operations. GE called its approach to planning and improvement as "workout sessions." Workout sessions were intended to make employees reexamine business problems and practices by taking apart the present business to improve its focus on the customer. These sessions fostered fiery debates, with the walls of a room becoming covered with sheets of paper showing captions and colored ink diagrams of how the business does and should operate. This hot debate about practices was a cultural change, as at first it could be viewed as a kind of hanging out of one's dirty laundry in public. In this perspective, workout culture was initially shocking to the new GE employees in Japan. For example Kohei Tanaka, president of the new Japan Leasing (also acquired by GE) commented that the language of workout and the open vigor of its debates were new to him. Business planning and workout improvement discussions provided a strategic methodology for GE to install a global GE culture in all its companies.

Case Analysis

This strategic acquisition of businesses and integration into GE corporate culture was a core strategic competency of GE. From 1988 to 1999, GE Capital acquired and integrated more than 300 financial-services companies. In this way, GE became the world's then largest non-bank financial institution, with $300 billion in assets. To succeed, GE leadership had to mastered the arts of identifying and acquiring potentially-successful-but-under-performing companies and training its newly acquired employees to turn these organizations into profitable moneymakers, through the strategic culture of formal practices of planning goals and deliberately improving operations.

PLANNING AND STRATEGY

As we saw in this case, it was through GE's taking the strategic opportunities of aggressive acquisitions of troubled financial services businesses in Asia in the 1990s, that GE Capital was strategically positioning itself to become number one in nonbank financial services in the world. We recall from Chapter 6 that Jack Welch, then CEO of GE, had instituted his strategic policy that all the businesses of GE were to be either the first or second in a large industry or be sold off from GE.

The point of his strategic policy was to provide a clear long-term direction for all the managers in GE with a strategic metric that was applicable and adaptable from industry to industry. Look at the size of the industry in which the business operated and measure who was first or second in market domination in the industry. If it was a GE business, then that manager was secure. If not, GE would sell the business, and the manager would go with it. Strategic policies in providing long-term direction told the business planners in GE how to play the game.

Strategic policies should properly define the rules of the game for business plans.

Another point in this case was that the first thing GE would do after an acquisition was to separate out profitable from unprofitable parts of the acquired businesses, keeping the profitable parts in a new GE business and disposing of the unprofitable parts as the old business. This organizational action provided the new GE manager with an opportunity to grow and not be burdened with past unprofitable decisions made before GE acquired the business. Under Welsh, a second GE strategic policy was that GE provided reasonable returns to stockholders. If after an business acquisition, managers spent capital and time trying to turn around a bad business decision, then it was a waste of GE's time.

Strategic policies should not only define the rules of the game but also provide proper measures of organizational performance.

A third point illustrated in this case was that after GE acquired a firm, intensive training introduced the new employees to GE's business culture. They made ninety-day, one-year, and eighteen-month business plans. These clearly set the business targets and ways to meet these targets for the new GE managers in the acquired firm. Work-out sessions were also implemented for manager and employee participation in identifying business problems and ways to improve operations to meet business plan targets. The time-frame of the business plans were short-term to underscore the period required for managers of the new acquisition to have the firm performing to GE strategic policies or it might not remain a GE firm.

Business plans should provide clear operational targets in a clear time frame, provided the targets are strategically realistic.

The work-out sessions were as important to the GE culture as was the business plans. Planning targets that cannot be reached merely make for employee despair and not incentive. If a plan requires improvement in productivity, profitability, or sales growth, then a management process to find ways to do these things is an important part of the planning process.

ORGANIZATIONAL CONTEXT OF PLANNING

Whether in the realm of the military or in the realm of business, strategy and planning are intended to guide future actions. Strategy and planning analyze an action as into means and ends. Strategy defines the long-term direction, and planning the short-term means and ends. What makes strategy and planning complex is that these cognitive activities are performed within the contexts of groups and organizations. No single mind of any individual in a group or organization cognates all strategy and planning. Individuals must depend on each other.

For example, in the case of D-Day, Roosevelt and Churchill, in consultation with their military advisers made the strategic decisions on a series of operations—invasion of North Africa, invasion of Sicily, invasion of Italy, invasion of France. These were the long-term strategic goals for the western Allies in Europe. Their generals planned the battles, the short-term goals. For example, Eisenhower planned the invasion of North Africa and the invasion of France. Montgomery planned the battle of El Alamain and the invasion of Sicily. Clark planned the invasion of Italy. (And incidently, Patton saved the invasion of North Africa, Sicily, and the invasion of France twice, once off the beaches of Normandy and again at the Battle of the Bulge in the Ardennes forest. Things don't always go as planned.)

For example, in the case of GE Capital's acquisitions in Asia, at the top executive level of GE, its CEO Welch had set the long-term goals of the strategy for GE to be in the industry of financial services and also set the near-term plans of aggressive acquisitions of Asian firms during the Asian recession. His managers in Asia planned the acquisitions and implemented the integration of acquisitions into the GE firm. At their level, their strategy consisted of deciding which distressed financial companies GE should buy. And at their level, their planning was deciding which businesses of the acquisitions to keep and how to improve their operations.

In the strategy and planning of action for groups, different executives and managers (leaders and generals) play different roles in formulating and implementing strategies and plans.

This is what makes both the ideas of planning and the processes of planning both critical and difficult in organizational contexts. All the actors involved have to strategize, plan, and act well for the whole organization to be effectively led and operated.

An organizational culture that fosters and rewards management for thinking strategically, planning well, and effectively implementing good strategy and plans provides the basis for the long-term and short-term success of business, and this is what strategic management is all about.

CASE STUDY: Rise and Fall of Osborne Computer in the 1980s

Next we turn from strategy and planning in operating businesses to the special case of planning a new business start-up. This case is special because in a business start-up, strategy and planning are inseparable and events are more surprising than in the case of an existing business.

For example, in the strategic planning of the D-Day invasion we looked at changes in operating organizations (e.g., war undertaken by existing governments in the middle of the twentieth century). Strategic planning for *changes in ongoing organizations* differs considerably for strategic planning for *starting a new organization*. In an ongoing organization, strategic planning need focus on *changes to operations,* whereas in a new start-up organization, strategic planning needs to plan a whole new organization and then quickly change plans as events unfold the dynamics of the new venture.

We recall (from Chapter 2) how the dynamics of a new business venture can be anticipated as a set of key milestone events that need to be successfully met. Now we look at how planning in a new venture needs to deal with such milestones and in the context of strategically planning a whole new organization. We will see that starting a new business, strategic planning and operational planning cannot be separated, because the short-term is so critical. We will next look at the appropriate planning format for a new business, which presents fully integrated strategic and operational plans. To see this, we now examine a case of how rapidly a new company can go from instant success to instant failure.

This example occurred in the early days of the new personal computer industry in the 1980s. As information technology progressed in the second half of the twentieth century, computer innovations provided several periods of innovation and new business start-ups. During the early 1980s, the innovation of the personal computer in information technology provided many instances of new businesses. Of these firms begun in the then new personal computer industrial sector, only Microsoft, Dell, Apple, and Compaq survived into the 1990s. Still, it is instructive to look back at some of the new start-ups and failures in a historical setting in order to clearly see the action and dynamics of action in new business ventures.

This case study is of Osborne Computer, long gone and forgotten, yet it still provides a historically interesting example of how swiftly a new company can rise and fall in times of hot innovative action. It is rare to find so clear and swift a case of rapid rise and fall of a new high-tech business. The case of Osborne Computers nicely illustrates the swift dynamics of strategy and action in new enterprises.

Osborne shipped its first computers in July of 1981. Only two months later, it had reached its first $1 million in sales. By the second year sales had grown to $100 million. Yet six months later into the third year, Osborne Computer went bankrupt. The rise was due to entrepreneurial strategy, and the fall was due to wrong product strategy (Osborne and Dvorak, 1984).

Historically, the personal computer market began in the late 1970s. Adam Osborne was one of many interested in the new information technology of the personal computer, which was based upon a new electronic component, the semiconductor chip central processing unit or microprocessor. This chip of a microprocessor put all the computational operations in a computer on a single transistorized silicon chip, thereby making relatively cheap computer possible. Osborne wrote and published books about programming microprocessor-based computers. We recall that Jobs began Apple in this same period.

By the summer of 1980, Osborne was working for McGraw-Hill, to whom he had sold his computer book publishing company. He decided to create his own computer because he saw an awkwardness in the early market of personal computers. At the time, computers were only being sold as components and separate software—separate computer, disk drive, monitor, printer, software. Osborne decided to package them as a portable computer and to sell it for less than competitors' equivalent component sets. (This product would grow into the laptop product line—but it was still too big to fit on anyone's lap.)

Osborne incorporated a new company and hired Lee Felsenstein to design the electronics for the computer. Then he presented his ideas for the new venture to Jack Melchor, a venture capitalist. Melchor invested $40,000. It was an instant success. Priced at $1,795, the Osborne was more conveniently packaged and more completely equipped at a lower price than any other personal computer of that time. Its sales soared and its distribution spread rapidly. By the end of the first year, sales of the Osborne personal computer soared to one third of the then leader in personal computers, Apple. Osborne had created a new market niche.

Raid production expansion was needed as sales grew so rapidly, and Adam Osborne managed this expansion by subcontracting for all components. He had planned only to assemble and test units. All components were purchased, and all printed-circuit boards constructed by subcontractors. The parts were purchased by PH Components, and the boards stuffed and tested by Testology.

There were tactical problems. Of course, one should expect problems in any plan and respond quickly and flexibly by solving tactical problems. The first problem was

> Osborne Computer Corporation quickly ran into problems in August and September of 1981 when Testology was unable to make a logic tester work, and consequently was stuffing and testing many boards but delivering very few. We were nevertheless still committed by contract to pay PH Components and even for those components that Testology could not make it work.
> —(Osborne and Dvorak, July 16, 1984, p. 55).

The second problem had to do with increasing production to meet the rapidly growing demand for the product, and this problem Osborne solved in time. However, a third tactical problem in the new venture was in the product design. This Osborne did not solve in time, and it would kill the company!

In a new venture, all tactical problems are dangerous and, if not solved in time, can kill the venture.

Here is how it happened. In the original design of the transportable PC, Osborne had chosen a very small 5-inch screen. This made it very difficult to read and write anything on the screen. It did not provide a standard 80-letter-wide writing space (like the width of standard typewriter paper). In the summer of 1982 Osborne's competitor Kaypro introduced a similar model with a 9-inch screen and 80-column width. Kaypro had corrected Osborne's display mistake, with a larger screen showing 80 characters. This small difference in features made a big difference in the application of the personal computer as a word processor. Kaypro sales soared. Osborne sales plunged.

This is a general lesson for innovative new businesses:

A new product concept can create a new market niche and/or alter existing market structures. Competitors enter the new market, focusing on obviously weak features of the innovative product.

Osborne had chosen a 5-inch display to minimize interfering radio noise inside the computer. Product design always requires trade-offs on desirable features. He later planned to enlarge the screen, but delayed the introduction of the larger-screen model too long.

Then another tactical problem occurred, which together with the tactical problem of the small-screen would doom the company. This was an unfortunate delay in the generation of additional capital through a public offering. Osborne had planned it for the summer of 1982 but put it off until early 1983. By then, the brokerage firm decided not to make the offer because of the sales slump that began the preceding summer. Potential Osborne customers decided to wait for Osborne to offer a new model with a larger screen, like the Kaypro screen.

Kaypro's competition and rumors that Osborne was preparing an improved model had created the sales crisis. Sales slowed in the fall of 1982 and by the following spring were nonexistent.

Under a sharp drop in income, the capitalization of a new business is the critical tactical factor of immediate survival. Improving capitalization and a new product were the immediate tactical problems for Osborne. He had begun planning for a public offering and a new president in the fall of 1982. Unfortunately, then the rapid drop in sales began.

Robert Jaunich was hired as the new president and started in January 1983. The Osborne board selected Saloman Brothers as lead underwriter for the public offering, and preliminary meetings between the underwriters and Osborne management also began in January.

Earlier, Osborne had started the company with $100,000 in capital, raising $900,000 more in February 1981. Yet by September 1981 (two months after shipping the first products in July 1982), production expansion required more capital. On Friday, September 4, 1981, the Bank of America informed Osborne that it would not provide the requested loans without additional equity. Over the weekend, another $1.6 million was raised from venture capital sources.

But financial trouble again appeared in December 1982, just before the new president and underwriting talks began. That month, the Bank of America refused to extend the line of credit for the company above $10 million. The troubles grew. The underwriters, after conducting due-diligence interviews issued an opinion that the time was not right for a public offering. Their reasons were that the transition to new management was incomplete, two new products were to be introduced which might be delayed, and the company's future success depended on how well the new products did, and there was continuing instability in financial forecasts.

The losses kept increasing. On April 6, 1983, the president told stockholders that the company would report a loss of $1 million for the most recent year, ended February 26, 1983. On April 22, the underwriters were told that the loss was revised to $4 million, On Monday, Osborne said that he was told the loss would be more like $8 to $10 million. An emergency meeting was held on April 29, as described by Osborne:

Here are the highlights of the April 29 board meeting:

1. Januich claimed he had no knowledge of possible poorer financial data until April 21.
2. Januich observed that the death of the Osborne 1 had occurred faster than anyone could have predicted.
3. Dennis Bovin of Salomon Brothers, on behalf of the underwriters, stated that they would work with the company to raise money providing top management remained and no evidence of fraud or misrepresentation was found.

—(Osborne and Dvorak, July 23, 1984, p. 47)

Another meeting was held on May 5, in which the board resolved that management should try to raise $30 million in new equity. At that time the equity base was $13 million. Existing venture capital investors added a new $12 million.

On April 17, the new model, Executive, was formally announced and shipped during the first week of May. Its price was set at $2,495, and the price of the Osborne 1 was raised to $1,995. The prices were too high for the immediate inflow of cash that was then needed: "Executives were not selling at $2,495 and Osbornes were barely trickling out the door at $1,995" (Osborne and Dvorak, July 23, 1984, p. 50). Competition from Kaypro continued to hold Osborne's sales down. For a company with small capitalization and large losses, an immediate upturn in sales was desperately needed. It didn't come. In September 1983, Osborne Computer Co filed for bankruptcy.

Thus after the summer of 1982, Osborne's sales decline had created a financial crisis. November and December passed without significant sales. January, February, March, and April passed without sales. The Osborne Company was going through a cash-flow hemorrhage. Bank loans could not be increased, and the public offering had not occurred. Osborne made several attempts to raise new capital privately, but there still was not enough cash to carry the company. In September 1983, the Osborne Company declared insolvency under Chapter 11 of the bankruptcy law. All the millions of equity on paper only a year before had vanished.

Incidently, in 1983 Kaypro, its competitor, had one of its most profitable years. Family-owned Kaypro offered 5 million shares to the public in 1983. They realized $9 million from the sale of part of their equity. Yet after another year, another new start-up called Compaq drove Kaypro out of the computer business. Compaq came out with a product that looked similar to the Kaypro, but it had the 8088 Intel chip and Microsoft operating system that the new IBC PC was using. Kaypro also failed to make a timely product change to meet new competition.

Case Analysis

We saw in an earlier chapter the kinds of milestones that new ventures must achieve for long term business success:

1. Acquisition of start-up capital
2. Development of new product and/or service
3. Establishment of production/delivery capabilities
4. Initial sales and sales growth
5. Production and distribution expansion
6. Meeting competitive challenges
7. Product improvement, production improvement and product diversification
8. Organizational and management development
9. Capital liquidity

Osborne Computer failed to successfully meet the milestone of matching competition. In new ventures, the point of planning is to anticipate milestone challenges and to be prepared for them. When they occur, unprepared, competition may not give a new business time to change.

What is striking to think about in this failure is that only two years later a company like Osborne was started that did grow to a major and long-term business success. It was a new hot computer business startup begun with a computer package that was 8088 a copy of the Osborne—only with Kaypro's larger sized screen, Intel's 8088 CPU chip, MS-DOS operating system software, and Phoenix input-output software to emulate the IBM PC. That company was Compaq, which became a huge success and was still selling personal computers and servers in the year 2000. With a few product changes, Osborne's company could have followed a path like Compaq's and might have become a major success instead of a classic example of a new venture meteor burning brightly in the commercial sky briefly and plunging to earth.

BUSINESS PLANS FOR NEW VENTURES

Planning for the start of a new business is different from planning for an ongoing business. In the new business start-up, everything must be planned de novo— from strategy to operations. In a continuing business, strategic planning is on *change to current operations*. Now let us look at format of a plan for a new business.

All new businesses need a business plan, and venture capitalists require one before considering an investment. The business plan expresses in detail the entrepreneurial vision and plan for a new enterprise system. The purposes of the business plan are:

1. To identify the complete set of policies and strategic assumptions of the business model of the new business
2. To chart the course and identify the resources needed for the new venture
3. To attract venture capital

To meet these purposes the format of a business plan for a new venture needs both to describe the strategic business model for the venture and to forecast the progress of the venture. Therefore, the name business plan is somewhat of a misnomer, for it is more a model of a new enterprise and forecast of action rather than a plan of an ongoing operation.

As we saw before, a strategic plan for an ongoing operations focuses on changes in strategic direction and planning the details of the next term of operations. A strategic plan for an ongoing operation does not contain the strategic business model, but is built upon it. Accordingly, the format of a business plan for an operating business using the categories of

1. Mission and stakeholders
2. Objectives and metrics
3. Scenarios and knowledge
4. Strategy and goals
5. Organization and resources
6. Tactics and competition
7. Budgets

In contrast, a new venture business plan should explicitly contain the strategic business model. Accordingly, the format for a new venture business plan differs from that of a strategic plan of an ongoing organization. In writing a business plan for a new venture, topics explicitly lay out the strategic business model for the new start-up. We recall (from Chapter 4) that a complete set of business policies for a strategic business model should include:

- Innovation strategy
- Product strategy
- Production strategy
- Marketing strategy
- Diversification strategy
- Organization strategy
- Information strategy
- Competitive strategy
- Finance strategy

We can use these categories to construct the format of a new-venture business plan, whose format covers the following topics:

1. *Executive Summary*. This is a one-page summary of the highlights of the business plan, written last and placed first. Its purpose is to generate sufficient interest by a potential investor to read the whole plan. The strategic concept of the new enterprise should be summarized as to how the business should provide functional capability to customers, who they are and what are their application needs. Then the concept should identify what kind of product or service the new business will provide for that customer and how it will provide value to the customer. About this assumption of adding value to the customer, it is important to be very explicit, for the pricing of the

product or service depends on how much value it provides for the customer. Translating "value-addedness" into price is one of the most critical assumptions an entrepreneur will make in a business plan.

2. *Innovation Strategy*. The innovation strategy should identify and discuss the new idea that is being brought into the enterprise to provide new kinds of products or services or to improve production or delivery of current kinds of products or services. When the innovation is in a product or service, the discussion should indicate how it will change customers' applications or create new applications. When innovation is in production or organization for an already existing and standard product or service, then the value-addedness must translate into significant cost reduction and production quality improvement for the technological innovation to provide an entering competitive edge for a new business going up against existing competitors.

   One needs to address the continuing progress in the innovation. What further research needs to be done, who will do it, and how will the new business acquire and implement it?

3. *Product Strategy*. The next section should describe the concept of the new product or service that the new business will produce and market. The technical specifications of the product or service should be detailed, along with the current state of the development and design of the product. If the product is still in development, then a development schedule should be given, and technical risks in the development identified and described. Careful attention should be paid to identifying the technical risks and schedule for two reasons. The first is to make sure that sufficient capital is raised to carry through the development and begin manufacturing and sales. The second reason is to protect the entrepreneur by due diligence from potential law suits by investors, if development fails or falls behind schedule.

4. *Production Strategy*. The business plan must also envision how the new product/service will be produced and the capital required to establish production. Production planning will require judgments about what parts and materials to produce and how to fabricate or assemble the product or service. The trade-off judgments here are capital and learning costs of establishing production versus loss of control over proprietary knowledge and costs through purchasing. The advantages of producing in-house are that costs and quality can controlled and a proprietary technology can used in design and/or in manufacturing, but this comes at capital costs. The advantages of outsourcing and purchasing parts and even fabrication is that this reduces capital requirements but cannot provide any competitive advantages over competitors (who can source the same parts and materials).

   It is also important to estimate the capital required to expand production. For a new venture that is quickly successful, the most common way for an entrepreneur to continue to dilute equity is to need second and third rounds of investments to expand production.

5. *Marketing Strategy.* The marketing strategy needs to identify the potential customers for the new product or service and the applications context in which these customers will use the new product or service. The marketing strategy should also identify the customer requirements for the product or service and the price bracket for which the customer may pay for the product or service. The marketing strategy should identify the distribution channels for getting the product to the customer, and the costs and problems in setting up or entering these distribution channels. The marketing strategy should also plan the sales force, how they are organized and rewarded. The marketing strategy should identify efficient and effective means of advertizing and distributing information about the new product or service to potential customers.

6. *Diversification Strategy.* It is also important in the business plan not only to identify the initial product or service but also a planned family of products and product lines and services that the business will evolve. It is rare that a single product will be sufficient to build a successful company. A product family and product lines are usually necessary for long-term commercial success.

7. *Organization Strategy.* Who are the initial management team? The experience and credentials of the management team for the new business should be described. This is very important because experiences of successful venture capitalists have emphasized that what investors are basically investing in is the management. The organization of the business and operating procedures should be planned, and how staff will be recruited and trained should be outlined.

8. *Information Strategy.* How will the enterprise use information technology? What will be its "bricks and clicks" balance? In what parts of the operation can progress in information technology provide a competitive advantage?

9. *Competition Strategy.* The competitive strategy should identify the way the new business intends to compete and what its planned competitive advantages are. It is important to benchmark competing products or products for which the new product or service may substitute. An important feature of such benchmarking is specifying technical performance and features of competing products and prices of these. The plan should show the rate of anticipated technology diffusion of the new product or service into the market, and critical assumptions which facilitate or hinder that diffusion.

10. *Financial Strategy.* The financial plan should be constructed to anticipate the cash flow of the operation through the critical milestones of a new venture. It should begin with a sales projection and planned growth and

penetration of the market over the first five years of operation. For these sales projections, the financial plan should then forecast income, expenditures, and profits for the first five years of operation. In addition, working capital and balance sheets should be constructed for each of these years. Additional needs of further financing should be identified and discussed. The financial plan should show projected return to investors as increase in equity. Finally, the financial plan should have a cash-out plan so investors and entrepreneurs can gain liquidity.

USING THE NEW-VENTURE BUSINESS PLAN

As we noted, one use of the new-venture business plan is to attract venture capital and to plan the adventure. For example, many business schools have established student competitions to write new-venture business plans. One of these was an annual student competition in business plans during the 1990s held by MIT. In 1996 and 1997, the MIT Enterprise Competition offered a prize of $50,000 dollars to the student submitting the best business plan for a new company.

The competition was limited to MIT students, but nonstudents could team with an MIT student to submit a business plan. The competition announcement nicely summarized the purposes of a business plan and its audiences:

- A business plan is a document that conceptualizes the totality of a significant business opportunity for a new venture
- Presents the organizational building process to pursue and realize this opportunity
- Identifies the resources needed
- Exposes the risks and rewards expected
- Proposes specific action for the parties addressed

Audiences for a business plan included the founding team, potential investors, and employees, customers, and suppliers or regulatory bodies (MIT, 1996).

Also, planning the new business adventure is important because all action requires a kind of completeness of means for successfully attaining an end. For example, one must produce a product before having something to sell, and one must sell the product to obtain revenue, and one must collect the revenue to obtain a cash flow, and one must pay for production and sales of the product—all of which together can create a profit on investment. This completeness of performing all the actions necessary for a successful business is what is crucial to plan. For if not planned, it may not happen—and then the business fails.

It is true that nothing ever goes according to plan, but without some planning nothing ever goes. The plan should lay out what must be done to make a new business go and what the assumptions are upon which the plan is based.

When nothing ever goes to plan, it is because either the plan was not fully implemented or some assumptions of the plan turned out not to be valid. One needs to know both these things for successful planned action:

1. What needs to be done
2. What assumptions need changing for revising planned action

Good plans control the completeness of a successful action and can identify the faulty assumptions in unsuccessful action.

EVALUATING PLAN FORMATS

As we discussed earlier, all planning is intended to provide guidance for future action. Therefore, the quality and quantity of planning needed is determined by the use of a plan to facilitate successful action. In this book, we are examining the full scope and complexity of the idea of strategic planning. We have shown how elaborate a strategic planning process can be with the full set of formal techniques of planning scenarios, business strategic models, strategic vision, and formal strategic and operational plans. How much of this set and how detailed the use of the set depends upon the complexity, risks, and uncertainty of action.

When the environments have not been changing and current operations are successful with high profits and steady growth, little formal planning needs to be done except scanning the horizon for changes in technology, markets, competition. When environments have been changing, new competitors are entering, markets are altering, and the old tried-and-true operations are no longer being successful, then detailed planning needs to be done and strategic change undertaken.

The amount and detail and extent of strategic planning required for successful action depends on the conditions of business.

As we saw in the case of the invasion of Europe, a massive amount of planning and logistics build-up was necessary to assemble and successfully land an army of a half million Allied soldiers on the beaches of Normandy in 1945.

Moreover, when businesses are new and beginning, not a great deal of planning is necessary—only as much to detail assumptions and identify critical milestones and convince investors to fund the new business. The important thing in a new business is not only planning but quickly adapting to new action—meeting challenges of product development and production, sales, production expansion, product changes, competition, keeping cash flow within working capital bounds.

In cases of brand new action with little prior experience, adaptability of action is more important than detailed planning, which soon is out of date anyway.

SUMMARY: USING THE TECHNIQUE OF PLANNING LOGIC

The plan format for ongoing businesses and new businesses differ, and the format of the strategic plan needs to be adopted as appropriate.

1. Choose the appropriate plan format
   - An ongoing, large business needs to focus primarily on changes in current operations and should separate the strategic plan from the operational plan to facilitate the focus on changes.
   - A new business start-up needs to use a format that does not separate strategic and operational planning and should focus upon the business model, assumptions, and the milestones of the new venture.
2. Keep the strategic plan as brief as possible
   - Ongoing organizations should focus on *change* as actionable items, targets, responsibility, measurable outcomes.
   - Even for a very large organization, when an annual strategic plan becomes too long to read in a brief time and too dull to keep interest in reading it, then the planning process is flawed.
3. Use the plan to monitor progress
   - If a plan cannot be used to monitor progress of activities through appropriate milestones and goals, then the plan can not help identify unwarranted assumptions and unanticipated risks in the strategic business model of the company.
4. When devising a plan, keep in mind why plans fail
   - Plans fail when future action is not well thought through.
   - Plans fail when unrealistic assumptions and targets are expressed in the plan.
   - Plans fail when competitors and markets do unanticipated things.
   - Plans fail when the form of planning overwhelms the substance of planning.
   - Plans fail when too much bad luck simply overwhelms action.

5. Use the planning process to know when and how to revise action

- The planning process should not be simply a formal exercise for management to show that they appear to know what they are doing. A planning process will eventually put the company at risk if it does not

—Seriously probe and question the current business conventions

—Effectively scope the horizon for important change

—Facilitate real and effective top-down-and-bottom-up communication about operations and change

For Reflection

Read books on the history of the Xerox company. Also search for Fortune magazine articles on Xerox from 1960 to present and read them. How did planning guide the company at first to great success, and then how did planning later fail at Xerox? If you were in charge of Xerox when the twenty-first century began, how would you plan Xerox's future?

CHAPTER 7

STRATEGIC VISION

PRINCIPLE

Strategic vision is an intuitive view of the future focusing upon desirable changes in strategic perception, commitment, preparation, and policy.

STRATEGIC TECHNIQUE

1. Prepare a strategic vision statement
2. Specify strategic perception
3. Specify strategic commitment
4. Specify strategic preparation
5. Specify strategic policy changes

CASE STUDIES

Sony Corporation
Musashi: A Samurai's Vision
Welch: A CEO's Vision

INTRODUCTION

From the interactions of the top-down and bottom-up perspectives in the strategic planning process, both a planning scenario and corporate strategic models are formulated. From these anticipated and needed changes, a strategic vision for the future direction of the corporation can be synthesized. We now look in detail how a strategic vision occurs and what it does and how it is expressed and used. We will find that a good strategic vision succinctly expresses the reason for change, the shareholders in change, the opportunities and challenges in change, and the strategic action items for change.

Basic principles of traditional strategy have always included the relationship of leadership to strategic vision and change:

1. Strategic vision is the fundamental responsibility of leadership, since top management only has the authority to make major changes in operating organizations.
2. Strategic change is only periodically necessary, but to be effective such change must be envisioned, anticipated, and planned.
3. Sources for strategic vision are either external in the environments of the organization or internal as opportunities developed within the organization.

This importance of effective top-level leadership in strategic vision has long been recognized. For example, the early writings of Chester Barnard emphasized that strategy was an important executive function (Barnard, 1938). Many other writers, such as A. Chandler (1962) and J. B. Quinn (1981), further emphasized the executive importance of rethinking strategy when structural change has made past strategy ineffective. John P. Kotter, for example, observed more than 100 companies trying to change themselves to become better competitors: "A few of these corporate change efforts have been very successful. A few have been utter failures. Most fall somewhere in between . . ." (Kotter, 1995, p. 59).

Kotter saw strategic vision of leadership as the vital role in change, suggesting that several steps were necessary to change an organization (Kotter, 1995, p. 59):

- Establishing a sense of urgency for change and forming a high-level management coalition for change
- Creating a vision of change and communicating that vision through the organization
- Empowering managers to act on the vision and planning for and creating short-term wins about change

- Consolidating improvements and institutionalizing new approaches for continuing change

The strategic problem is to anticipate innovative change.

For example, in historical hindsight, it is easy to pick out why and even when a new technology was substituted for an old technology. Initially, a new high-tech product will probably not perform a given function as well as an existing product. Its potential for substitution lies in a natural advantage in the progress of applied knowledge, but there is never any guarantee that a potentially large natural advantage will ever be realized. All new knowledge-based products or services require considerable incremental improvement in performance before their performance will surpass that of existing products. Moreover, while the incremental improvements to the new product are being sought, competitors are spurred to seek incremental improvements in their existing products. So the performance goal the new knowledge-based product is aiming for is a running target—it is a race between radically new knowledge and improvements in existing knowledge-based products.

Also we recall the transilience of innovations can impact a firm by either preserving or destroying competencies in production and/or in marketing; therein lies the challenge of strategic vision. Current corporate strategies can successfully deal with regular and niche-creation innovations, since the knowledge and skill base of the organization is not affected. (In the latter case, of course, marketing skills must be sharpened and new markets entered.) However, the revolutionary and architectural innovations require true *strategic reorientations*. Revolutionary innovations require reorientation of production competencies, and architectural innovations require orientations of both production and market competencies. We recall that we saw the importance of next-generation products in a previous case of Steve Jobs' experiences in strategy.

Next-generation products are always either revolutionary or architectural innovations, requiring restructuring of production and/or market competencies.

Also in reviewing strategy theory, we saw how in the literature on strategy, the principle of vision has been one of the most puzzling aspects of strategy to students of strategy. What constitutes vision—except that it derives from intuition based upon experience? Why is the intuitive vision of one leader judged in history really visionary (e.g., Jobs) and others who fail (e.g., other Apple

CEOs) are judged reactionary or laggard in vision? Can the capability of being visionary be learned by anyone, or are the truly visionary only born and not made?

Furthermore from Mintzberg's writings, we have seen that this scholarly puzzle about vision arises from the basic cognitive fact that vision is based upon the intuitive function of the mind and not upon the analytical function—intuitive thinking versus analytical thinking. Leadership vision is an intuitive activity of the top leadership of an organization and difficult to prescribe in the kind of analytical, rule-based procedures in which business education abounds and in which formal plans get expressed. Much of the problem about strategic vision of leadership in large organizations has to do with the difficulty in some management cultures of providing processes that encourage *intuition* in a group context as opposed to *analysis* in a group context.

CASE STUDY: Origin of Sony Corporation

We begin to look at the idea of strategic vision in a business by reviewing a historically famous and important case of vision in the origin and growth of Sony Corporation. This case provides a clear illustration of the role of strategic vision in the development of a globally successful firm in the second half of the twentieth century. Sony's founders were Akio Morita and Masaru Ibuka.

In 1944 in the last years of that war, Morita was finishing his education and developing his interest in the new technology of electronics. Morita was a university student studying applied physics under Professor Tsunesaburo Asada at Tokyo University. Asada's laboratory was performing research for the Japanese navy. Morita worked for him on electronics. Physics applied in the technology of electronics is an area of applied knowledge and in the interface between science and technology. It is in such science/technology interfaces that advanced technologies are created. Morita was being trained by Asada with a scientific orientation but motivated to create advanced technology. This provided a lasting influence on Morita as a basis for his future strategic intuitions in business.

We recall from the case of Mitsui how Japanese culture in embracing the challenges of industrialization encouraged an active interest in acquiring and innovating technology. Asada wrote a short weekly column for Tokyo newspapers, in which he explained the latest developments in research and technology. Sometimes when Asada was busy, he would let his student, Morita, would write these for him. One column Morita wrote was about the theory of atomic energy, in which Morita noted that the energy in the atom was so great that if a weapon could be made using the atom's energy it would be an extremely powerful weapon. But in 1944, no one in Japan knew that their then enemy, the Americans, had made such a weapon which soon would be used upon Japan.

Morita was trained to be attentive to learning the new advances in science and to imagining what technological use could be made of them. Morita viewed science as source of ideas for new technologies. Viewing science as a source for new technology was been the hallmark for successful vision of high-tech firms in the twentieth century.

As Morita neared graduation, he had not yet been drafted. But the war was still going on, and he learned that he could enlist as a career technical officer in the navy by taking an examination and then be assigned to a research facility instead of a ship (which would allow him to finish his education). In early 1945, he was assigned to the optics laboratory in Office of Aviation Technology at Yokosuka. There he worked on the problem of preventing static electric streaks on the films of aerial mapping cameras.

After graduating from the university, he underwent a four-month officer's indoctrination and training course and received the rank of lieutenant. He was assigned back to the optical division and began supervising a special unit working on thermal guidance weapons and on a night-vision gun sight. This was in a small town south of Kamakura on Sagami bay:

> We were based in a big old country house in Zushi. . . . The house was built on the Western style, faced with stucco, with a courtyard garden . . . The house was built at the foot of a cliff just above the beach, and I took a room at the nearby Nagisa Hotel. . . . It seemed incongruous because sometimes it was as peaceful as any beach resort, yet we were right under the return path of the B-29's."
>
> —(Morita et al., 1986, p. 29)

The American long range bombers, B-29s, were dropping incendiary bombs on major Japanese cities, such as Tokyo, destroying entire cities. Morita was worried about the American bombers but felt lucky to be at Sagami bay. During the July and August of 1945, the Tokyo-Yokohama area was bombed almost daily. Morita could watch the big silver B-29's passing overhead. Watching with him was a colleague on the special project, Masaru Ibuka, who owned his own electronic company. Morita and Ibuka became good friends (and after the war, they would found the Sony corporation). One of their worries was that the military might not give up the war no matter how badly it was going. An American invasion would likely occur there where they were stationed on the Miura Peninsula at Sagami Bay. They knew that this peninsula would become a bloody battleground, a last battleground for the fanatical Japanese military.

The next year in August of 1946, Morita took leave to visit his family at their ancestral home in Kosugaya. While he was there, the two atomic bombs were dropped on Nagasaki and Hiroshima; and grudgingly, the military did finally obey the Emperor's order to surrender. Morita recalled the occasion:

> I was shaken awake by my mother in the early morning . . . she said the Emperor Hirohito was going to make an announcement on the radio at noon. Even the announcement that the emperor would speak to the nation was stunning. . . . The emperor's voice had never been heard by the Japanese people . . . Because I was, after all, a naval officer, I put on my full uniform, including my sword, and I stood at attention while we listened to the broadcast. . . . the high, thin voice of His Majesty came through. . . . "The war was over."
>
> —(Morita et al., 1986, p. 34)

The post-war world was one of grim survival. Morita returned to his station on August 16. American occupation forces arrived without incident. Morita waited at the station for days without orders. Finally Morita received the order to close the project. He bartered equipment for railway tickets and sent the staff home:

> The new period of peace was strange. The bombers did not come anymore, but many cities looked as though there was nothing more to bomb. . . . The Morita family was fortunate because we had lost no one in the war and the company offices and factory in Nagoya and even our home, survived with no serous bombing damage.
>
> —(Moria et al., 1986, p. 41)

The Morita family business was a brewery run by Morita's father. Although elder sons, such as Akio Morita, were expected to succeed their father in the business, His father was still healthy and robust and did not yet need him. Besides Morita's interests were in advanced electronics, and he accepted a position to teach physics at the Tokyo Institute of Technology.

There he looked up his friend Ibuka. Ibuka was starting a new electronics company. Morita did not really want to teach. He wanted to create new technology. He decided to join Ibuka in founding a new company. He left the Institute and worked with Ibuka. His father invested in the new firm, which first they called Tokyo Telecommunications Engineering Company and later renamed Sony.

EXPLORATION AND REPETITION IN ACTION

We will pause in this case to discuss a basic idea to the concept of strategic vision—the differences between *repetitive action* and *exploratory action*—planning to go where one has been before and planning to go *where no one has gone before*.

Successful vision is based upon the same conditions that all successful intuition rests—an experiential base. Therefore, the basis of experience provides the grounds for and foundation of intuition. Repetitive action provides a detailed experiential base for detailed planning. Exploratory action provides a inexperienced base for only sketchy planning.

Repetitive action can be planned in operational detail, since one has gone there before and the route of the action has been well mapped. This is the common kind of planning performed in organizations for the short-term time horizons— quarterly and yearly planning. For example, if you plan to drive from New York to Portland in the United States, you can purchase detailed road maps and plan the route and times of travel precisely. However, the first U.S. citizens to explore the route of what came to be known as the Oregon Trail had no such maps and had to explore and find the route, using the help of American Indians living in the area at the time. Strategic planning is about exploration into an uncharted future.

As we see so far in the case of Sony, the experiential base for the founding of Sony was Morita's and Ibuka's (and all of Japan's experiences) in the defeat of the country in war. The future of Japan was forced to be a new exploration for survival, since the occupation of Japan by the United States was then determined to democratize the governmental institution of the nation and to forbid any immediate reestablishment of its former military might.

A plan is a systematic approach into the future, but known and unknown futures require different balances of analysis and intuition in the planning activity.

An example of this is a set of construction projects. To the extent that one has previously designed similar buildings and used similar processes, the new construction project will be a customization of such previously used designs. Such a construction project can be carefully analyzed and planned in detail and schedule, using the prior experience of the previous designs and projects. In this case, project planning tools (e.g., PERT) are extremely useful in laying out the scheduled map for the project.

To the extent a future is unexplored, unique, and nonrepetitive, analysis in strategy becomes less important and intuition more important.

In the cases of uniqueness, such as radically innovative projects, then the project is more of an exploration into the unknown than a similar repetition of the projects of the past. Such nonrepetitive project planning is more like being on an adventure into unknown territory. Although one may start with a plan, it will prove to be true that much more important than the initial plan is the preparation, supplies, and the ability to be flexible and adaptable and capable of rapid learning in completely new environments.

In exploration, the strategic vision is more important to success than is the initial plan.

Strategic visions are necessary to long-term planning because nothing in the long-term future ever really happens quite as planned. Also the farther into the future, the more it happens that one learns what the real goals are. It is a strategic vision that enables the explorer to effectively use a planning process as a real guide in any long-term action:

- The strategic vision allows the planner to exploit unplanned opportunities and to revise goals.
- The strategic vision allows plans to cope with unanticipated delays and problems.
- A delayed and altered plan can be rescheduled and redirected because the strategic vision will provide continuing long-term direction.

CASE STUDY: SONY Corporation, Continued

We continue the case of the origin of Sony, noting that the situation then facing Ibuka and Morita in starting their firm was one of a very uncertain future. They could not then have devised a meaningful and detailed plan since postwar Japan would be new and unchartered territory.

In starting the new firm, Morita and Ibuka were going into unexplored territory. Their new firm was a company with only a strategic vision about being high tech, but they had no specific product plan.

Morita's partner, Masaru Ibuka, had been born in 1908 in Nikko City, north of Tokyo. His father was an engineer, but he had died when Ibuka was only three years old. His mother was a graduate of Japan Women's College (Nihon Joshi Daigaku) and had taught kindergarten. Ibuka was inventive and had always been fascinated with technology. As a boy, Ibuka had liked radio and built one using three vacuum tubes. But since vacuum tubes then were expensive in Japan, Ibuka had fabricated his own tubes. Then Ibuka went to Waseda University and studied engineering. He patented the first of his many inventions while a student. (During Ibuka's whole life, he continued to invent and design, so that at the age of 83 in 1991, he had acquired 104 patents. He would design Japan's first transistor radio, transistorized television, a videocassette recorder for home use, the Walkman personal stereo, and a compact disc player.)

In 1946, when Morita and Ibuka formed their new firm with $500, they did not even have a product plan. They searched for a product to produce; and they considered producing radios. But Ibuka decided this was a bad commercial strategy, since he was certain the existing large Japanese electronic firms would soon produce radios and be unwilling to sell components to others. Moreover, the radio was then a standard technology and could not provide their new company with an innovative competitive edge. Ibuka was anticipating the competitive conditions in the industrial value-chain. Ibuka knew that they had to have a product for which they were the technology leaders. A small firm has a chance against bigger competitors only with strategic leadership in technology and knowledge.

But what competitive edge? What product? Immediate products were necessary for cash flow survival. At first, they made shortwave radio adapters to enhance the medium-wave radios that were widely owned in Japan. Shortwave could receive broadcasts from other countries, an important source of news in occupied Japan. Shortwave radios were in short supply, and this product began their business.

Next they noted that many Japanese households had prewar phonographs that needed repair. They began making new motors and magnetic pickups. American things were arriving in Japan, and the American swing and jazz records were very popular. But the parts business was not a future. They still wanted to produce a completely new high-tech consumer product—that was their strategic vision.

At that time they knew of the wire recorder, which had been invented in Germany just before the war. Ibuka found a company, Sumitomo Metals Corporation, that could make the special kind of small, precise-diameter steel wire for such a recorder. Ibuka decided to produce a wire recorder. But there was a problem. Sumitomo was not interested in a small order from a new, untried company; and they would not then be able to produce a wire recorder. Ibuka and Moria continued to keep the company afloat with their small parts business.

Meanwhile, the U.S. occupation forces had taken over the Japan Broadcasting Company, NHK, and needed new technical equipment. Ibuka was familiar with audio-mixing units and submitted a bid to make one for the U.S. forces. He received a contract and made the mixing unit. When delivering the unit to the NHK station, he saw a new American high-tech product, a Wilcox-Gay tape recorder, which the Army had brought from the U.S. It was the first tape recorder Ibuka had ever seen. He looked it over and could see immediately that it had technical advantages over wire recorders.

In the wire recorder, the wire had to pass over the recording and playback heads at very high speeds to obtain decent fidelity in the reproduction of sound. But the tape of the tape recorder with its wide size provided a much larger magnetic area for signal recording and therefore could be allowed to travel much more slowly in providing fidelity of reproduction. Also, to get a long enough playing time, wire reels had to be very big; whereas the larger magnetic area of a recording tape meant that tape reels could be much shorter for the same playing time. And the wire could not be spliced as tape could, in order to correct recording errors or to rearrange recorded sequences. For these reasons, Ibuka knew that tape would inevitably replace wire in audio recorders. This was the new high-tech product for which he had been searching! Ibuka understood what kind of electronics, physics, and chemistry would be required to copy and develop the new technology. Ibuka decided tape recorders was to be their product.

One sees here a strategic vision at work. Ibuka was committed to a high-tech consumer electronic product—he had no plan, just a commitment. He was looking for a new high-tech product with which to compete with the bigger but slower-moving established Japanese electronics firms. He saw an innovative foreign technology and decided to duplicate it. His decision was based upon the distinct advantages of the new technology.

There are many technical problems in innovating a new product, but Ibuka and Morita knew how to proceed—with systems analysis of the high-tech product and with new applied knowledge through the process of applied scientific research. The realized the critical part of the system which they did not know how to make was the recording tape. Earlier, they had worked on wire recorders as a possible high tech product and understood how to make the mechanical and electrical components of a recorder.

The tape was a subsystem of the new high-tech product, providing the recording medium. The tape would be composed of materials that were physical and chemical based. The rest of the system was based in electronic circuitry and mechanical processes. Ibuka and Morita wanted to produce the tape as well as the machine, in order to obtain a follow-on sales business. This commercial strategy to produce tape as well as tape recorders occurred immediately after they had made a decision to produce tape recorders.

Ibuka and Morita's first technical problem was the base material for the tape, but again there was a supply problem. The American firm 3M was then the source of the base tape material, but they were also the major tape producer. Ibuka and Morita thought they might not wish to supply tape to a small Japanese competitor then.

There were severe materials shortages in post-war Japan. Ibuka and Morita found they could not obtain cellophane in Japan, which was their first idea for a base. They tried it, even knowing it might prove inadequate. It turned out to be wholly inadequate. They cut it into long quarter-inch-wide strips and coated it with various ferromagnetic materials. The cellophane stretched hopelessly after only one or two passes as tape on the machine. They even hired chemists to improve cellophane but without success.

They needed another material. Morita went to one of his cousins, Goro Kodera, who worked for Honsu Paper company. Morita asked him if they could produce a very strong, very thin and very smooth kraft-paper for their tape. Kodera said he would try. He did, and it worked. They had a base material, which they sliced into tape, using razor blades at first.

The next problem was the magnetic coating to put on the paper tape. Nobutoshi Kihara, one of the young engineers they had hired, did empirical research by grinding up magnetic materials for the powder for the tape. At first the magnetic materials were too powerful for the system. Then Kihara tried burning oxalic ferrite into ferric oxide. That worked! Morita and Khara searched all the pharmaceutical wholesalers in their district of Tokyo, finding the only store that stocked oxalic ferrite. They bought two bottles of it, cooked it in a pan until it turned brown and black (the brown stuff being the ferric oxide they wanted and the black stuff being ferrous tetraoxide they didn't want).

Next they mixed the brown stuff with Japanese lacquer and tried to airbrush it onto the tape. That didn't work. They tried brushes. They found that a brush made from the very fine bristles from the fur of a raccoon's belly did the best job.

Still it was terrible tape, of very poor fidelity. They really needed plastic tape. Finally, they got a supply of plastic tape. With a supply of plastic tape and the ferric oxide coating, they had a recording-tape technology. This was the critical technology for their new product.

Tape became a major source of cash-flow for Sony. They continued to put heavy development into it. Many years later, in 1965, IBM choose Sony magnetic tape as their suppliers for IBM computer magnetic data storage.

But back to the year 1950. Ibuka and Morita's first tape machine, which used their new plastic-based magnetic tape, had turned out to be very bulky and heavy (about 75 pounds) and expensive (170 thousand yen). Next they began learning something about the marketing of new and expensive high-tech products. Japanese consumers simply wouldn't buy it. They had to look for another market.

At the time there was an acute labor shortage of stenographers, because during the war so many students had been pushed from school into war material production. Ibuka and Morita demonstrated their new tape recorder to the Japan Supreme Court and immediately sold it twenty machines. It was the breakthrough sale for their new high-tech product.

We see that from their strategic vision to produce leading-edge high-tech consumer products, they were learning some product and marketing strategy.

They redesigned their tape recorder into a medium-sized machine (a little larger than an attache case). They also simplified it for a single speed and sold it at a much lower price. They then sold their modified product to schools for English language instruction to Japanese students.

As their product and marketing strategy progressed, they next developed intellectual property strategy. To obtain a high-quality recorded-signal, they had purchased a license to an invention patented by a Dr. Knszo Nagai, which was a high-frequency AC bias system for recording. This demagnetized the tape before it reached the recording head, reducing background and prior recording noise. At that time, the patent was owned by Anritsu Electric, and they bought half-rights in the patent from them in 1949. Eventually when Americans imported U.S. made tape recorders into Japan using the AC bias system technology, the U.S. firms had to pay royalties to Sony. This encouraged Sony to be aggressive about intellectual property.

By 1950, the new company had products, tape recorders, magnetic recording tapes, and intellectual property. They were using their strategic vision, but they still did not have a strategic plan. That was to come next.

In 1952, Ibuka decided to try exporting their tapes and recording machines to the United States. He visited the United States to study its markets, and since he had earlier read about the invention of the transistor at Bell Labs, he also visited Western Electric in New York (then the patent holder on the transistor). Ibuka was impressed by the new technology. He wanted it. In the following year in 1953, Morita went to America to purchase a license to the transistor from Western Electric for $25,000 dollars—a big sum to the new company in those days.

Ibuka had appreciated the inherently great performance advantage that transistors potentially had over vacuum tubes. A transistor could operate at a fraction of the size and with a fraction of operating current. Ibuka and Morita knew that any business which made portable consumer electronics products would have eventually change from vacuum tube circuits to transistorized circuits.

This was the beginning of Sony's strategic plan—transistorized circuitry and miniaturization.

Tubes to transistors—so obvious! Now we might say that in hindsight. Yet ponder this strategic mystery. It is a historical fact that the U.S. consumer electronics industry, which in the 1950s was the greatest electronics industry in the whole world, had almost completely disappeared by 1980! And the reason was a failure of technology strategy. The U.S. consumer electronics firms failed to transform their products from tubes to transistors in a timely, committed manner. That little transistor—an American invention (and its follow-on key invention, the integrated circuit semiconductor chip)—was the technical key to the rise to world dominance of the Japanese consumer electronics industry and the corresponding demise of the American consumer electronics industry.

Ibuka and Morita then had a strategic plan which focused upon a core technology competency—transistorized electronics. But the transistor invention had to be improved to use it in a radio. They had to improve the frequency response of the transistor to a wider range.

The problem with the original transistor invented at Bell Labs was its poor frequency response. The original transistors were constructed out of two kinds of semiconductors, arranged like a sandwich, in which the middle slab controls the current flow between the outer two slabs. Since current in semiconductors can either be carried by electrons or by holes (holes are unfilled electronic orbits around atoms), one can design either hole-electron-hole carrier combinations (positive-negative-positive: pnp) or electron-hole-electron combinations (negative-positive-negative: npn).

The original Bell Labs transistor had a pnp sandwich of germanium-indium-germanium. Electrons (the negative carriers) inherently move faster through a semiconductor than holes (the positive carriers). The physical reason for this is that holes wait for an electron to put into its empty orbit from a neighboring atom before that empty orbit appears to have moved from one atom to another. This is inherently a slower process than a relatively freely moving electron passing by one atom after another.

So the first thing the Sony researchers had to do to make the new technology of the transistor useful was to speed up the signal processing capability of the transistor by using electrons rather than holes as carriers. The Sony researchers accordingly reversed the order of the transistor sandwich: from a positive-negative-positive structure to a negative-positive-negative structure (indium-germanium-indium). The development of the transistor by altering its phenomenal basis from hole conduction to electron conduction is an example of a knowledge strategy.

If Ibuka's and Morita's new electronics firm had been staffed with only electronics engineers and without any scientists, they would not have been able to understand the new physics of semiconductors. They would not have had the technical imagination to begin developing the transistor, knowing they could reverse the combination to seek a higher frequency response. Ibuka and Morita had established a firm with both an innovative electronics technical capability and also with an innovative applied physics capability. That's how Morita was trained. Applied physics underlay Morita's knowledge strategy.

The new firm had good research physicists. In fact, they were so good that during the course of the transistor research one of them, Leo Esaki, discovered a new fundamental phenomenum of physics—quantum tunneling (in which electrons can sometimes tunnel through physical barriers that would bar them, if they obeyed classical physical laws and not quantum physics). In 1973, Esaki won the Nobel Prize in physics.

The next problem the researchers faced was the choice of materials for the bases of the transistor and its impurities. Without adding a small quantity of different atoms "doping," neither germanium or indium conducts electricity. The doped atoms "impurities" make these materials semiconducting, as opposed to nonconducting. They decided to discard the indium used in Bell Labs' original version of the transistor. Indium had too low a melting point for use in a commercial transistor. They tried working with the combination of gallium with antimony as its doping atom. That didn't work well either.

Next they tried replacing the doping element of antimony in the gallium with phosphorus. At first, the results were not encouraging, but they persisted. Eventually they found just the right level of phosphorus doping. Then they had an npn transistor of gallium-germanium-gallium structure, with just the right amount of phosphorus atoms doping the gallium materials. Sony researchers had developed a high-frequency germanium transistor, which was commercially adequate for their pocket radio.

So the radio—the consumer product that Ibuka would not produce a few years back (since at that time the new firm had no technological competitive advantage)—would now become a second flagship product line, the pocket radio. In 1955, they produced their first transistorized radio in a small size, as a pocketable transistor radio. However, since the radio turned out to be just a little larger than a standard men's shirt pocket, they did sew a slightly larger pocket on the front of their salesmen shirts, when they went out to market the new product.

The development of the transistor for radio application is an example of the Japanese acquisition of a foreign-invented applied knowledge and the subsequent improvement of that knowledge for commercialization by their knowledge asset capability of applied research. This pattern of acquisition of foreign-originated knowledge and subsequent improvement of applied knowledge for commercialization was the common pattern in both early and later industrial development of Japan that led to its emergence first as a world-military power and second as a world-economic power.

With their new knowledge strategy of the transistor and the new product of the pocket radio, Ibuka and Morita decided to change the name of the new firm. They now had global aspirations, and they wanted a globally recognizable name. They changed from the Tokyo Telecommunications Engineering Company to Sony.

When Sony introduced its transistorized pocket radio into America, they discovered that Texas Instruments had independently innovated a transistorized pocket radio. But Texas Instruments had no strong commitment to the consumer market, and they soon dropped the product. Sony was committed to the consumer electronics market and began their climb to a world leader in consumer electronics.

This is the second element of a strategic vision—commitment. TI's commercial successes remained in the industrial and military markets, where its real heart was—its strategic perception and commitment.

Sony focused upon the consumer electronics market and became an innovative, high-tech, top-quality consumer electronics firm and a giant, global company. Sony also introduced the first transistorized small black and white television set. In color television, Sony innovated a single-gun, three-color TV tube. It innovated the Walkman series of miniature audio players. It innovated the first home video cassette recorder (VCR), after the industrial version had been invented in the United States.

Ibuka and Morita had imbued their new corporation with a strategic vision and strategic plan which searched for and focused new technologies on advanced consumer electronics products—with a corporate technology competence in transistorized and miniaturized products.

Case Analysis

We see in this case the starting and building of a major company on the founders' strategic vision. They had no operational plan, only a direction. The experiential base of their vision was the postwar situation with its overwhelming need to rebuild Japan upon an advanced high tech basis. Their other experiential base was an understanding of progress in the knowledge area of physics and the applied knowledge area of electronics. Upon these experiential bases, their vision was to build a new company upon an area of advancing applied knowledge. To do this, they committed to exploration for a new business. They found, acquired, developed, and applied new knowledge on electronics to build innovate consumer electronic products. They acquired new applied knowledge by licensing the new technology of the transistor. They then had a knowledge strategy to continue to develop the new technology through research. Their product strategy was then to apply the new technology in a stream of new miniaturized products, such as pocket radios, walkmans, VCRs, and so on. Their strategic vision created new markets through innovation.

STRATEGIC VISION: PERCEPTION, COMMITMENT, PREPARATION, POLICY

We recall in the review of the schools of strategy that there were two schools that especially emphasized the roles of cognition of learning in the formulation of strategy:

- **Cognitive School.** This school focused upon the cognitive base of strategy, including intuition. It emphasized the role of information and knowledge structures in formulating strategy and included a constructivist view of the strategy process that sees strategies as creative constructs of what reality could become.

• **Learning School.** This school viewed strategy as a kind of learning process in which formulation and implementation interact for the organization to learn from past planning and experience.

The lessons from these schools urged us to carefully examine how the cognitive function of intuition operates in strategy formulation and how intuition interacts with an experiential base for learning. To do this we will look at the key subideas in the concept of strategic vision: perception, commitment, preparation, and policy.

Strategic Perception

As we earlier noted, all visions are formed upon an experiential base. *Thus the first subidea in the idea of strategic vision is that of perception.* Perception is about what one sees in the world, experiences. In the case of Morita and Ibuka, the immediate and stark perception in Japan then was the loss of the war, the terrible devastation, and the desperate need for survival. The perception of the future that, Morita and Ibuka shared was the opportunities in consumer electronics. Their commitment was to begin a high-tech consumer electronics company. The preparation to do so required the search for a first high-tech product, which they found in the wire recorder. Their policy was to innovate, and their product would not be the wire recorder but an innovative tape recorder. Perception includes past experiences and projected future possible experiences.

In strategy to innovate for the future, the strategic perceptual space needs to include relationships between knowledge, product, and applications. Consider the sketch in Figure 7.1, wherein the ideas of applied knowledge, product, and application are sketched together as forming a kind of three dimensional space in which to perceive, see, and think about an innovation strategy in a high-tech business—as in the case of Sony.

We recall from the second chapter that in an industrial life cycle, the market for a new applied knowledge product does not begin until innovative new products (or services) incorporating the new knowledge are available to be used by a customer in an application. The customer must see the value of the innovative product/service sufficient to justify its price. Therefore, the focus of innovation strategy for the application of new knowledge is to identify an initial application and a product/service specifications of the customer requirements for the application.

Strategic Commitment

The second ideational component of a strategic vision is commitment. Of the visionary perception of the world, to what within the vision does one com-

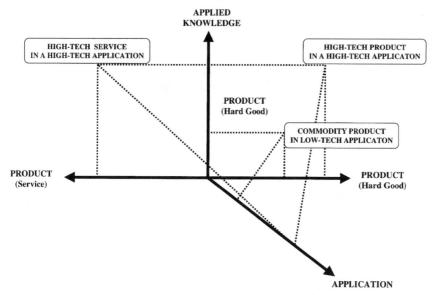

FIGURE 7.1   DIMENSIONS OF STRATEGIC PERCEPTION IN BUSINESS

mit? A strategic vision is not only a perception of the future but also a commitment to a direction of action in the future.

In the Sony case, Morita's and Ibuka's strategic perception centered upon the applied knowledge area of electronics (and later transistorized electronics), but their strategic commitment was to consumer electronics. Therefore, they had to match perception to commitment, which they did first with tape recorders and then with transistorized consumer electronic products.

Strategic Preparation

The third ideational component of a strategic vision is preparation for the future. Once a strategic perception and commitment is made in a strategic vision, one needs to prepare for future action in the direction of the commitment. Preparation usually requires acquiring knowledge and skills and resources for future action.

Morita and Ibuka prepared for the commitment to consumer electronics by licensing technologies and doing applied research to develop the applied knowledge into products. They built the resources necessary to pursue this first by manufacturing radio parts kits and then by manufacturing and selling the tape recorder.

Strategic Policy

Finally, the translation of strategic vision (which has strategic perception, commitment, and preparation) into organizational action is through the formulation of appropriate strategic policies for the organization.

For example, Murita's and Ibuka's strategic policies were to seek out new applied knowledge, develop knowledge asset capabilities in the new knowledge through applied research, design innovative new consumer electronic products, and market them worldwide.

Strategic perception, commitment, preparation, and policies are the four fundamental components of a strategic vision.

ROLES OF ANALYSIS AND INTUITION IN STRATEGY

Now with this understanding of strategic vision as involving strategic perception, commitment, preparation, and policies, we can better regard the proper roles of analysis and intuition in strategy:

- In strategy, intuition arises from an experiential basis as a strategic perception and commitment, which together creates a visionary goal.
- Analysis can then detail the preparation and policies necessary to carry out a strategic vision.

Strategic intuition is the synthesis of a committed perception about the world. Strategic analysis is a means of realizing a world shaped by a strategic intuition, through preparation and policies. We recall that of the many Western students of strategy, Henry Mintzberg has been a particularly strong advocate of the importance of the role of intuition in the strategic process (Mintzberg, 1990). And we recall that analysis and synthesis are complementary cognitive functions—analysis is taking apart a complex concept into simpler component ideas, and synthesis is taking discrete, apparently unrelated ideas and constructs them into a new coherent idea. Analysis divides; synthesis unites.

Intuition, which creates wholes, does not always create the correct totality, nor is intuitive strategy always successful. In business strategy, the market is ultimately the test of whether a strategic intuition is proper or improper, a success or a failure. The problem in business strategy is how to evaluate a strategic intuition with strategic analysis. A strategic intuition about visionary perception and commitment should then be strategically analyzed in preparation and policies for the intuitive vision. If the strategic preparation is possible and economic and the strategic policies are concrete and feasible, then the intuitive visionary perception and commitment have a real chance of being implemented.

Some firms have put elaborate procedures in place for a strategic plan and then been disappointed at the outcome, erroneously concluding that strategic planning is useless. Yet if they had not distinguished analytical strategy from intuitive strategy, they could not know whether the failure was one strategic analysis or of strategic intuition.

Although the market is ultimately the test, the evaluation of a correct strategy prior to implementation is not easy. Strategy is always a top management responsibility. One way that management fails is to blame every failure in strategy on implementation by lower staff. Actually every failure in strategy is also a failure by top management, because good strategy should include strategy for implementation.

One symptom of a failure in strategic vision is when a strategic planning process results in a voluminous planning report, which is then neither read nor used. Another symptom is when a large corporation creates an expensive corporate research laboratory that remains isolated from the firm's business units and is always being criticized by management as being unrelated to the firm's businesses. Another symptom of a lack of strategic vision in a diversified firm is when a strategic business unit's product development incorrectly implements research results from corporate research because the business divisions are looking sideways and backwards toward competitor's past technology and not forward toward technology that can leapfrog a competitor's capabilities. Other symptoms are when a firm's productivity lags and continues to lag behind competition, when new products not getting out in a timely manner, and when a firm is falling behind competitors in the race of competence.

An analytical-strategic plan of preparation and policy cannot substitute for an intuitive-strategic vision of perception and commitment but must follow it.

Thus what we have seen illustrated in the case of Sony was how the company was constructed and guided by a strategic vision with strategic components of the perception and commitment of their founders on the importance of advanced technology applied to consumer goods, their preparation in developing corporate competency in innovation, and their *policies* producing and marketing high-tech products for global consumer markets. Good strategic vision is thus a concrete ideational process in action—not some vague, nebulous thing—that "strategy thing." The components of strategic vision constitute very concrete elements for 'intuition' in strategy as formulating corporate goals, perception and commitment, followed by 'analysis' in strategy for preparation and policy.

In Western culture, the term *intuition* is one of those terms that is widely and frequently used, yet remains vague and poorly defined. Outside of the specialized areas of Jung-Myer-Briggs schools of personality and gestalt psychology and writers on creativity and "breakthrough" thinking, one finds in the literature very little concrete, prescriptive, profound, how-to-do-it specification of intuition and vision. How does one create totalities? How does one conceive conceptual wholes? How does one perceive and individuate objects? How does one synthesize?

Thus in Western management literature, while there are many books on creativity, there is no consensus as to how synthesis occurs or how to facilitate synthesis. For example, even in Mintzberg's emphasis on the importance of intuition in strategy, he himself offered only a vague description of intuition: "Intuition is a deeply held sense that something is going to work. It is grounded in the context in which it is relevant and based on experience of that context. I cannot be intuitive about something I know nothing about" (Campbell, 1990, p. 109).

And this negative judgment is probably the most certain thing one can say about intuition, from the Western literature on intuition:

One cannot be successfully intuitive if one lacks a proper experiential base.

Yet this is the nub of the problem of action. If one must be intuitive to perceive correct strategy for an action, how can one gain the experience which only action provides, but prior to the action? This is why strategic commitment in vision must be followed by strategic preparation and policy.

CASE STUDY: Musashi: A Samurai's Strategic Vision

As Mintzberg pointed out, many of the American and European business writers on strategy have had conceptual difficulty in pinning down the idea of vision. Why? I think it might be (to paraphrase a former U.S. president) that the "vision thing" is a "cultural thing." One finds in non-Western writing on strategy (e.g., Chinese or Japanese writings), a more direct approach to intuition and strategy. In this case, we will examine an older version of strategy, written by a samurai warrior, Minyuomoto Musashi (from a way of life, which as we reviewed in an earlier chapter, the present traditions of Japanese management culture sprang—samurai to chonin to manager).

One can conjecture that the roots of the cultural difference between the European/American and Chinese/Japanese management thought about strategy lies in their different historical roots of philosophy. In medieval Europe, the literate caste was a priestly caste, and the warrior caste was generally illiterate. Whereas in feudal Japan (after the Tokogawa *shogunate*), the warrior caste was generally literate as were the priests.

Of course, priests have little experience with action, whereas action is the business of warriors—actions of battle and war. Primarily the experience of priests lies in contemplation and reflection, and this is what they teach their students. The medieval universities of Europe arose from cathedral schools and drew their faculty heavily from clergy. They, of course, taught what they knew and thought deeply about in their own experiential context—the experience of contemplation and reflection. Thus the classical philosophies of Europe arose as contemplative philosophies on existence. A famous example of this was the dictum of the sixteenth century philosopher, Renes Descartes, who (after explicitly rejecting religious dogma) premised: *Cognito, ergo sum.* (I think, therefore I am.)

This is not to criticize European philosophy, for after all, it was the origin of natural philosophy, and from natural philosophy, came modern science. Still, it is important to understand that there do exist philosophies focused on the problem of action as opposed to philosophies focused on the problem of existence.

Now compare that existential philosophical orientation in the West to the action orientation of some Eastern philosophies, most notably Japanese Zen Buddhism. Here the contemplative philosophy of traditional Indian Buddhism was transformed into a way to sharpen intuition for action.

A famous example of the adoption of Zen Buddhism for the way of action was Minyuomoto Musashi's *A Book of Five Rings.* This book has been long popular with Japanese management and became very popular with U.S. readers in the 1980s (Mushai, 1982). Musashi was a literate and highly cultivated warrior—but he was a warrior. Before he turned to writing, Musashi claimed he fought sixty sword duels without a single defeat. Paraphrasing Musashi, we can list nine of his admonishments to warriors:

This is the Way for warriors who want to learn my (Musashi's) strategy:

1. Do not think dishonestly.
2. The Way is in training.
3. Become acquainted with every art.
4. Know the ways of all professions.
5. Distinguish between gain and loss in worldly matters.
6. Develop intuitive judgment and understanding for everything.
7. Perceive things which cannot be seen.
8. Pay attention even to trifles.
9. Do nothing which is of no use.

How do these principles constitute a kind of philosophy of action? First, if one thinks dishonestly, one cannot be honest with oneself; and without self-honesty one cannot think clearly.

Second, the path to the attitude, the Way, must be in training if the attitude is one of action; for without training, one cannot act with skill.

Third and fourth, action always occurs within a complicated, holistic context—a system, in which many skills or arts and professions may be required to carry out an action. (This is why, for example, in innovation teams are necessary including personnel with technical, marketing, production, and financial skills.)

Fifth, action always requires some resources, tools, or supplies, and it results in outcomes that either deplete or replenish and/or improve previous resources. Thus all action must be strategically judged in gain or loss of resources.

Six and seven refer to the importance of the intuitive facility in the heat of action. Actions are complex, contradictory situations of flux and motion in which totalities must be instantly synthesized and comprehended. Hence a warrior's (or a manager's) intuitive ability is as important as his or her analytical ability. Musashi's precepts here argue that in the heat of battle, intuition and perception are critical cognitive functions, and hence must be trained and exercised before battle.

In the eighth precept, Musashi emphasized that in action, a detail can sidetrack or defeat a whole project (e.g., bugs in software programs can destroy a whole company's product reputation; in another industry, a poorly mixed antifreeze solution caused thousands of new automobiles to be recalled, harming the company's reputation for quality).

Finally the ninth precept emphasizes that in action, economy is important, for action consumes resources. Action must be focused, disciplined, and economically executed.

In summary, we see from this example, that the precepts are sensible ways to prepare for and behave in battle—they are efficient and effective guides to action—that is, a philosophy of action.

STRATEGIC VISION AND CORPORATE CULTURE

A corporate strategic vision resides both in its culture (its experiential past) and in the philosophy of its top leadership (its experiential present). We recall that one of the schools of strategy particularly emphasized the importance of culture to strategy:

Cultural School. This school emphasized the role of culture (as opposed to power) in the formulation and implementation of strategy.

Also we recall also that the strategic corporate model is a kind of model of the totality of a business. It rests upon a set of assumptions, the fundamentals of the business. As Lowell Steele nicely expressed this:

> Every business is based ultimately on a few simple ideas, principles, or even assumptions. They address the fundamentals of the business. . . . In the aggregate these fundamental features could be termed *the concept of the enterprise* . . . They are often so deeply internalized that they become invisible . . . The concept of an enterprise . . . can be thought of as a kind of internal guidance system that keeps a business on trajectory."
>
> —(Steele, 1989, p. 71)

Any given corporation may or may not be conscious of its culture, its basic concept of its enterprise. But if it is conscious and has procedures to inculcate culture, then it could be said to have a "way" (using a term from the oriental cultural writings on strategy). Strategy, as "way," is a philosophy of action, requiring perception, commitment, preparation, policy. The perception is the continual awareness of the greater scheme of things during all moments of action. The commitment is the plan at the moment of action. When Steele's business conventions are articulated as maxims or precepts for management, they form the "way" of that company's culture. (As for the example, the formal policies for the acquisition and integration of companies into Cisco became a strategy "way" at Cisco in order to continue to stay ahead in information technology.)

Where do such conventions arise for management? They arise from past successful past business experience, evening going back to the beginning.

> These beliefs and conventions are not so much taught or inculcated as they are absorbed. Many of them so deep in the bones that they are not even evident to those who live by them. They may persist for decades and literally go back to the foundation of a company.
>
> —(Steele, 1980, p. 71)

Perception and commitment arise from an experiential base—past business success. These then ground the future expectations of the business of the firm—management's concept of the enterprise. Together, perception and commitment form the base of the intuitive corporate culture, corporate "way." We recall that Steele listed several kinds of assumptions that form the way of a business, including;

1. Common understandings and even rationalizations about what business you are in.
2. Shared assumptions about the way you gain competitive advantage.
3. A joint sense of the manner in which the company grew, how it got to where it is.

4. Shared conventions about the quality and extent of information needed for decision making.
5. Shared or perceived conventions regarding the guidance and operational control of the enterprise and the way it will evaluate its performance.

—(Steele, 1989, p. 70)

Thus intuitive business vision is the shared concept of the enterprise of a leadership team. This concept of the enterprise is the set of assumptions about business practice and strategy that constitute the abstractions and generalizations (theory) of the "existential" basis of leadership in an organization. It implicitly and explicitly summarizes a management team's (and its predecessors) experience in succeeding to grow and maintain a competitive organization. The team's shared beliefs and conventions may not all be capable of being articulated at any time by all its members, but they, in aggregate, do form the critical assumptions behind strategy of how business is successfully practiced in that firm.

Steele also nicely summarized the evidence of this existential basis of prior business experience in generating the enterprise concept:

Clearly, a company's past history strongly influences its character. This is particularly apparent in companies that have played a powerful role in the growth of their industries. . . . Goodyear, Du Pont, General Electric, General Motors, Caterpillar and IBM. . . . Not surprisingly, that history of success also contains the seeds of trouble.

—(Steele, 1989, p. 82)

This is also an important point about vision and change—vision is partly based upon past, successes of the past, but change in the environment may not allow that past practice to be successful in the future.

Strategic vision is the difference between the successful enterprise of the past and the successful enterprise of the future.

CASE STUDY: Welch: A CEO's Strategic Vision

The above philosophical discussion may strike some as almost too ephemeral for a hard-headed corporate strategy, yet we should remind ourselves that strategy is one of the principal responsibilities of top management of a corporation. An example of this was the management philosophy of Jack Welch when he was CEO of the General Electric Corporation in the 1980s and 1990s. About midpoint in his tenure, Stratford Sherman wrote an article on Welch's strategy:

His (Welch's) ideas are simple: Face reality. Communicate clearly. Control your own destiny. But put together, they could rewrite the book on how to run a big company."

—(Sherman, 1989, p. 39)

Such precepts are really elements of a philosophy of action—part of Welch's personal philosophy for the manager.

At the time, Welch's management philosophy attracted many observers. For example, Noel Tichy and Ram Charan also discussed Jack Welch's philosophy and its implementation:

Jack E. Welch, Jr., chairman and CEO of General Electric (in 1989), leads one of the world's largest corporations. It is a very different corporation from the one he inherited in 1981. GE is now built around 14 distinct businesses, including aircraft engines, medical systems, engineering plastics, major appliances, NBC television, and financial services. They reflect the aggressive strategic redirection Welch unveiled soon after he became CEO.

—(Tichy and Charan, 1989, p. 112)

Sherman described Welch's reputation after that strategic redirection:

Neutron Jack, as he is sometimes called, is widely regarded as one of the world's most ruthless managers. The truth is more complex. Some of his actions are indeed harsh, and he antagonized people inside the company and out by fixing something they didn't think was broke. What is becoming clear only now is how those moves fit into a larger plan to strengthen the enterprise and to make its remaining employees more secure."

—(Sherman, 1989, p. 39)

Jack Welch graduated with a Ph.D. in chemistry from the University of Illinois. He went to work for GE as a chemical engineer in its plastics business at Pittsfield, Pennsylvania. GE had invented a new thermoplastic called Lexan, which has an exceptional structural strength. It was a technical innovation that could be a large business for GE, but at the time Lexan had no markets and few sales. Welch saw its potential and pushed its commercialization. At age 27, he gained managerial responsibility for GE's plastic business (as a profit-and-loss center). He remained in Pittsfield for seventeen years, increasing the plastics business at an average earnings growth of 33 percent a year.

In 1977 he was promoted to GE's corporate headquarters as a senior vice president for consumer products. The CEO of GE then was Reginold H. Jones. Jones had become GE's president in 1970 and had turned GE from a state of chronic shortage of cash to financial strength by 1977.

In the 1960s GE had taken on, at the same time, three major new technological areas: mainframe computers, nuclear energy, and commercial jet engines. All three had gobbled up money for research and development, but GE had succeeded commercially only in jet engines. IBM beat them in the computer market, and the nuclear industry turned environmentally and politically sour (Banks, 1984). Jones closed down the computer business and nuclear industry business to get GE's financial health back in shape.

Jones had a financial background, and in 1980 he chose as his successor the technically trained Welsh. But it was also important to Jones that Welch had risen through the ranks of production. Jones thought production experience would provide the understanding of how businesses should be run. Jones had viewed the strategic problems of GE as alternately requiring strategic focus upon financial knowledge and upon technical knowledge, with an appreciation that both must be strong. As the new CEO, Welch continued change at GE:

> In 1981, Welch declared that the company would focus its operations on three "strategic circles"—core manufacturing units such as lighting and locomotives, technology-intensive businesses, and services—and that each of its businesses would rank first or second in its global market. GE has achieved world market-share leadership in nearly all of its fourteen businesses. In 1988, its 300,000 employees generated revenues of more than $50 billion and net income of $3.4 billion.
>
> —(Tichy and Charan, 1989, p. 112)

While Jones had put GE back on its financial feet, the GE that Welch was to lead then was typical of the large financial conglomerates that many managers assembled in the 1960s and 1980s—diversified but without any integration or synergy between businesses. Welch rationalized its businesses into fourteen.

Welch believed that a strategic attitude for a conglomerated company must have not only a financial strategy but also a market and technology strategy.

> He (Welch) sees global markets inevitably coming to be dominated by fewer, ever more formidable players—steamrollers like Phillips and Siemens and Toshiba. To prosper in this world, Welch believes, GE must achieve competitive advantages that allow it to rank first or second in every market it serves. So often is this simple concept repeated around GE, people express it as a single, seven-syllable world: "number-one-and-number-two."
>
> —(Sherman, 1989, p. 40)

In 1989, in the U.S. market and in the world market, GE was first in aircraft engines, in circuit breakers, in electric motors, in engineering plastics, industrial and power systems, locomotives, and medical diagnostic imaging: "Welch loves big, complex businesses with only a few competitors" (Sherman, 1989, p. 41).

When Welch was asked what he thought made a good manager, he replied:

"I prefer the term *business leader*. Good business leaders create a vision, articulate the vision, passionately own the vision, relentlessly drive it to completion." (Tichy and Charan, 1989, p. 113)

You see this is an expression of a manager with a strategic vision—an intuitive philosophy of action. The traditional functions of management are to plan, acquire resources, organize, implement, control, and supervise. Welch is emphasizing that even before planning is vision. Vision derives from philosophy of action which derives from perception and commitment upon an experiential base.

Sherman (1989, p. 50) summarized the six rules of Welch's strategic philosophy:

- Face reality as it is, not as it was or as you wish it were.
- Be candid with everyone.
- Don't manage, lead.
- Change before you have to.
- If you don't have a competitive advantage, don't compete.
- Control your own destiny, or someone else will.

Case Analysis

It is interesting to compare Welch's aphorisms to those of the samurai from Chapter 4:

1. Do not think dishonestly.
2. The Way is in training.
3. Become acquainted with every art.
4. Know the Ways of all professions.
5. Distinguish between gain and loss in worldly matters.
6. Develop intuitive judgement and understanding for everything.
7. Perceive things which cannot be seen.
8. Pay attention even to trifles.
9. Do nothing which is of no use.

One sees the similarity in strategic philosophy—although Welch was a manager of the twentieth century and Musashi a warrior of the eighteenth century. What they had in common was a philosophy of action.

The "Tao," the "Way," the "Strategy" is the perception, commitment, preparation, and policies before any battle and before all battles—the strategic philosophy of action.

STRATEGIC VISION AND COMPETITIVENESS

The environment for competition continues to change as technology alters and as the world industrializes. Many students of modern economic change have noted the rapid pace of technological change and economic change. For example, Kim Clark (1989) emphasized some of the changing features of competition, among which are

- A continuing and growing worldwide dissemination of scientific and technical knowledge
- An increasing number of global competitors competing in different national markets
- At the same time, the fragmentation of mass markets into market niches and rapidly changing customer preferences as a wider variety of products are offered
- A continuing revolution in computer and communications technologies that provide corporate capabilities of more rapid responsiveness and greater flexibility
- The proliferation of the number of technologies that may be relevant to any given product, including mechanical, electronic and software technologies and choices of materials

In the light of the above changes, Clark proposed five precepts for corporate strategy:

- Managers should understand the technological core of a business and envision that as a strategic advantage.
- Managers should take a broad, worldwide view of technical competence, seeking out the best technology wherever it can be found.
- Managers should focus upon time as the critical factor in using innovation for competitive advantage.
- Managers should discipline business function around the function of production (in production the technical knowledge of the company is focused into a value-adding activity to the customer).
- Managers should integrate all business functions through the information system of the firm.

We see that Clark's precepts for a strategic vision require deepening manager's concerns for technology, widening their horizons on technical change, focusing their attention on timeliness, and integrating technical activities around the science of manufacturing. Clark emphasized that management's fundamental responsibility with regard to the technical competence of an organization is to deliberately build such a competence.

As another example, T.G. Eshenbach and G. A. Geistauts (1987, p. 63) offered precepts for engineers. They argued that the perspective of engineers should be broadened to view their companies as kinds of socio-technical systems:

- Think of the firm as a total system.
- Focus on the interaction between the firm and its environment.
- Concentrate on the firm's most fundamental questions and issues, including the basic mission, definition of the business and goals.
- Be explicit about value judgments in technology assessments and R&D cost/ benefit analyses.
- Emphasize anticipatory adaptive control for the firm to optimize long-run performance in the face of inherent uncertainty.
- Articulate a philosophy of management that represents a permanent commitment to integrative, systematic long-range planning.
- Develop an ongoing planning process, wherein strategy is continuously re-examined.

We note in this strategic philosophy the emphasis upon perceiving the business in a systems view, envisioning the firm as an economic value-adding transformation. In addition they emphasized that engineers should focus upon the interaction between technology and business goals. They also advocated an emphasis on anticipatory attitudes, formalized in a planning process. Technology planning processes in firms should be designed to foster a business strategic attitude in their research scientists and engineers to enable them to collaborate strategically with business managers.

As a third example of strategic precepts for managing innovative firms, Lowell Steele (Steel, 1989, p. 345) offered what he thought a "technologically effective" enterprise should be capable of:

- Taking a systems view of technology
- Being aware of the dynamics of maturation of technologies and industries
- Being explicit about how the enterprise uses technology for a competitive advantage
- Articulating a clear sense of is what are the businesses of a firm
- Knowing who are its competitors
- Being aware of the changing nature of competition
- Being relentless in its pursuit of excellence
- Effectively dealing with uncertainty and ambiguity

For all the above examples of commentators on corporate strategic vision, their precepts provide the elements for their philosophies of action, their way. Clark's "way" drew from the perspective of the manager and emphasized the need for a manager to be aware of technology and commitment to globalization, time and production. Eshenbach and Geistauts' "way" drew from the perspective of the engineer and emphasized the need for an engineer to be aware of the business system and commitment to adaptive control and long-range planning. Steele's "way" drew from the interface of research and business and emphasized need for a balance between business and engineering views on an enterprise, with commitments to competitiveness and excellence.

Which "way" is best? They are all best, depending upon one's experience of the world, perception, and commitment to action. Each "way" sees the world from the particular experiential base of action. For technology strategy, the management personnel require a "way" to be more aware of and attendant to technology as a competitive factor. Technical personal require a "way" to be more aware of and attendant to the business implications of technology as a part of the enterprise. Formulating strategic attitudes in the corporation requires bringing each group into a "way," in order for all groups to cooperate in the total business enterprise.

Management's strategic precepts (about the nature of the world and about the values for action) choose the focus of corporate perception, commitment, preparation, and policy—strategic vision.

STRATEGIC PRECEPTS

A precept is a command or principle intended as a general rule of action. In the cases of Welch's and Musashi's expressions of their principles of action, we see they are stated as precepts.

Strategic visions are expressed as precepts.

We can summarize precepts we have seen in this chapter about strategic action in the following categories:

A. On Intuition
  - Do not think dishonestly.
  - Face reality as it is, not as it was or as you wish it were.
  - Perceive things which cannot be seen.
  - Pay attention even to trifles.
  - Deal with uncertainty and ambiguity.

- Develop intuitive judgment and understanding for everything.
- The Way is in training.

B. On Action
- Time is the essence.
- Develop an ongoing planning process, wherein strategy is continuously reexamined.
- Emphasize anticipatory and adaptive control in the face of inherent uncertainty.
- Control your own destiny, or someone else will.

C. On Business
- Distinguish between gain and loss in worldly matters.
- Think of the firm as a total system.
- Focus on the interaction between the firm and its environment.
- Articulate a clear sense of what businesses a firm is in.
- Be relentless in pursuit of excellence.
- Discipline technical functions around the science of production.
- Integrate operations around the information system.

D. On Competition
- Know who your competitors are.
- Be aware of the changing nature of competition.
- If you don't have a competitive advantage, don't compete.

E. On Technology
- Become acquainted with every art.
- Know the ways of all professions.
- Take a systems view of technology.
- Take a global view on the distribution of technical competence.
- Be aware of the dynamics of maturation of technologies and industries.
- Be explicit about how one uses technology for a competitive advantage.
- Know the technological core of a company and link it to strategic intent.

Strategic precepts about action instruct on how to act in cases of exploratory action, as opposed to repetitive action. Strategic vision of a company needs to be expressed in strategic precepts that focus upon the change needed for the company's future prosperity and survival. Precepts that merely reinforce current practices provide no strategic guidance beyond the precept of "continue to do as you are doing."

Strategic vision should be expressed as a brief set of strategic precepts as to what kind of and how change should be implemented.

For example, in the case of Welch's leadership of GE, his major strategic precepts expressing his strategic vision for GE's businesses were to be in big markets and be number one or number two dominant player in the market.

As another example, in the case of Sony, Morita's and Ibuka's strategic vision for Sony was to innovate new high-tech consumer products and to lead in transistorized and miniature products.

SUMMARY: TECHNIQUE FOR USING STRATEGIC VISION

The strategy process uses strategic vision as the result of the strategic thinking of preparing planning scenarios and strategic corporate models as depicted in Figure 1.1. Formulating strategic vision in an organizational context can be facilitated by the following procedure:

1. Prepare a strategic vision statement
   - After reviewing the planning scenario and strategic business model, a strategic leadership team should prepare a brief strategic vision statement that focuses on changes in strategic perceptions, commitments, preparations, and policies.
   - Mission statements in operational plans are not the equivalent of a strategy vision in strategic planning because mission statements focus upon a continuity, whereas strategic vision focuses on changes to continuity.
2. Express strategic perception
   - To what significant changes need to be attended in the future environments and businesses.
3. Express strategic commitment
   - What changes in operational directions and commitment of resources will be necessary?
       What new efforts will be required as, new products, services, operations, markets, businesses, and so on?
   - Which measures of performance will be primary in addressing successful action on the changes.
       Leadership should be careful about what they really wish in performance (as opposed to what they say) for they are closely observed by their followers, who distinguish the real commitment of their leaders from their public statements.
4. Express strategic preparation
   - What kinds of preparation are necessary to prepare for the future as, research, investments, acquisitions, training, and so on.

5. Express strategic policy
   - What changes in business policies and business practices need to be revised?

For Reflection

Read some of the classic books on strategy (such as, Sun Tzu, *The Art of War*; Karl Von Clausewitz, *On War*; Niccollo Machiavelli, *The Prince*). From these, what precepts about strategy are useful in the domain of business strategy? Why?

CHAPTER 8

MARKETS AND INNOVATION

PRINCIPLE

Linear change in markets is due to innovation.

STRATEGIC TECHNIQUE

1. Scan progress in new knowledge
2. Anticipate new functional capabilities from new knowledge
3. Identify next-generation innovative products and/or services
4. Identify current markets that will change under the innovations and those that may not change
5. Write market-change scenarios under such innovations

CASE STUDIES

Innovation of the Internet
FreeMarkets
General Motors Loses Market Share

INTRODUCTION

In developing a strategic business model, we should understand the basics of innovation strategy to market strategy. Markets are stable or markets change. These alternative conditions provide different strategic challenges. In the case of stable markets, the strategic challenge is to analyze the market properly and provide products or services carefully designed for the right segments of the market. In the case of changing markets, the strategic challenge is to determine the forces for change and to anticipate the right kind of products or services and approaches to exploit the changing market.

Since everyone takes marketing courses that deal with the complexities of markets and planning marketing, we will instead focus here not on marketing strategy but on the reasons for market change. Changes in markets are strategic issues that always need to be addressed in strategy because such they dramatically alter a business's competitiveness.

Markets provide a structure of trading to exchange goods and services between sellers and buyers. Changes in the trade structure can change the market. In business history, *innovation* has created market changes that are linear in effect. This is to say the new form of the changed market never returned to the old form of the market structure.

One can see this impact of innovation upon the linearity of change in market forms even today, particularly by looking at older traditional forms of markets. Traditional markets can still be found in some parts of the world where societies are still agricultural. The economic historian Fernand Braudel has nicely described these:

> In their elementary form, [traditional] markets still exist today. Survivals of the past, they are held on fixed days. And we can see them with our own eyes on our local market-places, with all the bustle and mess, the cries, strong smells and fresh produce. In the past, they were recognizably the same: a few trestles, a canopy to keep off the rain; stall holders, each with a numbered place . . . a crowd of buyers and a multitude of petty traders . . . bakers selling coarse bread, . . . butchers with displays of meat, . . . wholesalers selling fish, butter and cheese in large quantities, . . . straw, hay, wood, wool, hemp, flax. . . .
>
> —(Braudel, 1979, p. 29)

The traditional market developed in agricultural societies as a way of getting the surplus country produce to the cities for exchange for the crafted products produced by the artisans and merchants in the city:

> Markets in towns were generally held once or twice a week. In order to supply them, the surrounding countryside needed time to produce goods and to collect them; and it had to be able to divert a section of the labor force to selling the produce.
>
> —(Braudel, 1979, p. 29)

The country-to-city market is the historical origin of all markets with the bartering of goods or services as the basic units of exchange. When money was introduced into a market as the basic unit of exchange, the reach of the market was extended over space and in time. Money allowed exchange of goods and services without an immediate barter. So it is that markets go back to the beginnings of agriculture, to the origin of cities, and to the dawns of civilizations. Innovations changed market structures from local bartering markets to money exchanges markets to distant markets to global markets. In the 1990s, the innovation of the Internet added new electronic market structure to the world. Electronic markets are a new way to perform the age-old function of matching demand to supply.

Yet new market forms, like traditional market forms, still ultimately perform the same function of economically connecting society to nature—in that all food is ultimately generated from hunting and gathering, agriculture, fishing, and manufacturing activities that acquire food and materials and products from nature. This is the essential image in Braudel's expression that the "surrounding countryside needed time to produce goods and collect them."

All forms of markets are basically ways to economically connect society to nature.

This basic truth about all markets is particularly important to keep in mind in addressing the new strategic challenge of the Internet and electronic commerce. It creates new markets that succeed only as effective ways to connect society to nature.

A second market truth we also need to keep in mind in addressing the new forms of markets is that:

Markets change when the ways of exchanging goods and services change in an economy and/or when the kinds of goods and services change.

Progress in information technology has impacted markets in both ways. The ways in which exchanges occurred was altered as customers could be accessed through Internet services. Value was added through new products and services or through sales of existing products/services through the Internet. Lifestyle changes in communication through the Internet developed rapidly. E-mail became a major means of communication among all age groups, and chat rooms and bulletin boards were spawned. Web-streamed radio, music, and television developed. By the year 2000, e-commerce was rapidly developing into different kinds of markets, including retail commerce, business-to-business commerce, auctions, entertainment, and education.

CASE STUDY: Innovation of the Internet

When we began, we noted that two essential strategic capabilities for a company were the capability to *prosper* and the capability to *change*. Prosperity and change are also essential in a whole society, one special kind of change has created prosperity in society—innovation. The innovation of the Internet is one in a sequence of basic innovations in the twentieth century that created strong economic development and prosperity.

The Internet is both an idea and an implementation. As an idea, the Internet consists of the technical information knowledge of how to connect computers into networks—information technology. As an implementation, the Internet is a system of hardware, connections, and software that enable attached computers of different organizations and people in different locations to communicate—Internet system. These two ideas of a technology and of a system are essential features in the innovation of any new major economic capability in society.

The historical setting of this case was in the last three decades of the twentieth century, when progress in the applied knowledge of computers was stimulating some researchers to envision communication from computer to computer. The origin of the Internet was an earlier computer network called ARPAnet. ARPAnet's origin can be traced back to Dr. J.C.R. Licklider. Licklider served in 1962 in the Advanced Research Projects Agency (ARPA), which funded advanced military research projects for the U.S. Department of Defence. He headed research in ARPA on how to use computers for military command and control (Hauben, 1993). Licklider began funding projects on networking computers in the newly created ARPA research program, Information Processing Techniques. He also wrote a series of memos on his thoughts about networking computers that were to influence the computer science research community.

About the same time, a key idea in computer networking derived from work done by Leonard Kleinrock at MIT. He had been working on the idea of sending information in packaged groups, or packet switching. He published the first paper on packet switching in 1962 and a second in 1964.

In 1965, Lawrence Roberts at the Massachusetts Institute of Technology connected a computer at MIT to one in California through a telephone line. This was one of the first prototypes of computer communications (it would later be called a wide area network (WAN) of computer communications). In 1966, Roberts submitted a proposal to ARPA to develop a computer network for protection of U.S. military communications under a nuclear attack. This was called the Advanced Research Projects Administration Network, or ARPAnet, and it would eventually become the Internet.

By this time, Robert W. Taylor had replaced Licklider as program officer of ARPA's Information Processing Techniques Office. Taylor had read Licklider's memos and was thinking along the same lines of the importance of computer networks. He funded Robert's ARPAnet project: "The Internet has many fathers, but few deserve the more than Robert W. Taylor. In 1966 . . . Taylor (funded the) idea for Internet's precursor,the ARPAnet." (Markoff, 1999; p. C38)

Earlier, Taylor had been a systems engineer at the Martin Company and then a research manager the National Aeronautics and Space Administration (NASA), funding advances in computer knowledge. Next he went to ARPA and became interested in the possibility of communications between computers. In his office, there were three terminals time-sharing computers in three different (research) programs that ARPA was supporting. He watched how communities of people built up around each time-sharing computer:

> As these three time-sharing projects came alive, they collected users around their respective campuses . . . [but] . . . the only users . . . had to be local users because there was no network. . . . The thing that really struck me about this evolution was how these three systems caused communities to get built. People who didn't know one another previously would now find themselves using the same system.
>
> —(Markoff, 1999, p. C38)

Taylor was also struck by the fact that each time-sharing computer system had its own commands:

> There was one other trigger that turned me to the ARPAnet. For each of these three terminals, I had three different sets of user commands. . . . I said. . . . It is obvious what to do: If you have these three terminals, there ought to be one terminal that goes anywhere you want to go where you have interactive computing. That idea is the ARPAnet.
>
> —(Markoff, 1999, p. C38)

In 1965, Taylor proposed to the then head of ARPA, Charlie Herzfeld, the idea for a communications computer network using standard protocols. Next in 1967, a meeting was held by ARPA to discuss and reach a consensus on the technical specifications for a standard protocol for sending messages between computers; and these were called the Interface Messaging Processor (IMP). Using these to design messaging software, the first node on the new ARPAnet was installed on a computer on the campus of the University of California at Los Angeles. The second node was installed at the Stanford Research Institute, and the ARPAnet began to grow from one computer research setting to another. By 1969, ARPAnet was up and running, and Taylor left ARPA to work at Xerox's Palo Alto Research Center.

As the ARPAnet grew, there was the need for control of the system, and it was decided to control it through another protocol, called Network Control Protocol (NCP). This was begun in December 1970 by a committee of researchers called the network working group.

The ARPAnet grew as an overall interconnected, independent multiple sets of smaller networks. In 1972, a new program officer at ARPA, Robert Kahn, then proposed an advance of the protocols for communication as an open architecture accessible to anyone, and these were formulated as the Transmission Control Protocol/Internet Protocol (TCP/IP). These were to become the open standards upon which later the world's Internet would be based.

While the ARPAnet was being expanded in the 1970s, other computer networks were being constructed by other government agencies and universities. In 1981, the National Science Foundation (NSF) established a supercomputer centers program, which needed to have researchers' computers throughout the United States able to connect to the five NSF-funded supercomputer centers (in order for researchers all over the country to use these supercomputers). NSF and ARPA began sharing communication between the networks, and the possibility of a truly national Internet became envisioned. In 1988, a committee of the National Research Council, was formed to explore the idea of an open, commercialized Internet. They sponsored a series of public conferences at Harvard's Kennedy School of Government on the "Commercialization and privatization of the Internet."

In April 1995, NSF stopped supporting its own NSFnet backbone of leased communication lines, and the Internet was privatized. The Internet grew to connect more than 50,000 networks all over the world. On October 24, 1995, the Federal Network Council defined the Internet as

- Logically linked together by a globally unique address space based on the Internet Protocol (IP)
- Able to support communications using the Transmission Control Protocol/Internet Protocol (TCP/IP) standards

Case Analysis

One can see in this case that the innovation of the Internet was motivated by researchers seeking ways to have computers communication with each other—a new kind of functional capability in computation. The invention of the computer networks required the creation of several technical ideas.

The first technical idea was that computer messages should be transmitted in brief, small bursts of electronic digital signals rather than a continuous connection used in the preceding human voice telephone system. Computers could talk to each other in bursts of digital bits, different from how humans talked to each other in continuous streams of analogue sounds. Thus the physical basis of computer communication, packet switching, was different from the technology of the human phone system, continuous connection.

The second technical idea was that formatting of the digital messages between computers needed to be standard in format to send message packets, and these open standards became the Internet's (TCP/IP) standards.

The third technical idea was that a universal address repository needed to be created, so computers could know where to send messages to one another. This became the Internet directory (where at first all addresses ended in .edu for universities, .com for businesses, .org for other organizations).

Finally, in addition to these technical ideas of the Internet, we saw that a physical structure of the Internet needed to be constructed. The architecture of the Internet system which evolved is partially described in Figure 8.1.

Therein two kinds of contexts are shown connected to the Internet. There is a home context connected through modems and telephone lines (or cable or DSL connections) to an Internet Portal Server (e.g., AOL), which connects with its server and router to the Internet. Also there is a business context with an internal local area network (LAN) connected to the Internet with its own server and router.

Also shown on this figure are two of the important technical ideas of the

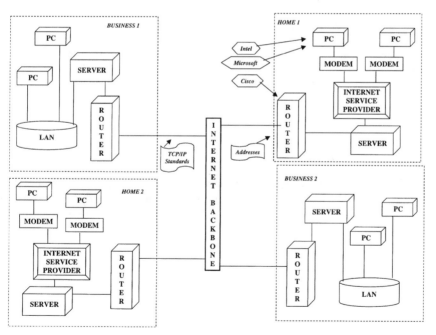

FIGURE 8.1 ARCHITECTURE OF INTERNET

Internet system, the message standards and the registered address system. The Internet standards (TCP/IP) provide the formats for all messages exchanged on the Internet. Each router accesses a universal directory for registered addresses of all sites.

Also, it is interesting to note that in this Internet system, only three information technology companies have built dominant positions: Intel in PC CPU chips, Microsoft in PC operating system software, and Cisco in routers.

FUNCTIONAL CAPABILITIES IN AN ECONOMY

The Internet is an example of a broad class of systems that provide the capability of an economic function in society. In this case, the economic function is communication, and the Internet is a subsystem of a larger set of communication systems in an economy.

The oldest communication system in the U.S. economy was the United States Postal Service, established when the nation was new. Following the postal system, other innovations in communications were the telegraph, telephone, radio, and television. Computer networks and the Internet are only the most recent innovation in the economic function of communications.

To appreciate the strength of innovation as a force for change in economies, we need to understand how new functional capabilities are created in a society's economy. We saw in the innovation of the Internet, that this innovation resulted from research to develop the technical ideas of computer networks and commercial introductions of products and services to establish the capability in the U.S. economy.

Technological Innovation

All new functional capabilities in an economy are created through research to create the new technical ideas and sales of new high-tech products and services based on these technical ideas to establish the capability in the economy. The first part, new technical ideas, is called a technology. The second part, sales of high-tech products and services, is called commercial innovation. Together, the two parts for creating and establishing a new functional capability in an economy is called technological innovation.

> *Technological innovation is the invention and introduction into the market place of new functional capabilities of an economy.*

In the case of the technological innovation of the Internet, a U.S. government agency, ARPA, was charged with the governmental mission of performing research for new military capability of the country, which conceived of and funded the development of the technical ideas for the Internet. Two of the key technical ideas were packet switching for packaging how messages were sent in computer networks (as discrete packages) and protocols for formatting the transmission and reception of data packaged as packets of information. Technical ideas such as these provided part of the information technology for the Internet.

The introduction into the marketplace of this functional capability required first transfer of the technology from the defense research agency, ARPA, to a civilian research agency, NSF. NSF funded the purchase of new equipment by universities to connect their computers together into a national Internet. This purchase of equipment and services by universities established a first market for the introduction into the market place of the new functional capabilities of computer network communications. Simultaneously, research laboratories of companies, working with universities and in defense, began to buy the same kind of equipment to connect their computers into the national research network fostered by NSF.

Along the way, Tim Bernes, a European researcher connected into this network, invented software for transferring data, a network browser, and a university graduate student at a computer research center funded by NSF added a graphical user interface to this browser. Then a venture capitalist funded a new business to commercialization and sell this browser as Netscape.

So the innovation of the Internet did take this common pattern of first research and invention of new technical ideas and then development and commercialization of new products and services embodying these new ideas—technological innovation.

The innovation of the Internet was a new functional capability in the world's economies, a new information technology innovation.

Economic Functional Systems

In this case we saw how a new technology was innovated that had a very forceful impact upon industry and commerce. The innovation of the Internet made electronic commerce possible, which immediately had a huge impact on the stock market of the late 1990s.

In general, all new industries have originated from innovations so basic as to provide to an economy an entirely new and important functional capability. Examples of historically important innovations of economic functional systems are listed in Figure 8.2.

As we can see in this partial list of some of the world's most significant inventions, innovation has played a major role in the development of human civilization. And since the Industrial Revolution in England in the late 1770s, innovation has been the major driving force for change in business and the economy.

An economic functional system is a means of providing a basic and useful technical ability in a society.

INNOVATION	FUNCTION	DATE
TOOLS	TECHNOLOGY	PRE-HISTORY
POTTERY	MATERIALS	STONE AGE
BRONZE	MATERIALS	2500s BC
WRITING	LITERACY	2500s BC
IRON	MATERIALS	1500s BC
GUN	WEAPONS	1300s AD
PRINTING	LTERACY	1400s
TELESCOPE	SCIENCE	1500s
MICROSCOPE	SCIENCE	1700s
STEAM ENGINE	POWER	1700s
POWERED MACHINERY	PRODUCTION	1700s
RAILROADS	TRANSPORTATION	1830s
TELEGRAPH	COMMUNICATIONS	1850s
CHEMICALS	MATERIALS	1850s
STEAM SHIPS	TRANSPORTATION	1860s
CAMERAS	IMAGES	1860s
TELEPHONE	COMMUNICATIONS	1880s
ELECTRIC LIGHTING	ILLUMINATION	1880s
ELECTRICAL POWER	POWER	1880s
BICYCLES	TRANSPORTATION	1880s
AUTOMOBILES	TRANSPORTATION	1890s
AIRPLANES	TRANSPORTATION	1900s
PLASTICS	MATERIALS	1900s
MOVIES	COMMUNICATIONS	1910s
ELECTRON TUBES	ELECTRONICS	1910s
RADIO	COMMUNICATIONS	1920s
RADAR	SENSING	1930s
SPACE ROCKETS	TRANSPORTATION	1930s
NUCLEAR FISSION	WEAPONS	1930s
TELEVISION	COMMUNICATIONS	1930s
COMPUTERS	COMPUTATION	1940s
TRANSISTORS	ELECTRONICS	1940s
SATELLITES	TRANSPORTATION	1950s
INTEGRATED CIRCUITS	ELECTRONICS	1950s
COMPUTER NETWORKS	COMMUNICATIONS	1970s

FIGURE 8.2 HISTORICALLY IMPORTANT INNOVATIONS

Understanding the importance of innovation to strategic change is fundamental to understanding strategy, for the major challenges to the successful implementation of good strategy are the innovative contexts of a business and of an industry. The profitable opportunities for new businesses and industries in new functional-capability systems occurs in providing products and services to establish and run the functional-capability system. For example, the innovation of the Internet provided business opportunities for new information technology companies to provide goods and services for the Internet. Examples of successful new Internet information companies were Cisco, which provided routers to connect the Internet, AOL, which provided portal services to connect to the Internet, and Netscape, which provided browser software to explore the Internet.

In basic technological innovation, the major opportunities for the origin of new industries and businesses and the growth of new markets occur in providing products and services for customers to use new functional-capability systems for their applications.

Application Systems

So new businesses and industries spring up to provide goods and services for a new societal economic functional system. A new functional capability system in an economy provides the opportunity for new applications of this functional capability—customer application systems. This also provides new business opportunities through the new goods and services needed to enable customer applications of a new economic functional system.

New high-tech products and services for these customer applications also create new industries.

For example, early uses of the Internet included several kinds of customer applications—e-mail, retail business transactions, business-to-business transactions, entertainment, information, distance learning, and so on.

Innovation of a new functional-capability system in a society alters the market structures of the economy of the society.

Thus successful strategy must focus not only upon the kind of innovation and the industrial dynamics due to innovation but also on how a customer uses the products and services of the new functional capability—the customer's application system.

For example in the late 1990s, a consumer's application system to use the Internet consisted of a personal computer, a phone modem connection to an Internet portal service (e.g., AOL), and browser software (e.g., Netscape). And the way customers use an application system of the Internet was for different kinds of tasks, such as e-mail, shopping, information searching, distance learning, and so on.

An application system may be complex and consist of many devices, software, and services, such as:

1. A major device system and all the technologies embodied in the device (e.g., Cisco's router in computer network application systems)
2. Key peripheral systems and all the technologies embodied in the peripherals (e.g., LANs and WANs that connect to the Internet)
3. Strategies, tactics, and control technologies for using the major device system and peripheral systems in the application (e.g., AOL's Internet service provider strategy)

A customer's application system to use a functional-capability system consists of the set of high-tech products and services required to use the capability.

These concepts of the functional-capability system and the application system are important for strategy because they provide different kinds of industrial markets in which industries can originate, grow, and mature.

For example, the information technology businesses stimulated by the new Internet of the 1990s lay in businesses providing products and services for the performance of the functional capability of the Internet (e.g., routers, servers, switches, telephone services, etc.) and for the performance of applications of the Internet (e.g., personal computers, browser software, portal services, etc.)

A major source of new industries and new business ventures has been new basic technological innovations of new functional capabilities.

CASE STUDY: FreeMarkets

Next we look at change in markets that is linear—changed markets that can never return to prior forms. Linear change in markets occurs from the impact of innovation. We examine a historically interesting case of how innovation altered an existing market structure by creating a new market form: business-to-business electronic markets (B2B). The time of the case was in the decade of the 1990s when the commercial applications began on the Internet. Glen Meakem was a pioneer in altering the business marketplace of manufacturers purchasing parts and supplies from supplying vendors with his new company, FreeMarkets.

In 1993 Meakem first pitched his vision for a new kind of market to his then employer, General Electric:

> The idea was as simple as it was radical. Meakem proposed to make suppliers compete for manufacturers' orders in live, open, electronic auctions. No more golf-course schmoozing, no more haggling, no more sealed bids. The market for semifinished goods from circuit boards to packaging materials would become

as efficient as the market for stocks, and prices would drop to levels purchasing managers previously could only dream of.

—(Tully, 2000, p. 132)

In his pitch, Meakem claimed that GE would save billions of dollars in its own manufacturing divisions. In addition, Meakem foresaw that GE could become a trading market facilitator for all American manufacturers, collecting billions more in transaction fees.

Meakam got his idea, while working at General Electric. After graduating from Harvard Business School (and also having served in a reserve unit in the Gulf War) Meakem joined GE in 1994. At GE, he joined purchasing:

On his second day in Fairfield, Meakem joined a conference call that would change his life. A manager at GE's transportation division was describing an exotic, GE-sponsored event at a Marriott in Pittsburgh. In one ballroom, dozens of suppliers surveyed samples and drawings for machined metal components that GE wanted to order In another ballroom, GE managers manned a line of flip charts. Suppliers who'd viewed the equipment in one ballroom hustled around the corner to scribble down bids. As they received new low bids, the GE scribes crossed through the former low bid on the flip chart and wrote the new "best price" below it."

—(Tully, 2000, p. 133)

Meakem then had the idea that if GE held an auction like this electronically, GE could build a kind of commodity market for industrial supplies. In 1993, the Internet was still primarily a research tool. At that time, GE had a private electronic network for supply management, and Meakem proposed using it to try his electronic auction concept. His boss, Gary Reiner, told him to try it out at an experimental level. Meakem did this, holding his first GE auction on their private network, which resulted in a deal for circuit-boards from a new supplier in India that saved GE 45% in cost of parts supply.

Next Meakem proposed to GE's headquarters in Fairfield, Connecticut, that GE should start a parts auction business. Meakem told his boss that the new venture would cost $10 million, but it would change the world. But at the time, GE's leadership worried about the cost and risk of the new business Meakem had proposed. His boss decided that Meakem should continue to experiment with his idea with the GE Information Services group. They thought the idea had promise but preferred to have Meakem try it out in GE Information Services.

Meakem knew he was so right that he had to pursue his vision. He quit his job at GE and started his own business, FreeMarkets. It was an early success with a large market capitalization ($7 billion in 2000), and Meakem had become a multimillionaire. The early success of FreeMarkets came from servicing the needs of manufacturing customers, thorough lowering prices by electronically increasing supply to the manufacturing customers demands. Large companies, such as General Motors, United Technologies, Raytheon, and Quaker Oats, used FreeMarket. They saved 15% on the average in buying parts and materials.

Previously, a manufacturer would send out a request for quotation for a part. But this request didn't spell out a lot of information, such as delivery schedule, how much part inventory the supplier needed to hold over time, who the supplier might be, how innovative the supplier, and so on. The electronic communication format of the Internet provides for a lot more communication and discussion of information between manufacturer and parts suppliers than was possible in the old paper and mail format. So in the old mode, most manufacturers choose the easy way, going with previous suppliers whose performance they already knew.

In the new open system, the supplier offers to supply the part with specified schedule, payment terms, inventory arrangements. The lowest price is then found in the electronic auction.

To prepare for an auction, FreeMarkets acts as a consultant to show their clients how to detail all requirements for their requests, and FreeMarket provides expertise on how to find qualified suppliers. The auction requires only about a half an hour, where sellers see competitor's bids in real time. For example at an auction for industrial diamonds, the bidding for the first black diamonds appeared initially at $738,000; but quickly bids lowed the price to $612,000. Finally at the 20 minute deadline for the auction, the price declines to $585,000.

As of the 2000, FreeMarkets was not yet profitable, but its revenues were growing, with sales rising between 25 and 50% each semester. In 1999, FreeMarkets had handled $2.7 billion of the $3 billion exchanged at industrial auctions held that year.

But as a new kind of market is created, other competitors enter to share profitability of the market. As the commercial applications on the Internet evolved, other providers of electronic catalogs and markets emerged. For example, Web sites such as 3-Steel, MetalSite, and PlasticsNet.com created global spot markets for standard processed materials like steel, chemicals, and plastics. Auctions for the most basic commodities (e.g., wheat and fuel oil) had long existed as futures exchanges.

FreeMarkets' biggest competition came from some of their former clients. In 1998, General Motors was FreeMarkets' second-biggest customer, but in the fall of 1999, GM took a stake in B2B software provider Commerce One. In January 2000, GM left FreeMarkets, and General Motors, Ford and DaimlerChrysler announced that they would combine forces to create a single Internet marketplace for their supplier purchasing.

As in any new marketplace, the strategic issue was who would make the most money in the new electronic markets: "No one doubts that there is tremendous money to be made helping companies do business with one another over the Internet. And judging by the enormous multiples that investors are paying, They seem pretty sure which kinds of companies are going to make that money. But are they right?" (Oppel, 2000, p. 3)

The strategic issue was how the profit pie in industrial electronic markets was to be shared: "Leah Knight, a principal analyst at the Gartner Group, said the automakers must have realized that they could dictate terms, and "if they want to take a larger percentage of the transaction, that comes with the turf." (Oppel, 2000, p. 3)

Another example in 2000 of this kind of big manufacturers joining together to use customer-market clout in B2B markets were four major defence contractors joined forces to establish an supplier electronic-market:

> The world's four largest military contractors will announce on Tuesday the formation of a joint Internet commerce exchange for buying supplies and selling products. . . . Lockheed Martin Corp., Boeing Co., Raytheon Co., and BAE Systems PLC (formerly British Aerospace) have come together . . . , partnering with the facilitator Commerce One Inc.
>
> —(Schneider, 2000, p. E1)

There were several ways to generate exchange revenue in electronic auction markets: from recurring subscription and maintenance fees from participants, from a small commission on each transaction, and from equity stakes in the new exchanges. The amounts could be very large. In 1999, the B2B transactions were $145 million, with projections to grow to $7.3 trillion by 2004. If exchange revenue generated only a fraction of the volume, the revenue could be very large (for example, a transaction cut of 0.25% on $7.3 trillion still generates $20 billion).

As the B2B activity began to evolve, its market structure also refined:

> The B2B market is still in its infancy, and its structure and players remain in rapid flux. . . . Most B2B activity to date has centered on on-line exchanges and auctions, and most observers have assumed that these electronic marketplaces would come to dominate the B2B landscape . . . However, most Internet exchanges are floundering. The suffer from meager transaction volume and equally meager revenues, and they face a raft of competitors . . . The hard truth is that few of these exchanges will ever create the liquidity needed to survive.
>
> —(Wise and Morrison, 2000, p. 86)

Case Analysis

When the twenty-first century began, information technology had impacted markets in all three ways:

1. Customers were accessed through Internet services and information interactions.

2. Value was added through new products and services or through sales of existing products/services through the Internet.

3. Life style changes in communication through the Internet developed rapidly. E-mail became a major means of communication through all age groups with chat rooms and bulletin boards spawned. Web-streamed radio, music, and television developed.

By 2000, e-commerce had rapidly developed in several sectors, such as consumer retail commerce, B2B sales, and auctions. Functional systems in an economy provide the functional capabilities to satisfy societal needs of the economy. Radical new innovations that provide improvement in functional systems have large impacts upon economic development and competition.

STRATEGIC MARKET ANALYSIS

While the niching of a market does begin along dimensions of price and performance, still market analysis for stable markets is not quite so simple. Even in these broad categories, further differentiation of markets can exist in just how people differently use a product/service in applications. For example, in the automobile markets, different applications of autos as trucks, commuting cars, family cars, sports cars, recreational cars divided the auto markets so that there were low performance, high-performance, and luxury markets in all these different application segments.

It is therefore very important in terms of strategically analyzing a market of examining the applications in which a customer uses a product/service and the kinds of tasks performed in these applications. In the late 1980s, a term used about the proper design of products or services for a market became popular called the "face of the customer." This phrase was used to remind the product designer that the ultimate success or failure of any product depends on how it appears to the face of the customer—what value the customer sees in the product. So the challenge of product or service design is envisioning the face of the customer in the market niche for whom the design is intended.

Accordingly, strategic market analysis must go down further into the segmentation of the market (beyond price and performance) into the face of the customer—seeing directly the customer's use of a product in the customer's applications and tasks:

- An economic functional system is a means of providing a basic and useful technical ability in a society.
- A major source of new industries and new business ventures has been new basic innovations of new functional capabilities.

- Innovation of a new functional-capability system in a society alters the market structures of the economy of the society.
- A customer's application system to use a functional-capability system consists of the set of high-tech products and services required to use the capability.

Thus the connection of innovation to the market is through the *application system* of the customer, which uses the new high-tech products and services of the innovative economic functional system.

We can illustrate this connection of a high-tech business and its products/services to the market of customers and their applications in Figure 8.3.

A useful technique to view interactions between business systems and customer systems is to use a set of overlapping Venn diagrams, as in Figure 8.3. A Venn diagram sketched as an oval symbolizes the relationships between different sets of things. One oval is one set of things (e.g., the set of all the elements in a business system, the set of all the elements in a product system, or the set of all the elements in a customer system). When two Venn diagram ovals overlap, this overlap expresses that the two sets share mutual set members (e.g., the business system and the customer system share elements of the product system). Where there is no overlap of set circles, then these sets share no members in common.

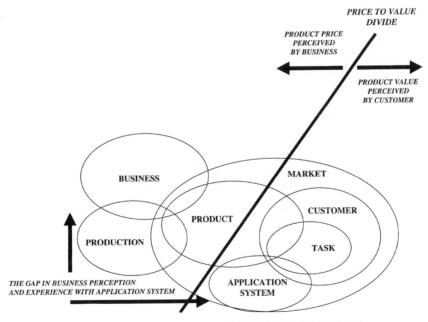

FIGURE 8.3 SEEING THE FACE OF THE CUSTOMER

In a business system the basic transformation is the value-adding activities of producing products from supplies and selling products to customers. A business system interacts with its customers through the sale of its products/services to the customer. In the value-adding operations of its business system, a company develops, designs, and produces products and services for sale to customers.

A product (or service) needs to be thought of as a system, product system, because all products/services need to be envisioned as a set of connected elements that produce transformations in product use or service delivery. For example, a computer system is a product that transforms data input into analytical output. An automobile system is a product that transports people and materials over land. An airplane is a product that transports people and material through air.

The production operations of a business also should be viewed as a system, production system. Generically, all production systems can be described as sets of unit production processes connected by materials-handling subsystems and by production-control subsystems. Examples of production systems include chemical production systems, integrated-circuit chip production systems, telephone systems, airline service systems, Internet service delivery systems, and so on.

The market system encompasses the sets of customers who purchase the products/services sold by a business system. In a market system the basic transformation is the exchange of goods and services between sellers and buyers.

How customers use products also needs to be viewed as kinds of systems—application system, tool system, and task system.

The application system describes the functional capability which a set of products can provide a customer (e.g., the Internet). A tool system describes the means a customer can make of an application system (e.g., Web browsers to explore the Internet). A task system is the set of activities in which a customer uses the application and tool systems to provide valuable benefits to the customer. Customers often use the same product in different applications, such as the use of a home computer for education, entertainment, business, and so on. The transforming operations, systems, of the different tasks provide different requirements on the application and tool systems and upon the success of products/services intended for the systems.

CASE STUDY: General Motors Loses Market Share

We now look at the strategic techniques for analyzing stable markets. As a case of strategy in stable markets, we look again at General Motors strategy when the twentieth century ended. We recall from the previous case of Sloan at GM (in Chapter 3) that early in the twentieth century in the United States, General Motors (GM) overcame Ford's innovative lead and gained market share to dominate the U.S. automobile market after the middle of the twentieth century. In the 1950s and 1960s, GM grew to have about 54% of the automobile market, but then began to slip with poor design and engineering to the rebuilding auto industry of the word, particularly in Japan and Europe. By 1995, GM's share had declined to 34% and continued declining to 29% by the year 2000.

That continuing decline from of the last five years of the twentieth century was particularly puzzling to GM management because they had been using tried-and-true management techniques for marketing:

> GM launched a sweeping overhaul of its marketing back in 1995. That was just after Ronald L. Zarrella was recruited from Bausch & Lomb Inc. to head North American sales and marketing. He promised to restore the auto maker's declining fortunes by applying the brand-management techniques used by consumer-products companies.
>
> —(Welch, 2000, p. 213)

The concept of brand management has proven successful in consumer products in stable markets for products wherein the performance of the product has not changed for years. Brand management uses:

1. consumer research to analyze an existing market
2. Provides authority to a brand manager
   - to alter products of the brand to match the analysis of consumer preferences
   - to change advertizing to reach the consumers

GM had some success with brand management, particularly in its Pontiac division. Pontiac's market share held steady by targeting younger buyers with sporty styling. But overall, brand management had not stopped or even slowed GM's declining fortunes, as GM's share of the U.S. Automobile market continued to decline to 29% from the 34% (at the time its brand management campaign began).

Getting the market strategy for brand management is not always easy. For example, Oldsmobile brand managers had tried for five years to recast the division toward younger car buyers, particularly with its Aurora model but without great success. Finally GM closed the Oldsmobile brand and division.

In addition to trying to get the advertizing campaigns right, there were other problems:

> Misguided campaigns are just one of the problems eating away at GM's brand-building. Another is the company's constant push to hike sales via rebates. . . . An even bigger problem is the company's inability to differentiate or dump its

similar lines. . . . With eight brands and 80 vehicles, GM is by far the most extended carmaker in the businesses. Sculpting brand images for that many cars is a Herculean task.

—(Welch, 2000, p. 214)

Case Analysis

This case illustrates that marketing and product design are strategically inter-linked. A product needs to be differentiated and proper for the market niche for which the product is intended. Brand management can tweak a product for better focus upon a market, but the product has to have been designed roughly correctly in the first place for the kind of customer in the market niche.

The case also illustrates that the range of product brands need to be rationalized to cover the market niches. We recall that in the 1920s after Sloan began to head GM, one of the first product strategies he implemented was to rationalize GM's brands as to prices and features to cover the automobile market from the lowest to highest priced cars. GM needed to go back and rationalize again its product line. Its 8 brands and 80 vehicle model were not properly covering the market niches and or were overlapping and confusing.

Moreover, the major problem for GM during the period of the 1980s and 1990s had not been just brand management. It had more to do with quality in auto design and manufacturing. Since the 1970s, some other auto makers simply better designed and manufactured better cars. Japanese automobile manufacturers ex-celled in manufacturing and design skills over American manufacturers and used their quality advantage to expand their share of the U.S. automobile market.

The rapid decline of GM's market share was due to the expansion of Japanese auto sales in the U.S from the late 1970s. By the year 2000, foreign-based au-tomobile firms (i.e., Japanese and European) had captured over a third of the U.S. market—primarily at the expense of GM's lower quality. This was the major cause of GM's lose of dominance, poor quality in auto manufacturing and design. It was not a problem that brand management alone could solve, as this case still showed.

MARKET NICHES

No market is homogeneous. Customers in the market for a product good or service have different needs and applications for the product and different price sensitiv-ities and different aesthetics. Thus in analyzing an existing market into its seg-ments of more homogenous classes of customer, it is important to know how to design the right kind of product at the right price for a class of customers, or market niche.

For any kind of market, the most general segmentation occurs along the dimensions of price and performance, as illustrated in Figure 8.4. Some customers prefer a product or service that has high performance or advanced features and are willing to pay prices in a market and create a high-performance market niche. Some customers are so price sensitive that they are willing to purchase a product of lower performance but at a much lower price and create the low-performance market niche. Some customers can differentiate quality products with advanced performance when offered at a low price and create a quality market niche. Finally, some customers will accept low-performance and pay high prices for fashionable products and create a *fashion market niche*.

A quality market niche is the hardest niche to serve and differentiate from a low-performance market niche because a producer for this kind of market must excel competitors in manufacturing skill and efficiency to produce at low cost. In the previous case by 1980, Japanese auto makers had gained a decisive lead quality over U.S. automakers and used it over the next two decades to grow a substantial market share in the U.S. market. This market niche is available only to a competitor who gains and maintains a distinct and clear advantage of quality over competitors.

Some kinds of changes do occur in stable markets, and these are what we will call "circular" change because this kind of change recurs periodically. The sources of circular changes in a stable market are

- Further market niching
- Fashion
- Demographics

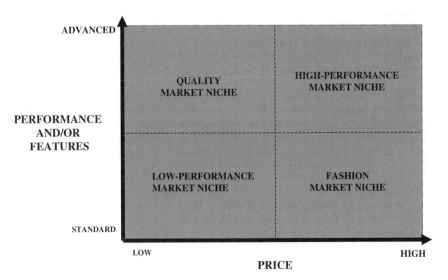

FIGURE 8.4   MARKET NICHES

Market structure is composed of markets niches in which different features and/or performance of a product or service is valued by different segments of customers. The corporate capability of serving different market niches depends upon the cost of production and the responsiveness of the market niches to price differentials. Refinement of market niches is possible as production flexibility improves and costs of production drop. Anticipating changes of market niche structure is very important to market strategy.

Fashion differentiates commodity-like products or services without differences in performance or features. Fashion changes are cyclic over the long-term as they depend upon aesthetics within market groups. Change in fashion tends to be lead by some groups and design-leaders in these groups. Anticipating fashion change requires identification at a time of the fashion leader groups and design leaders.

Demographics can alter markets as the size of the population grows or shrinks and as the relative percentage of the age-groups in a population change. Over the twentieth century, life span in industrialized countries continued to lengthen, with a resultant increasingly larger proportion of the population growing into the old-age category. This dramatically increased consumer spending upon medical care, retirement communities, and nursing homes.

In addition, prosperity in the second half of the twentieth century in industrialized countries led to increasing consumer spending by youth, leading to expansion of youth-oriented markets.

PRODUCT DESIGN

For a business to design a commercially successful product or service to sell to a market niche of customers, the business needs to know specifically the needs of the customer as niche market—the face of the customer (in terms of the kinds of applications, tools, and tasks in which the customer will use the product). The technical risks in designing a product or service for this face arises from knowing exactly and precisely these needs.

The technical risks arise from uncertainties about what the product or service can really do for a customer in applications, how well does it do it, how much resources it will consume, how dependable is it, how easy is it to maintain and repair, and how safe is it:

- Functionality
- Performance
- Efficiency

- Dependability
- Maintainability and repairability
- Safety and environment

Functionality of a product, process, or service means for what kind of purposes can it be used. For example, different industries are often classified by purpose: food, transportation, clothing, energy, health, education, recreation, and so on. The goods and services within these industries satisfy these different purposes. Furthermore, within a purpose are usually different applications. For example, in transportation, there are applications of travel for business, vacation, and personal travel.

The performance of a good or service for a function means what is the degree of fulfillment of a purpose. For example, different food groups provide different kinds and levels of nutritional requirements.

The efficiency of a good or service for a level of performance of a function means what are the amounts of resources consumed to provide a unit level of performance? For example, different automobiles attain different fuel efficiencies at the same speed.

Dependability, maintainability, repairability indicate how frequently a product or service will perform when required and can be easily serviced for maintenance and repair.

Safety has both immediate and long-term requirements. Safety in the performance and safety from aftereffects over time. Environmental impact of a product or service includes the impacts on the environment from production, use, and disposal.

In contrast to technical risks in trying to design products for the face of the customer, there also are commercial risks, from uncertainties about just who are the customer's, how they use the product, what specifications are necessary for use, how they gain information and access to the product, and what *price* they are willing to pay:

- Customer type
- Application
- Specifications
- Distribution and advertising
- Price

The customer, application and specifications together define the market niche of a product, process, or service. The distribution and advertising together define the marketing of the product/process/service.

The price set for a new product/process/service needs to be acceptable to the market and provide a large enough gross margin to provide an adequate return on the investment required to innovate the new product or process service.

Solving the technical variables correctly is both necessary and costly (they constitute the research and development costs of developing and designing a product). Even if successfully accomplished, the commercial variables must be correctly solved (they constitute the production and marketing costs). Despite how much a business may learn of a customer or a set of customers in market analysis, still the variability of the desired balance of these technical and commercial factors will vary across a market and even a market niche. There is always a range of technical variables possible in the design of a new product, process, or service. Which variables will turn out to map correctly to the future required set of commercial variables is never clear initially but only in retrospect. And that is why market analysis is necessary but never sufficient for commercial success (as we saw in the case of GM).

After a good market analysis for designing a product, there must always be some market leadership kind of commitment in gambling on a product design to create a competitive advantage for the product. This is why brand management alone is not capable of guaranteeing business success. Quality leadership in design and production is also necessary.

Neither brand management alone nor quality leadership alone can win, only the combination succeeds.

Why integrating brand management and product leadership is not easy arises from the difficulty of a business system to directly observe the 'face of the customer' even with good market analysis, as we saw in Figure 8.3.

A business system uses its knowledge in designing products and services and in designing production systems. Then the interaction between the business system and the customer system occurs in marketing and sales activities of the business. The overlapping Venn ovals in Figure 8.3 emphasize that

1. A business system can directly see a market system through its marketing and sales activities.
2. A business system can directly see its products (services) through its design and production activities.

However, the nonoverlapping Venn ovals also emphasize that

3. A business system seldom directly sees each customer system.
4. Nor does a business system directly see each customer's applications and tasks.

What this says about the problem of the face of the customer (i.e., customer needs and requirements) is that a business system can never (unless it has only a very few and intimate customers) see directly all its customers and the details of the customer's use of its product in customer's applications and tasks. And even if it does indirectly collect good information about these intimate details of customers, still there will be variability of customer uses and desires across applications and tasks. Accordingly, establishing the precise value to the customer is difficult and inherently ambiguous.

A business's knowledge of the exact value of a product or service to a customer is always only approximately known.

This is why the brand management approach illustrated in the case of GM's losing market share was not immediately successful. It takes time to interact with customers to improve the accuracy of understanding a market of customers valuing of a product/service. And this is why Pontiac at GM had been successful in brand management, for it had then started doing this in 1981 and not lately in 1995.

Accordingly, as illustrated in Figure 8.3, the overlapping interaction between the business system, product system, and customer system occurs in the price of the product sold by the business to the customer. The value of the product to the customer occurs in the overlap between the customer, product, and application systems. The utility of the product occurs in the overlapping interaction between the customer and application systems. The performance and maintenance of a product in application occurs in the interaction between the product and application systems.

The importance of viewing these system interactions is in making clear in which systems interactions the experience of the concept occurs:

Utility between the customer and application systems

Value among the customer, application, and product systems

Price among the business, product, and customer systems

This is basically why the economic/business concepts of utility, value, price are not identical (although they are sometimes treated as synonyms in some economic writings). This is important because it explains *where the difficulty lies* in successful commercialization of an innovative new product or service.

Look now at the dark line showing the divide between the business system and application system considered in Figure 8.3. This indicates that the direct experience of the business system with the application system often does not occur because the business system and application system do not overlap. This creates a kind of divide in perception between the business systems perception of the value of a product as the product price and the customer systems perception of the product as product value in its contribution to utility in the application system.

The experiential divide in perception of product price and product value is inherent as the missing experience of the business system with the application system. And this why commercial success from only market analysis or brand management seldom provides commercial success.

> *The problem for commercial success of a product/service arises from a lack of detailed understanding and knowledge about the value, utility, performance and maintenance of the product within a customer's application—and hence a gamble of quality leadership must always be undertaken.*

The firm that best matches product price to product value succeeds in the competition:

- When products or services are differentiated in performance, a higher-performing product can be priced at a higher price than a lower-performing product when it delivers higher value to the customer as more utility in the application (e.g., Porter's product-differentiation strategy).
- When products or services are not differentiated in performance, commodity-type products, then the lower-priced product succeeds because commodity-type products all deliver the same value to a customer.

> *The importance of using information technology in the process of commercializing innovative new products and/or services is to get correct the match between product/service price and value through both market analysis and quality leadership.*

SUMMARY: TECHNIQUE OF ANTICIPATING MARKET CHANGE

Forecasting market change is needed in both the planning scenario and in constructing a strategic corporate model:

1. Scan progress in science for new knowledge
   - In the planning scenario, use the model of societal structures to examine progress in science within the culture sector in the model
   - Relate new scientific knowledge to relevance of advancement in technology in the economy sector of the model
   - Distinguish incremental change in technologies from discontinuities

2. Anticipate new functional capabilities
   - Forecast progress in technology as impacting changes in application systems
   - Judge value of change in application systems and tools to customer systems
3. Identify next-generation product systems
   - For dramatic progress in technologies, imagine new features of application systems and tools that would allow new tasks
   - Imagine what the face of the customer will demand in these changes in new or changed tasks
4. Identify current markets that will not experience functional change
   - Analyze types of groups and fashion leaders for guiding cyclic change in these markets
   - Reexamine market niche structure to see if further refinement is likely
5. Write scenarios for market change in the competitive discontinuities
   - Imagine how next-generation products or services can alter requirements for product-lines and services

For Reflection

Identify a major innovation and trace its subsequent product lines and markets. What were applications in these markets? What were its first markets? What were eventually its largest markets? Were the firms that eventually dominated the markets, technology leaders or market leaders or both?

CHAPTER 9

COMPETITION AND STRUCTURE

STRATEGY PRINCIPLE

Competitive strategy is implemented in the core capability of product or service design

STRATEGIC TECHNIQUE

1. Depict the value-chain industrial structure of a business
2. Forecast changes in technologies within the structure
3. Analyze impact of these changes upon business competitiveness
4. Use information technology to improve design capabilities
5. Plan future products/service

CASE STUDIES

Vertical Integration in the Automobile Industry
Ford's Taurus Project in the 1980s
Design Process of a Computer Firm
Motorola's Product Roadmap

INTRODUCTION

In developing a strategic business model, we should understand the basics of strategy in competition and in industrial structures as an input to strategic thinking. Competition occurs directly as businesses provide products and services and compete with other businesses for customers, by the availability, quality, and price of the product or service. Both the context of how these businesses compete and how they design products and services are key factors in this competition. Therefore, competitive strategy focuses upon the factors of competitiveness:

- The competitive context
- Product/service design capability of the business

We first examine strategic issues of the context of competition and the product/service design capability.

COMPETITIVE CONTEXT

We recall that in the strategy literature, the positioning school emphasized strategically analyzing the conditions of competition:

Positioning School
This school emphasizes strategy as general positions selected from analyzes of industrial situations. The role of analysis in specifying the industrial situations uses techniques such as value chain analysis, game theoretical structuring, and so on.

In the early 1980s, Michael Porter emphasized the importance of looking at the competitive situation in any industry to understand the complexities of what makes up competitive advantages in strategy, and his five forces model of competition became a popular way to think about strategy (Porter, 1985). Figure 9.1: summarizes Porter's model of a competitive situation:

1. Business selling a product/service
2. Competitors selling similar products/services
3. Customers as buyers of the product/service
4. Suppliers of resources to the business and its competitors
5. Potential substitute products/services
6. Potential new entrant sellers into the industry

The smaller dotted box of Figure 9.1 delineates the boundary of the competition situation for the business. Porter argued that the struggle for power of the participants determined relative competitive advantages.

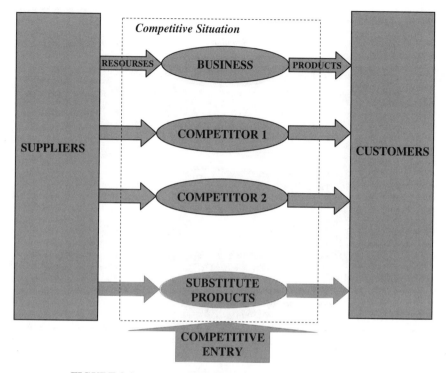

FIGURE 9.1   COMPETITIVE SITUATION OF A BUSINESS

For example, we recall the case of Amazon versus Barnes and Noble (from Chapter 1), in which Riggio began his business as a new entrant seller in the book retail industry, first in college text books and then second in trade books. Later Riggio used junk bond financing to acquire many of the competing bookchains. Suppliers to Barnes and Noble were book publishers, of which there are many and who produce many books. Substitute products or services in other industries were magazines, journals, movies, television, and so on.

Any strategy for a competitive advantage needs to consider the traditional situation of the five different participants (competitive forces) within an industry.

Now this model of the competitive situation that bounds a business in an industry is bounded itself by the state of the knowledge in the business. Thus we must add to Porter's model a larger contextual bound of the state of industrial knowledge, as shown in Figure 9.1A as a larger dashed box encompassing the dotted box of the competitive situation.

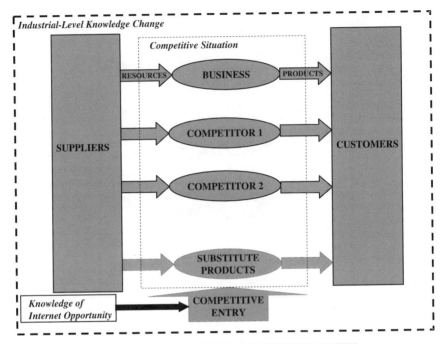

FIGURE 9.1A INTERNET COMPETITIVE ENTRY

In the case of Amazon versus Barnes and Noble, Amazon entered the established book retailing business through online use of the Internet. Jeff Bezos used his knowledge of the new Internet system to envision a new business opportunity. The change in knowledge at the industrial level (as the new information technologies and Internet system) provided Bezos with the competitive entry opportunity of becoming a new competitor through his *knowledge of internet opportunity.* Thus the competitive situation was changed for Barnes and Noble, and they had to respond by setting up their own Web business.

Information technology alters the traditional five forces model of competition in that all competitive situations are bounded by a knowledge structure of the industry. Change in the knowledge structure alters the competitive situation.

In a modern competitive situation model, one must indicate explicitly that the boundary of a business competitive situation is encompassed in a larger boundary of change in industrial-level knowledge.

New knowledge can alter competitive situations by providing new strategic business opportunities to those who have the new knowledge at the industrial level and can envision business opportunities in the new knowledge.

A model of the competitive situation is important as it indicates both the knowledge boundaries of the competitive context and the business boundaries:

- Change in the knowledge boundaries can alter the entire industrial structure through new kinds of competing businesses.
- Change in the business boundaries of the industry (due to the relative power among these different participants) limits the potential profit margins and profitability of businesses in the industry.

Riggio selected a lower profit margin as a strategy to enter the trade book market from his base as a textbook seller. Amazon.com used the cost advantage of not having bricks-and-mortar stores to reduce its online cost of selling books. However, because Amazon's pricing strategy was still limited by competitors such as Barnes and Noble, it was not able to price for both growth and profitability. At the time of this case study, Amazon had not yet had a profitable year but was financing its operations and growth from capital.

Industrial context of a business includes both the competitive boundaries of the business and also the boundaries of knowledge in the industry.

CASE STUDY: Vertical Integration in the Automobile Industry

The positioning school of strategy not only looked at the competitive position of a business in its industrial sector but also how that sector was structured. We next review the strategic concept of an industrial structure. We look at the historical case of how Toyota assembled of automobiles in 1983.

The production process then is sketched in Figure 9.2. Toyota purchased both manufactured components and materials for Toyota's own manufacturing processes from various suppliers. Toyota purchased components from suppliers, such as electrical parts, bearings, glass, radiators, batteries, and so on. Toyota also purchased processed materials, such as steel sheets and rolled steel, nonferrous metal products, oils, paint, and so on from other suppliers. Purchased components and materials were subjected to acceptance inspection.

Next, materials went through various production processes to be formed into parts (e.g., forging, casting, machining, stamping, plastic molding). In addition, some of the purchased components also went through further processing to be finished as components (e.g., heat treatment or additional machining).

Materials, components, and parts were eventually used for three subassembly systems in fabricating the automobile:

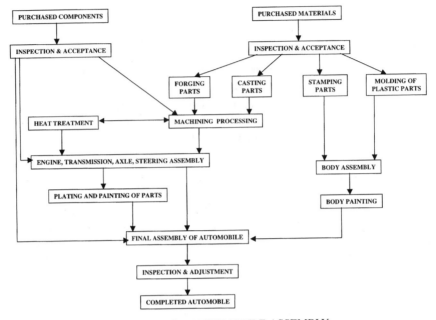

FIGURE 9.2 AUTOMOBILE ASSEMBLY

1. Power subsystems. Engine, axles, transmission, steering assembly, etc.
2. Chassis subsystems. Frame, suspension, brakes, etc.
3. Body subsystem. Body, seating, windows, doors, etc.

Various plating and painting processes prepared the power and chassis subsystems for final assembly, and the body was painted for final assembly. Finally, the three major fabrication subsystems were put together. After adjustments and inspection, the product emerged as a completed car.

In such a production system, Toyota was one of the most efficient automobile manufacturers than in the 1980s. "During the early 1980s, a dozen assembly plants turned out Toyota automobiles at the combined rate of more than 800 per hour" (Cusumano, 1985, p. 262).

At that production rate, much coordination with suppliers was required:

> Toyota managers were responsible for coordinating deliveries of components and subassembly manufacturing with the schedules of final assembly lines, where workers quickly joined engines, transmissions, steering components and frames with body shells . . . to manufacture a small car from basic components (excluding raw-materials processing) through final assembly, Toyota and its subcontractors took approximately 120 labor hours. . . .
> —(Cusumano, 1985, p. 262)

In the same period, one major difference between Japanese automobile manufacturers and American manufacturers was a much lower vertical integration of the Japanese automobile industry:

> Japan's 10 major automakers were more like a collection of manufacturing and assembly plants for bodies, engines, transmissions, and other key components than they were comprehensive automobile producers. From the mid-1970s through the early 1980s, Nissan and Toyota accounted for only 30 percent of the manufacturing costs for each car sold under their nameplates;
> —(Cusumano, 1985, p. 187)

In manufacturing, there is usually a choice of which components and parts to purchase from suppliers and which to produce internally. This is called the "degree of vertical integration." Cusumano (1985, p. 192) had figures that compared vertical integration between some companies in 1979:

- Nissan then produced 26 percent in-house
- Toyota 29 percent in-house
- GM 43 percent in-house
- Ford 36 percent in-house
- Chrysler 32 percent in-house

The reason that Japanese managers choose low vertical integration was partly historical:

> Subcontracting to subsidiaries or other firms reached these high levels in the Japanese automobile industry after demand expanded rapidly beginning around 1955. Managers decided that it was cheaper, safer, and faster to recruit suppliers rather than to hire more employees or invest directly in additional equipment for making components."
> —(Cusumano, 1985, p. 192)

Case Analysis

In this case we see an example of industrial structure in which manufactured products were produced from a combination of suppliers and in-house manufacturing and assembly processes.

INDUSTRIAL VALUE CHAIN

The kind of industrial structure that allowed Toyota to purchase materials and parts from suppliers is called an "industrial value chain," and its general form is sketched in Figure 9.3.

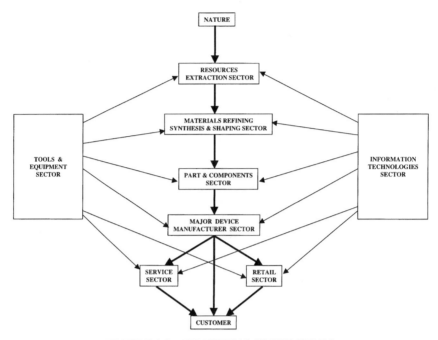

FIGURE 9.3   INDUSTRIAL VALUE CHAIN

Firms can be classified into industrial sectors whose relations to one another are as a kind of chain of "value-adding" transformations beginning with the extraction and processing of natural resources through the production of products and applications for final customers of the chain of transformations:

1. Nature. All physical goods begin with material created in the natural world.
2. Resources extraction sector. Businesses must first extract raw materials from nature (e.g., mining, petroleum, forestry, etc.).
3. Materials refining, synthesis, and shaping sector. Other businesses processed raw materials into materials products (e.g., steel, chemicals, lumber, etc.).
4. Part, component, and subsystems sector. Other businesses construct parts and components from these materials products (e.g., wheels, tires, windows, IC chips, etc.).
5. Major device manufacturing sector. Other businesses assemble parts and materials into major kinds of devices (e.g., automobiles, airplanes, houses, furniture, computers, etc.).
6. Retail sector. Retail businesses may sell manufactured devices to customers (e.g., automobile dealers, furniture stores, electronics stores, etc.).

7. Service sector. Service businesses may purchase devices to provide services to customers (such as bus lines, airlines, hospitals, etc.).

8. Tools and equipment sector. Parallel to this chain of value-adding, other businesses provide tools and equipment to the different sectors (e.g., machining equipment, dies and molds, chip fabrication equipment, etc.).

9. Information technologies sector. Also parallel to the value-adding chain, other businesses provide computers, software, and networking systems (e.g., personal computers, application software, LAN connections, etc.).

All of these industrial sectors interact to form a chain of value-adding activities to provide the necessary transformations for adding value from natural resources to the final product.

To understand a complete industrial system, one should envision the topology of producing systems as a flow of materials from natural resources through to customer applications.

COMPETITIVE CONDITIONS WITHIN AN INDUSTRIAL STRUCTURE

The strategic importance of describing the competitive situation of a business within an industrial value chain is that it facilitates understanding the competitive conditions of a business. The competitive context of the previously described five-forces competitive situation sits within a larger structural context of an *industrial value chain*. Competitive conditions differ in each sector of an industrial value chain:

- The competitive conditions of firms within an industrial sector are directly or indirectly impacted by changes in applied knowledge in all parts of their value chain.
- The kinds of knowledge which provide assets to a business depends upon the business's location in an industrial structure.

We next look at each kind of industrial sector in terms of its competitive context.

Resource Extraction Sector

For a business in the resource extraction sector, sources of raw materials are located and collected, mined or extracted from nature (e.g., the asset value of oil production firms can be measured in the estimated barrels of oil in their reserves; and the most proprietary knowledge they have are oil exploration techniques). In this sector, a business competes by finding and owning the rights to natural resources and by efficiently producing raw materials from these natural resources. The extraction industry firms provide raw materials to their customers, and the availability and quality and cost of these materials influence sales to customers.

In resource extraction industries, the most important assets are both material and knowledge based:

- Access to sources of raw materials
- Applied knowledge in the extraction of raw materials

The strategic competitive factors that especially discriminate among firms producing similar products within a resources extraction industrial sector include

1. The effectiveness of resource discovery techniques
2. The magnitude and quality of their discovered resources
3. The efficiency of the extraction knowledge
4. The use of capacity
5. The cost and efficiency of transportation for moving resources from extraction to refinement

The customers of the resource extraction sectors are the firms in the different materials refining, synthesis, and shaping sectors.

Information technology in sensing and modeling geological structures had revolutionized the resource industry in the 1980s. In the 1990s, information technology and the Internet began revolutionizing the markets for the materials and energy trading.

Materials Refining, Synthesis, and Shaping Sector

Businesses in the materials refining, synthesis, and shaping sector produce material products from the raw materials. The material products are processed and shaped into materials products that can be used in parts manufacture or fabrication. Examples are steel and aluminum industries and chemical industries.

In the materials refining, synthesis, and shaping industries, the most important strategic assets are physical plant and applied knowledge of synthesis or processing and of new applications for processed materials.

Strategic competitive factors for firms in the materials refining, synthesis and shaping industries include

1. Patents and proprietary knowledge about the creation of materials products
2. Patents and proprietary knowledge about the processes and control in production processes

3. Quality of produced materials
4. Cost of produced materials
5. Timely delivery of produced materials
6. Assisting customers in developing applications for material products

The customers of the materials refining, synthesis and shaping industries include firms in the parts and components, major device system integrator, applications system integrator, and tools, machinery, and equipment industries.

In the 1990s, information technology and the Internet began revolutionizing the markets for the materials industries.

Parts and Components Sector

In the parts and components sector, businesses manufacture parts and components and/or fabricate subsystem assemblies for their customers who are producers of major device systems. Examples are integrated chip producers, disk drive producers, automobile wheel producers, battery producers, and so on. For these kinds of businesses, the most important assets are both physical and knowledge based:

• Unique equipment for part production
• Production control
• Proprietary knowledge of part design

Strategic competitive factors for businesses in the parts and components industries include

1. Patents and proprietary knowledge about the design of products and components
2. Patents and proprietary knowledge about the processes and control in production processes
3. Quality of produced materials
4. Cost of produced materials
5. Timely delivery of produced materials
6. Assisting customers in developing applications for parts
7. Concurrent design capability with customers

These kinds of businesses constitute the largest numbers of firms in the manufacturing sectors of industry (and also the greatest diversity of businesses). Generally, they divide into electronics and mechanical type production. Electronic parts suppliers include IC manufacturers and other electronic parts, printed circuit board manufacturers and assemblers, and electronic subsystem designer and assemblers. Mechanical parts suppliers include mold and die makers, fabricators, and subsystem designer and assemblers.

Parts and components tend to be unique in technology or a kind of commodity product. In the former superior performance and in the latter costs and quality of production provide the principal means of competition.

In the 1990s, information technology as the Internet was revolutionizing the relationships between the parts and component businesses and their customers of the major device manufacturers (as we discussed in Chapter 8).

Major Device Manufacturer Sector

In the major device manufacturing sector, businesses fabricate and assemble a product which provides the major device for a generic class of customer applications. Examples are automobile manufacturers, airplane manufacturers, computer manufacturers, or construction firms.

For businesses in the major device systems integrator industry, the most important strategic assets are both physical and knowledge based:

- Production facilities and equipment
- Proprietary knowledge of design and production control
- Brand name and access to market distribution channels

Brands and distribution distinguish the asset value of device systems industry from parts industry. Competition early in the technology life cycle of an industry usually depends upon proprietary technology. But after the industry matures, businesses compete predominantly by brand recognition, production quality, price, luxury features, fashion, and customer service and cost to the customer of device replacement.

Competitive factors for businesses in the major-device manufacturing industries include

1. Patents and proprietary knowledge about the design of the major device system and about key competitive components and subsystems
2. Patents and proprietary knowledge about the processes and control in production processes
3. Performance and features of major device system
4. Costs of purchase and maintenance of major device system
5. Dependability and cost of repair of major device system
6. Availability and cost of distribution channels and timely delivery
7. Availability and nature of peripherals to complete a major device system into an applications system

8. Assisting customers in developing applications systems around the major device system

These kinds of businesses can vertically integrate backwards into their parts and materials suppliers to gain cost and competitive advantages (as did automobile makers in the early days of the industry). Or they can deverticalize at other times to gain cost and competitive advantages (as did automobile makers in the later days of the industry).

In the 1980s, information technology in design and manufacturing began revolutionized the production of major devices.

In the 1990s, information technologies and the Internet began revolutionizing the relationships between major device manufacturers and their customers in the retail sector.

Retail Sector

Strategic competitive factors for businesses in the retail sector (in which we include wholesalers) are:

1. Location
2. Brand franchises
3. Inventory control and logistics capability
4. Price sensitive advertising
5. After sales service
6. Point-of-sale information systems
7. Computerized communication with customers

Competition between these businesses focuses upon location, prices, brand product lines, and customer service. Information and knowledge strategies in these firms usually focus upon

1. Development of customer service applications
2. Development of logistics and scheduling technologies

Firms in the wholesale and retail industrial sectors include appliance, food, apparel, household furnishings, automobile dealerships, and so on. Generally, each sector specializes around a functional group of products (e.g., food, apparel, automobiles, etc). Grouping of products in general retail establishments, such as department stores and large grocery stores, may occur, or boutique shops in shopping malls may become established.

Information technology in the 1980s revolutionized inventory control techniques in retail sectors. The innovation of the Internet changed dramatically the competitive contexts and structures for retailers beginning in the late 1990s.

Service Sectors

Businesses in the service industries divide principally into service delivery firms and professional services firms. Service delivery businesses include airlines, buses and railroad lines, and telephone and communication firms. The major professional services firms are medical and legal businesses and engineering consultant businesses.

When the twentieth century ended, the service sectors had become the major sources of employment. For example, James Brian Quinn, Jordan J. Baruch, and Penny C. Paquette called attention to the importance of the service sector:

> The service sector has grown steadily in its contributions to the U.S. gross national product and now [1988] accounts for 71 percent of the United State's GNP and 75 percent of its employment. Far from being a negative development, today's service industries create major markets for consumer goods, lower virtually all manufacturer's costs, provide strong, stable markets for capital goods producers, and enhance overall U.S. competitiveness.
>
> —(Quinn et al. 1988, p. 45)

They emphasized several points about the service industries:

1. That they are capital intensive
2. That service companies are often of large scale
3. That some service technologies improve manufacturing systems.

In the 1990s, information technology, such as the Internet, was also revolutionizing the services industries.

Tool and Equipment Sector

Businesses in the tool and equipment industry supplies production equipment for all producing sectors of an industrial value chain. Firms specialize in types of equipment for different value chains, such as equipment for the ferrous materials industry, nonferrous materials industry, chemical industry, electronics industry, and so on. Examples are the mechanical machine tool industry, the industrial control industry, electronics equipment industry, chemical equipment industry, and so on.

Strategic competition between businesses here focuses upon performance capability and capacity of the equipment, price, and production system integrability. Strategic competitive factors include

1. Patents and proprietary knowledge about the design of the production equipment and tools
2. Proprietary knowledge about control in production processes
3. Performance and features of equipment and tools
4. Costs of purchase and maintenance of equipment and tools
5. Dependability and cost of repair of equipment and tools
6. Availability and cost of distribution channels and timely delivery
7. Availability and nature of peripherals to complete production equipment and tools into a production process system
8. Assisting customers in automating and controlling production processes

Information Technologies Sector

Businesses in the information technologies sector provide computational capabilities for all the businesses in an industrial value chain. They produce computers and networks and software.

Competitive factors for firms in the information technologies industries include

1. Proprietary tools for chip design
2. Patents and proprietary network interconnect hardware and software
3. Patents and proprietary computation hardware
4. Professional expertise
5. Copyrighted software

Businesses here focus their information and knowledge strategies on

1. Tools to improve chip and information system design capability
2. Interconnecting hardware and software
3. Computation platforms
4. Data and tools to improve consulting services capability
5. Software design capabilities

STRATEGIC INDUSTRIAL REORGANIZATION

An industrial value chain structure is changeable. Any industrial sector may integrate vertically, either backward into their supplier businesses or forward into their industrial customers' businesses.

For example, we recall that in the 1920s in the U.S. automobile industrial value chain, Durant assembled General Motors (GM) by vertically integrating backward in the auto supplier chain by buying Delco (an electrical auto parts company) and by buying the Fisher brothers' auto body production business. In the 1990s, GM deverticalized by selling off Delco.

Basic innovation may obsolete businesses within an industrial sector. For example, the invention of the transistor obsoleted the electron-tube-producing businesses (e.g., Sylvania) and the invention of the IC semiconductor chip obsoleted transistor producing businesses (e.g., Fairchild).

Dramatic changes in a strategic technology can alter the organization of an industrial value chain by

1. Altering vertical integration in the chain
2. Creating new product line variations in segments of the chain
3. Obsoleting product lines in segments of the chain
4. Providing substituting-technology products in segments of the chain
5. Fusing two different industrial value chains together or making obsolete an entire industrial value chain with a substituting value chain

Within an industrial value chain, the upstream sectors are viewed as suppliers and the downstream sectors are viewed as customers. Upstream technical change can impact downstream applications systems in several ways:

1. Lower cost to move applications into lower priced market niches
2. Improve quality to improve substitution into current applications systems
3. Improve performance to increase substitutions
4. Simultaneously improve, cost, quality and performance to increase substitutability, move into new market niches
5. Lower cost to provide multiple copy ownership of products in a market niche
6. Improve performance to adapt technology to new application systems (and new markets).

CASE STUDY: Ford's Taurus Project in the 1980s

Now we turn to the second important factor in competitiveness, design capability. To understand design capability and its competitive importance, we look at a historically interesting case where changes in the design procedures of a large firm was then essential to its business survival—the case of the Taurus design project in the Ford Motor Company at the beginning of the 1980s.

Times do change, and if change is not anticipated by management, a business crisis can occur. This happened extensively in the business world in the late 1970s. Changes in the political control over the supply of oil in the world with the establishment of an international oil cartel caused a dramatic rise in inflation. For the automobile industry, there was a sudden and unanticipated demand for fuel-efficient cars, which created a major crisis for the U.S. automobile industry. They found their current models uncompetitive, and they also realized that their design and manufacturing capabilities had also become uncompetitive.

1981 was a year of change or die for all three of the major U.S. auto manufacturers, Ford, Chrysler, General Motors. Chrysler asked for and received a U.S. government loan to retool for energy efficient models. General Motors undertook nearly a decade-long investment in new technologies and processes (and started a new car division, Saturn, and partnered with Toyota in the production of one car model). Ford's response was to redo its design procedures.

This case looks at Ford's re-engineering of its car design-and-development procedures, which they then called current engineering design (first used in the design of the Ford Taurus model). As the leader of the design project, Lew Veraldi, later talked about the project:

> We're very honored by all the attention about Team Taurus and by what it says about Ford Motor Company. But you must stop and ask yourself, why should a large company like Ford, which has been developing new cars for over eighty years, decide to change the way it does business?. . . . the need to change was 'survival'. That gets everyone's attention.
>
> —(Veraldi, 1988, p. 1)

In the economic turbulence of the 1970s due to the rapid rise in oil prices, Ford raised product design to the attention of senior management. But this only added another six months to an already five to six year product development cycle, when the Japanese firms could design and introduce a new car in less than four years. Veraldi described Ford's condition at that time:

> Remember when Taurus began, it was 1979–1980. Ford's image for quality was not very good. In addition, we were in the process of losing over $1 billion for two years in a row. That's a record that I believe still stands.
>
> —(Veraldi, 1988, p. 1)

Ford's product development process was essentially linear, as was then usual in the U.S. auto industry: beginning with (1) Concept Generation, then (2) Product Planning, next (3) Product Engineering, and finally (4) Process Engineering—after which (5) Full-scale Production would begin (Clark and Fujimoto, 1988). In this kind of linear product development process, first designers designed the shape of the new model and turned this shape over to engineers to design the mechanics. Next the design was given to manufacturing to figure out how to produce the design. Each group, model designers, design engineers, manufacturing engineers worked in isolation of each other.

The problem with that linear development was the number of design changes (and redesigns) it encouraged and consequently the long time it took to complete the development with all those changes. By the time the design reached manufacturing, many changes would be required to alter the design so that it could be manufactured with quality and low cost. This recycling of the design back and forth between manufacturing, engineering, and design took time and money. Finally, when marketing was allowed to see the design, it would be too late for them to make changes they thought the customer would prefer.

Changes in Ford's management philosophy began only after Henry Ford II retired in 1980:

> Ford, founded in 1903 by Henry Ford, was one of the very few large U.S. corporations where the top management position was traditionally held by a descendant of the founder. . . . At Ford, the bottom line was important, and the company culture was not people oriented. . . . Ford's style, as in many other companies, remained authoritarian. . . . As one former engineer put it, "When I was with Ford Manufacturing building the 1970s Mustangs, we didn't see the car we were going to make until eight or nine months before production was to start. And designers didn't want our ideas, either!"
>
> —(Quinn and Paquette, 1988, p. 1–2)

Representing the controlling interest in the company by the Ford family, Henry Ford II went through three presidents in trying to make Ford more competitive and next appointed Donald Peterson president, and Peterson's predecessor, Philip Caldwell, became Chairman of the Board. Earlier in 1973, Caldwell had been head of international operations and was assigned the task of developing in Germany the Ford small car called Fiesta. The development of the Fiesta cost $840 million and was then the most expensive car development project at Ford. Caldwell had chosen a design engineer to lead the project, and he was Lew Veraldi.

Veraldi had worked in Ford's design systems for twenty-five years. He had experienced the frustration of taking a design to manufacturing only to be told that it wasn't designed right to be manufactured. In the Fiesta project, he called in the manufacturing people at the beginning of the design and asked them: "Before we put this design on paper, how do you, the manufacturing and assembly people, want us to proceed to make your job easier?" (Quinn and Paquette, 1988, p 3)

Veraldi had tried team concurrency between design and manufacturing in the Fiesta project and liked the benefits of the experience. He had understand the importance of early and close cooperation between product design and manufacturing and saw that he could make that happen. Veraldi was then convinced of the real advantages in designing a car simultaneously between Product Engineering and Process Engineering and Marketing, rather than in sequential isolation. It had avoided many late changes or desirable changes that were identified too late to be made. And it saved money in the product development of the Fiesta.

Later in 1979, the new senior management under Caldwell and Petersen decided to replace Ford's mid-sized cars with innovative new products. The Taurus project was to be the first of these models, originally conceived as a five-passenger car with four-cylinder engine. But then it was a high risk development project, and essential to its success would be a change in management philosophy:

> The second ingredient on what it took to change (was) commitment of upper management. (It took) courage, I believe, of Mr. Philip Caldwell and Mr. Don Petersen to take a risk—in many ways, to bet the Company—to spend $3 billion when we were losing $1 billion per year."
>
> —(Veraldi, 1988, p. 1)

In 1979, Caldwell assigned Veraldi the job for the initial designs for the Taurus/Sable cars. In the summer of 1980, Veraldi and his group presented concepts for the new car to top management. After Veraldi's previous experience with concurrent engineering in the design of the Fiesta, senior management agreed to create a management team (later to be called Team Taurus) to try out the concurrent engineering ideas on the Taurus development project.

DESIGN PROCESS

We pause in this case to describe generally the design process of a business as a strategic capability. The way products or services are designed in an organization, the design process, is the key to the future product offering of the company— and so is a strategic process in the business. The policies that govern the process constitute an important part of the product strategy. In this case, a strategic change in design process was adopted by Ford in order to survive the competition with then automobiles designed and produced by firms in Japan. In 1980, after a long dominance in the world's automobile industry, American firms recognized that their competitors in Japan had created a competitive advantage in how they designed cars. The Ford strategic implementation of concurrent engineering design practice was one part of a way to catch up with competition.

In the design of new products or services, several areas of a firm's business functions need to be involved in the design process, from the beginning and throughout the design. These include

- Product design engineers
- Production design engineers
- Manufacturing managers
- Marketing and sales managers
- Finance managers
- Product and production researchers

A concurrent-engineering product-development team requires forming multi-functional teams for the management of complete development of a new product, and the product-development project-management team should consist of representatives from the different relevant activities.

The job of this product-program management group is to formulate the design requirements and specifications, to initially take into account considerations of manufacturing and marketing and finance and research into the product design as early as possible.

The product-program management group first conducts a competitive-benchmarking of competing products in all price-categories and establishes a list of best-of-breed performance and features for the product system. They next establish a product-development schedule and early prototyping goals and means. Then they identify early sources of supplies and draw upon their expertise and suggestions about part and subassembly design. This group also consults and solicits suggestions from dealers, from service firms that repair and maintain the product-systems, and from insurers for suggestions for product improvement and feature desirability. The group also consults representative samples of customers and solicits and analyzes reactions of these customers to current and competing products and possible prototypes of the new product.

After this, the product-program group compiles a want list for a new product—as a list of desirable product performance, features, and configurations. The group next analyzes these into priorities of must-have, very desirable, and desirable if possible within costs. Finally, the group continues to manage the product development process to encourage teamwork and cooperation in developing a product rapidly and of highest attainable quality and lowest cost

> *Concurrent engineering practices can facilitate the rapid and correct design of new products by improving communication and cooperation between research, product design, manufacturing, marketing/sales, and finance.*

In 1991, Boston University held a conference with manufacturers to summarize lessons that were being learned in speeding the product to market (based upon a project conducted on the subject by Stephen R. Rosenthal and Artemis March,

1991). The conferees particularly emphasized that how important were the early phases of the product development on cost and quality of the product design. The conferees also discussed lessons about collaboration in concurrent engineering:

- The selection of the team is a critical and complex problem and crucial to eventual project success.
- Senior management should concentrate their involvement in the early stages of defining the product concept but that they should resist calling all the plays, eliminating choices for the team.
- Attention must be paid to overcoming the barriers to communication in the "separate thought-worlds and meanings" that the different cultures of specialists bring with them.
- Special attention must also be paid to the difficulties in creating cross-functional team collaboration in organizational settings that are multisite or organized by functional specialization and that have older, long-established traditions of sequential responsibility in product development.
- After learning to collaborate effectively with each other, the team must still build credibility and cooperation and support from their home departments.

CASE STUDY: Ford's Taurus Project in the 1980s, Continued

Returning to the Ford Taurus case, the Team Taurus group consisted of a car-program-management group, headed by Lew Veraldi and consisting of key players, with its purpose to promote in the whole design process continuous interaction between Design, Engineering, Manufacturing and Marketing (also adding in top management, legal, purchasing and service organizations). The overall coordination of the design project was by a Car Product Development group.

The importance of organizing this kind of initially inclusive participation was to bring downstream judgments early into the design process. The car-program management group set initial management goals for the project:

- Best-in-class performance for the car
- First prototype to contain 100% of prototype parts to completely test the manufacturing prototype
- Improve the focus and timing of key decisions for rapid product development
- Reduce the complexity of the product to improve quality
- Eliminate late and avoidable design changes by bringing in manufacturing considerations early in the design
- Control changes, to as few as possible

To target the best-in-class goal, the team benchmarked competing products, examining the best vehicles in the world. They evaluated over 400 characteristics and derived a set of best objectives for the Taurus team to meet in its design. These feature provided the design standards for the features of the Taurus, and Veraldi thought the team was successful in meeting many of these standards. Another thing the group did was to create a want list from all relevant constituencies including:

> *Internally.* Ford designers, body and assembly engineers, line workers, marketing managers and dealers, legal and safety experts;
> *Externally.* Parts suppliers, insurance companies, independent service people, ergonomics experts, and consumers.

As an example of this process, Veraldi described the Taurus team's visiting manufacturing personnel in Atlanta. There they spoke to the personnel who would build the car and talked to them about the design. They received feedback, such as the assembly workers telling the team that to achieve consistent door openings and tight door fits, a one-piece bodyside was necessary. In response, the Taurus team reduced the number of components on the bodyside from 6 to 2.

BEST-OF-BREED DESIGN STRATEGY

We pause again to review the strategic concept of competing with the best features in a product model. For competitive strategy, the customer must see that a product has at least the same or better quality than equally priced competing products. The term "best-of-breed" was used by much of U.S. industry to compare product quality, and the technique for product comparisons was called "competitive benchmaking."

A product or a service is a kind of system, a product system or a service system. The product system embedding not only the principal transformational function of a generic technology-system, but also other technology-systems as product subsystems and features—the product system. Product designs are always compromises between performance and cost.

> *It is simply competitively foolish to design a product system that has any feature distinctly inferior to any feature of a competing product in the same price class.*

Moreover, there is a competitive advantage in providing, in a lower-price class product, the quality of features of product-systems in higher-price classes.

The systematic identification and listing of product-system features and their highest quality expression in a product-system of any price class is called "competitive benchmarking."

Competitive bench-marking allows a designer to see what are the best features in a current product-line but does not prevent surprises that competitor may have in store in their next model. In performing a competitive bench-marking effort, the design team should also use technology forecasting to guess what competitors might be able to add in the next model. For competitive purposes, those features whose quality are most noticeable by the customer are those which should take priority in design for providing best-of-breed product features.

The design into a product-system (of a given price class) the highest quality of features of all product-system price classes is called a "best-of-breed" product design.

Of course, given cost constraints in a lower-cost product system, one can seldom attain all the best-of-breed features. But here is a good place to use innovation in a product design.

Innovation that allows best-of-breed designs in lower-price classes adds product differentiation competitiveness to price competitiveness.

CASE STUDY: Ford's Taurus Project in the 1980s, Continued

Returning again to the case, one of the key changes in design specifications that came from this inclusive approach of Ford's new strategic product design process was a decision in April 1981 to refocus the Taurus as a six-passenger, six-cylinder family car (after gasoline prices began to stabilize). At the time, much of Ford's future was depending upon project and aiming the car just right for the family market was a critical strategic target. To this end, the design goal of obtaining a complete prototype early was important. The team conducted a consumer research project in Florida. A focus group of potential customers were selected based on demographics and asked to evaluate the prototype vehicles by driving them. Their opinions provided important evidence for the final design. For example, they confirmed the team's decision to make the ride and handling of the Taurus closer to European models.

These product-design techniques brought the Taurus project in on-time and below-budget. The introduction of the Taurus car of the Ford Division (and its variation Sable at Mercury Division) cost $250 million less than their design budgets. Both products were successful in the marketplace. In 1988 Lew Veraldi was promoted to a Vice President of Ford, in charge of the Luxury and Large Car Engineering and Planning.

Case Analysis

In this case, we see illustrated several of the important strategic policy concepts to improve the speed and correctness of the product development cycles:

- Concurrent engineering
- Competitive bench-marking and best-of-breed designs
- Rapid product prototyping

STRATEGIC MANAGEMENT OF THE PRODUCT DEVELOPMENT CYCLE

The strategy of developing new products or services includes the policies that govern management procedures for the development and design of new products and services. Since new products or services are continually be developed and designed to keep up with or ahead of competition, this process is a kind of cyclic process, over and over again, and so it is commonly called the product-development cycle.

Strategic change in the management of the product development cycle is a critical issue in competitiveness.

For example in 1991, a committee of the U.S. National Research Council reported on the importance of managing the product development process, particularly with the new computerization of design aides, then one of the new information technologies (NRC, 1991). The committee on Engineering Design Theory and Methodology reviewed modern computer-aided design aides for product development and argued that attention should also be paid to how these aides were used in engineering management of product development. They emphasized that in addition to the modern technologies of computer-aided-design and computer-aided manufacturing, the management of the production introduction process also required:

1. Continuous and incremental improvements in manufacturing
2. Appropriate project management techniques and design practices

In addition to reports such as the above, other studies in the twentieth century emphasized the importance of proper management of the product development process in the context of progress in information technologies. For example, then Robert Cooper studied a sample of 203 new product projects in 125 firms. He interviewed senior managers who identified a typical success and failure product in each firm:

> Success and failure were defined from a financial standpoint. For each project, the project leader and some team members were personally interviewed. . . . The final sample consisted of 125 successes and 80 failures.
>
> —(Cooper, 1990, p. 28)

Cooper (1990) identified several factors that financially successful product-development projects had in common:

1. Superior product that delivered unique benefits to the user were more often commercially successful than me-too products.
2. Well-defined product specifications developed a clearly focused product development.
3. The quality of the execution of the technical activities in development, testing, and pilot production affected the quality of the product.
4. Technological synergy between the firm's technical and production capabilities contributed to successful projects.
6. Marketing synergy between technical personnel and the firm's sales force facilitated the development of successful products.
7. The quality of execution of marketing activities was also important to product success.
8. Products that were targeted for attractive markets in terms of inherent profitability added to success.

One can see that, in these factors, the first and last have to do with the relationship of the product design to the market—a superior product aimed at a financially attractive market provides both competitiveness and profitability.

Also factors five and six have to do with marketing activity. Sales efforts are helped by products that fit well into an existing distribution system and by properly managed marketing activities that test and adapt a product to a market.

Factors two, three, and four have to do with the quality of the technical development process. Good management of the technical activities in product development help produce good products. And good project management includes having a clearly focused product definition, managing well the different phases of the product development process, and having the proper technical skills to execute the project.

The success of a new product introduction requires not only a good product design but good management of both the product-development process and marketing process. The product development cycle should be managed to provide:

- Innovation in product performance
- Innovation in productivity
- Responsiveness to market changes
- Competitive advantages

The impact of new information technology was to help compress the time of the development cycle as well as improving design quality. In product development process management, information technology would be used to facilitate the

1. Performing of design activities in parallel and making continuous improvement in manufacturing processes
2. Using computer-aided design, modeling, and simulation—designing in virtual time.

The concurrent engineering process, by performing activities in parallel, ensures that important design considerations that may be experienced later in the product development process are appropriately considered early in the design stage. Doing product development right the first time and in virtual time requires technologies for designing and simulating the product in the computer. In addition, it also requires rapid prototyping of a physical aspects of the system, since computer simulations will usually be limited in total reality. Product development times can be significantly shortened by prototyping parts and subsystems and the product system as early as possible. This allows testing for fit and performance and durability, serviceability, and fashion. Changes in design can then still be made before volume manufacturing has begun.

CASE STUDY: Design Process in a Computer Firm

We next look at how information strategy can provide an improved procedure for design capability. Design capability began to be impacted by progress in computer technologies in the 1980s. This case looks at the situation of using information technology to improve the design process and is an example of the type of information flows in a product-development process in 1990. Dundar Kocaoglu, M. Guven Iyigun, and Chuck Valceschini described the product development process of a computer firm. They summarized the information flows of the firm in the design process, Figure 9.4 (Kocaoglu et al., 1990).

There we see that a product idea should begin with marketing and research

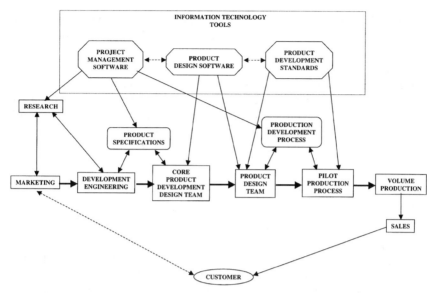

FIGURE 9.4   INFORMATION FLOWS IN PRODUCT DEVELOPMENT PROCESS

interacting with development engineering, together jointly conceiving an idea for a new product with a customer and defining system requirements for the new product.

Next, a core team is assembled, and it interacts with development engineering to refine the product-system requirements into a set of product specifications. Then the core team is expanded to a product team and performs the product development and any pilot production process changes need to produce the product. After that, high-volume production could begin.

During their research on the product development process with the computer firm, Kocaoglu et al. suggested improving the information flow with computerized design aides. First they suggested that computerized project management tools could facilitate the planning, scheduling, and interaction between research, development engineering, and the core and product teams. Next they suggested that computer-aided engineering tools could assist in the product specifications and designs performed by development engineering and the core and product teams. Then they suggested that the analysis of previous product designs could be codified to provide some company standards for best design and development practices—product development standards. This would be useful in order to continue to improve the management of the design and development process. Finally, Kocaoglu et al. recommended that the whole set of computer-aided tools for design and development be formulated and maintained within a design knowledge transfer system that would coordinate the use of these tools and provide access to information form one tool to the next.

What we see in this example is that the improvement of the service-technology in a product development process (through using the technologies of computer-aided-engineering design-software) also provides an opportunity for expanding the information flows and cooperation between research and product development and manufacturing and marketing to improve the overall coordination and effectiveness of product development.

LOGIC OF PRODUCT/SERVICE DESIGN

Strategic management of product design and development procedures centers upon a generic logic of design. All logic of design depends on the intellectual dichotomy of function and form. Design is creating form (morphology and logic) to perform function. For example, the design of a piece of hardware, such as a hammer, has the function of driving nails into wood in a customer's construction application. The form of the hammer consists of a heavy metal head and a long wooden or fiberglass handle. The head may have one flat surface for driving nails and a clawed surface for extracting nails. The combination of the weight of the head and the length of the handle facilitates the force the customer can apply when using the hammer.

All products, whether hardware software or services, can be so analyzed as to function and form in their design, and these provide the criteria that distinguish a good design from a bad design. A good design provides an appropriate form for a function. A bad design provides a function that does not perform adequately or safely or at the right cost. Designs of products that provide poor performance, are unsafe, and are high in price result in products that do not sell.

For a product design, the logic of design can be divided into several phases:

1. Customer requirements
2. Product specifications
3. Conceptual design
4. Preliminary design
5. Detail design
6. Product prototype
7. Testing
8. Final design

All phases of product designs create opportunities and risks in product design. For example, who is the customer and what are the product requirements for a customer is a critical judgment. This involves understanding the potential of market niches and application systems in these niches.

Setting the specifications requires translating customer needs to engineering specifications and is a procedure never very clear. In fact, creativity and innovation in this translation often results in higher quality products than a more plodding and literal translation. In addition, a given product will be part of a product family in order to cover market niches. Thus product design will occur within a broader activity of product family strategy. Within this strategy, product architecture and generic product platforms are critical decision decisions for profitability and competitiveness.

COMPARING HARDWARE, SOFTWARE, AND SERVICE DESIGN

From a strategic perspective, it is important to understand the differences in the design challenges of hardware, software, and services.

Modern products vary from hardware to software to combinations of hardware and software. It is important to understand the differences in design logic emphasis that occurs between hardware products, software products, and services. These differences lie in the relative complexity of the morphology or schematic logic of the core technologies of the product:

- In hardware design, physical morphology is complex and schematic logic relatively simple.
- In software design, physical morphology is relatively simple and schematic logic complex.
- In service design, both physical morphology and schematic logic are complex.

Hardware Design Strategy

Hard goods are physical products that can be used to satisfy physical needs, transform materials, carry out activities, or perform services. All hard goods are embodied in material manifestations, requiring materials and power resources. The physical aspects of hard goods require geometric and material design and manufacturing techniques.

The schematic logic of hard goods are in the control subsystem of the product. Traditionally, technologies for products and technologies for services had been relatively distinct, but with modern computer and communication technologies, they are increasingly sharing technologies for information and control in both product and service systems. Modern products now often embed control technologies in hardware and software. Technical innovation can be embedded in a product/service design in either:

1. New physical forms or materials or power sources of the product
2. New control systems in schematic logic in the design of the product's operations

Software Design Strategy

In contrast to hardware design, software design is principally concerned with complex schematic logics. Software design is the creation of schematic logics for a specific application. The elements of the schematic logic in software are conceptual linguistic primitives such as nominal terms (names) and relational terms (operations).

Thus it is important to software design to understand the nature of language. Language is the basic tool and form of thinking. We think in language, involving either an internal dialogue using language when we think to ourselves or as an external dialogue using language when we think with other people or computers. Thinking with other people or computers is usually called "communication" or "information." We learn to think in language as we interact with our parents as young children and acquire language. (Rare but poignant studies of children who grew up in isolation from human interaction have shown that language acquisition and thinking ability are interactive and must be acquired when very young.)

Principally, language

1. Sharpens and refines perception
2. Abstracts and generalizes experience from one specific event and context to another
3. Facilitates social cooperation and conflict

Therefore the development of language affects the nature of perception (how and what we see in the world), the nature of thinking (how and what we abstract and generalize of the world), and how we interact with one another.

Software design is the development of specialized languages and linguistic tools to facilitate thinking, computation, information, communication, cooperation, and competition.

A language is composed of a set of nominal and relational terms and a grammatical structure. The set of nominal and relational terms of a language constitutes the "dictionary" of the language. The grammatical structure of the language constitutes the architecture of the language. A linguistic logic is a kind of language to talk about the architecture (i.e., grammatical structure) of a language. Linguistic logic is a one-step linguistic regression of a language. As one constructs a linguistic logic as a language about language (first linguistic regression), one can also construct a language about a linguistic logic (second linguistic regression); this is often called a "meta-logic." Software design can thus be regressed into the design of the software language, software logic, and software meta-logic. This is why software design is complex in logic but simple in physical morphology.

The software language is what is coded in the software development. The software logic is the architecture of the software development. The software meta-logic is the boundary and assumptions about the software architecture.

In summary, in the designing software, coding tasks must include the expression of:

- System architecture. For example, a word processing program has an architecture of sentences, paragraphs, pages, and documents.
- Primitive linguistic operations. For example, in a word processing program linguistic primitives express fundamental editing operations such as delete, insert, erase, move, spell check.
- Data input and output. For example, in a word processing program, features must provide for inputs and/or outputs from several sources such as keyboard, storage disks, optical scanning, and electronic transmission.
- Coordination of activities. For example, in a word processing program, precedence ordering allows some operations to be called only after other operations (e.g. one must open a file before typing into it).

Software design therefore begins with a system approach to charting the kinds of allowed information and operations and possible flows among kinds of information and/or operations. Teams of programmers cooperate to write sections of the code and integrate the sections into an overall program. Debugging is a critical activity of software production.

Quality in software divides into bugs and defective disks or transmission. Failure of software to run properly is called a "software bug." Failure of software to run from a particular storage medium, disk or transmission is called a "hardware failure." Software bugs are the most serious problem for a software producer.

Software bugs arise from several sources:

- Poor programming
- Complexity of the application
- Newness of the application

The quality of programming depends on the skills of the programming team. The complexity of an application determines the complication of the architecture, coordination and number of lines of code. The more complex the application, the more bugs will occur simply from the inability of a team to comprehend the totality of the program operation in application. The more innovative and newer the application, the less experience will be available to determine the scope and details of the application. Thus bugs will arise from users trying to do something in the application for which programming was not planned.

Because of these kinds of sources of problems, the rule-of-thumb for large software programs is that bugs will never be entirely eliminated. Software producers must therefore depend on determining quality by the rate of the occurrence of bugs and not the absolute number of bugs. For this reason, software producers need to have marketing policies to quickly upgrade software versions and replace older versions when bugs are found by customers.

Accordingly, technological innovations that provide new functionality, extend functionality over more applications, or find bugs faster provide powerful competitive edges to software producers.

Service Design Strategy

Services are activities that provide value-added transformations or transactions or communications for a customer. Examples of service industries include banking, rental properties, medical services, accounting services, advertizing services, retail services, and food preparation services. As a sector of the economy, services have been and growing to provide the majority of areas of employment for a developed economy.

Service technologies are the tools and procedures used in the development and/or delivery of services.

Services can be internal to a productive organization or external as one of the products of the organization. Internal services provide activities necessary to the productive operations of the organization, such as engineering, personnel, marketing, manufacturing, and finance. External services are the products sold to customers or the assistance provided to customers.

Innovation of a new external service is the creation of a set of activities that can be sold to a customers. Innovation in service technologies are frequently dependent upon either new hardware or software. New schematic logics are important sources of innovation in service technologies.

Technologies for the service industries (or for internal services within a firm) use products and devices from manufacturers but in a procedural system that requires information and communications. Therefore what is unique in a service application is the information and communication and control procedures. The design of software for information handling, transactions, communication, and control are essential to services.

SUMMARY: TECHNIQUE FOR ANTICIPATING COMPETITION CHANGE

The strategy process requires the construction of a strategic business model, which incorporates a strategic issue of anticipated and planned changes in the conditions of competition. A systematic way to explore possible changes in competition is to use the following strategy technique:

1. Depict the value-chain industrial structure of a business
   - For each business in the company, construct and analyze the competitive conditions of all the sectors in the industrial value-chain of the business.
2. Forecast changes in technologies within the structure
   - Examine the progress in all the technologies within the value chain to determine possible impacts of progress on competition within the chain
3. Analyze impact of these changes upon business competitiveness
   - Formulate what kinds of products, services, quality, and cost will be necessary under changes
   - Foresee forecast possible changes in industry structure
   - Formulate opportunities for altering the structure to the competitive advantage of the business
4. Use information technology to improve design capabilities
   - Formulate design process strategies necessary to competitively design new products or services
   - Upgrading the quality of information technology in design tools and procedures is essential to prepare a business for continuing future competitiveness
5. Plan future products/services
   - Formulate product and/or service strategies for the future of the business

For Reflection

Examine the products of four different industries (one in hard goods, one in services, one in software, and one in entertainment). In an industry, what differentiates the high-end and low-end products? If you were asked to finance a new competitive entry into each business, in what kind of product strategy would you be willing to invest and why?

CHAPTER 10

OPERATIONS AND CONTROL

PRINCIPLE

Operations strategy should focus upon:
- Improving quality
- Lowering cost
- Increasing flexibility
- Adding e-commerce

STRATEGIC TECHNIQUE

1. Model a business in detail as a strategic enterprise model
2. Construct a three-plane description of the current operations structure
3. Examine wherein changes in operations are needed
4. Formulate strategic projects for operational change

CASE STUDIES

Ryanair

Apple Computer's Operations in 2000

General Electric's Refrigerator War in the 1980s
Outsourcing Manufacturing to Guadalarjara

INTRODUCTION

In developing a strategic business model, we should understand the basics of strategy in operations and control as an input to strategic thinking. Businesses create prosperity and survive through the effectiveness and efficiency of their operations in providing value to customers. Production processes can be service delivery systems for service businesses or manufacturing production systems for hard good production businesses.

Strategy for operations improvement involves strategic tradeoffs between the investments in operations improvements and the returns to investments from improved operations. In the short term, operations changes require capital investment, while in the long term, profitability is improved.

Operations deliver and produce the services and goods that a business may sell to a market. In formulating a strategic model of a business in the future, it is important to address the issue of what kinds of changes need to be made in business operations to prepare for a competitive and successful future. Strategic improvement in operations in a business involves improvement not only in the design processes for developing and designing products or services but also in the production systems. In the previous chapter, we reviewed strategic changes in the design processes; and in this chapter we will focus upon changes in the production operations (manufacturing or service delivery).

We recall that operations describe how current business activities are performed, and guiding operations are procedures. Procedures specify how the processes and activities of operations are to be performed, and guiding procedures are policies. Therefore, changes in operations require changes in the procedures and policies that guide operations. It is the change in policies for future operations that is strategic.

Operations changes need to be made for

1. Improvement of existing production processes (on service delivery)
2. New production processes (on delivery of new services)

Strategic improvement in current delivery/production processes always needs to be made to continuously improve service or product quality and to lower service or product cost. New delivery/production processes may be needed when new kinds of services or products are developed and designed.

CASE STUDY: Ryanair

We begin by looking at the strategic issues of improving service operations, operations of service firms. This case looks at the rise of a then new low-fare airline, Ryanair, in Ireland in the 1990s. The airline industry around the world had been regulated until the 1970s, when deregulation of the industry in some countries began altering competitive conditions, as Thomas Lawton summarized:

> Beginning in North America and spreading more recently to western Europe, the airline passenger market has witnessed a growing intensity in price-based competition. This intensified competition has been facilitated by policy deregulation initiatives.
>
> —(Lawton, 2000, p. 573)

In the United States, the most successful of the new airlines begun this way was Southwest Airlines, and in Europe the most successful was Ryanair:

> The largest and most successful of Europe's low fare airlines is the Irish operator, Ryanair. It is also the longest established, having first commenced scheduled services in June 1985, operating a fifteen-seater aircraft between Ireland and England. The market entry of this independent, privately owned airline, symbolized the first real threat to the near monopoly which the state-owned Aer Lingus had on the routes within Ireland and between Ireland and the U.K.
>
> —(Lawton, 2000, p. 574)

Ryanair provided a simplified service. No meals were served, seats were not reserved, and no restrictions were placed on the tickets. To meet the competition, Aer Lingus eventually had to reduce its ticket price, and the lower prices increased the volume of passengers:

> Ryanair's arrival helped precipitate a growth in the total air travel market, particularly between Ireland and the United Kingdom. This growth occurred primarily in what has been described as the "visiting friends and relatives" traffic.
>
> —(Lawton, 2000, p. 574)

From 1985 to 1995, the number of air travelers between the United Kingdom and Ireland grew from 1.8 to 5.8 million annually. And Ryanair's leadership in low costs allowed its low fares to provide excellent profits:

> Costs have fallen faster than yields within Ryanair, allowing profits to rise consistently. . . . This (has) translated into steadily increasing operating profit margins . . . going from 10.3 percent in 1994 to 17.6 percent in 1997.
>
> —(Lawton, 2000, p. 577)

Ryanair's strategy was not only to compete with low prices and low costs but also to open new routes. For example, it was first to offer services between Dublin and Bournemoth and between Dublin and Teeside.

Case Analysis

In this case, we see that the new-entry airline service competed against an established competitor with lower prices on existing routes, and it could do this through having lower costs. It expanded its markets by introducing new routes that had not had regular service previously.

Since the cost of the major device, the airplane, which allowed the service was the same for all competitors, the lower cost leadership of Ryanair had to be focused upon other aspects of the service delivery system (e.g., costs of food service, seat reservations, etc).

SERVICE SYSTEMS

Service systems provide a functional capability to a customer. For example, airline service provides the functional capability of travel by air from city to city in the world. Other services provide different functional capabilities, such as:

- A bus service provides a capability of land travel from city to city
- A phone service provides a capability of voice communications
- An Internet service provider provides a capability of global computer-to-computer communications
- A medical service provides the capability of dealing with medical problems.

The competition in providing a service depends on having the ability to deliver the service's *function* (e.g., operating planes, telephone lines, routers, doctors, etc.) and upon the other service factors in providing function, such as:

- **Effectiveness**. How well the function provision meets customers' needs (e.g., Ryanair increasing effectiveness by adding new air routes).
- **Efficiency**. How much resources are consumed in delivering a service (e.g., the fuel efficiency of airplanes purchased by Ryanair and Aer Lingus).
- **Capacity**. The service provider's capability of delivery services to many customers (e.g., the addition of Ryanair's services increased the total flight traffic between Ireland and England).
- **Price**. The service provider's valuing of the service to the customers (e.g., Ryanair's lower prices compared to Aer Lingus's prices).
- **Staff**. The competency, dedication, and responsiveness of the service provider's employees in serving customers (e.g., Ryanair's well-trained and dedicated employees).
- **Costs**. The cost of providing a service (e.g., Ryanair's no-frills and low cost operations).

- **Margins**. The difference between the price and costs of a service (e.g., Ryanair's good margins through lowering costs even at low prices).
- **Reputation**. The customers' perceptions of the reliability and quality of the service provider (e.g., the safety records and on-time services of Ryanair and Aer Lingus)

These service factors (function, effectiveness, efficiency, capacity, price, staff, costs, margins, reputation) provide the operation's sources of differentiating competitive factors between service providers. For example, Ryanair differentiated itself from its competitor primarily through operations of low prices, low costs, dedicated staff, and new routes.

MODELING OPERATIONS OF SERVICE SYSTEMS

Since the operations in a service are complex, it is very useful to have a strategic operations model of the system. Strategic operations models allow one to sort out the complexity of a system and identify exactly what are the critical issues that need to be addressed in improving and adapting a system's operations for the future. All services are kinds of managed systems, and all businesses are kinds of managed systems.

A managed system is any system of functional transformation that is managed for effective functional capability.

The operations of a managed system can generally be described by three related areas (or planes) of:

1. Activities of support of the transforming activities
2. Activities of the functional transformation
3. Activities of control of the transforming activities

These are illustrated in Figure 10.1 as the three planes of support, transformation, and control. Support Plane 1 describes the activities that support the production or service delivery activities of the Transformation Plane. The middle Transformation Plane 2 describes the operations of the system that directly provides the function of the service or production. The bottom Control Plane 3 describes the activities that control the service or production delivery activities of the Transformation Plane.

Sorting the operations into transformation activities, support activities and control activities (depicted on the three separate planes) helps to understand the complexity of the system. All managed systems require three areas of descriptions:

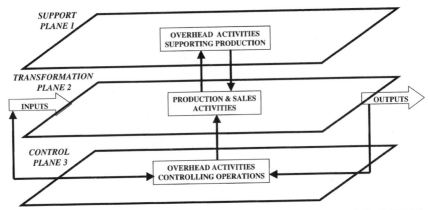

FIGURE 10.1   GENERIC OPERATIONS MODEL OF ANY MANAGED SYSTEM

1. Supporting activities of operations
2. Direct transforming activities of operations
3. Controlling activities of operations

Example: Service Systems Model of Airlines

To illustrate how this generic operations model of a managed system can be used to model airline services, we depict an airline service system operations model in Figure 10.2.

Therein, the transformation plane shows the city to city routes of the air transport system of a territory served by airlines. Air travelers (and freight) are inputs into the service to be flown from one city's airport to another city's airport. The output of the transformation of the functional capability of flight is the transportation of travelers and freight from city to city.

Support activities for these flight transformations include:

- Airplanes and fueling and maintenance systems
- Land transportation systems for travel to and from the airports
- Passenger handling systems for passenger check-in and loading and unloading passengers onto airplanes at flight gates in the airports
- Baggage handling systems for accepting and loading and delivering baggage and freight to and from the airplanes
- Food and beverage handling systems for refreshing passengers during flights.

Control activities for the flight transformations include:

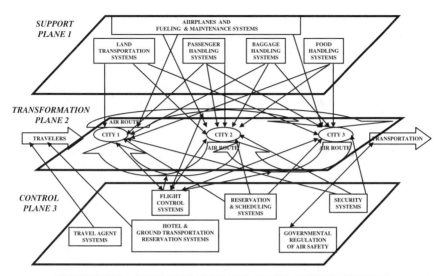

FIGURE 10.2   OPERATIONS MODEL OF AN AIRLINE SYSTEM

- Airport flight control systems to control air traffic
- Travel agent and reservation and sales systems for passengers to purchase air tickets and make flight reservations
- Airline reservation and scheduling systems
- Airport security systems
- Hotel and ground transportation reservation systems
- Governmental regulation of air safety

On the top support plane, support activities interact with producing activities on the transformation plane in a two-way interaction. For example, one support activity of Ryanair was a plane maintenance activity, which periodically takes a plane out of service and performs routine maintenance and/or repairs on the plane. The number of hours of flying time of a plane would be information that would trigger the maintenance activity of that plane, after which the plane would be returned to flight service (or a problem of a plane would be reported by a pilot to repair personnel, who would in turn diagnose and repair the problem).

On the bottom control plane, control activities acquire information of the service's outputs and use this information to control the producing activities and respond to inputs. For example, one control activity of Ryanair was to receive reservations and ticket purchases from traveler's for flights and then schedule flights to serve these travelers.

Activities on all three planes are all operations essential to a service business, although not all activities may belong to a single business in the service.

For example, in the airline service business of Ryanair, the airport flight control activities were essential to the taking off and landing of Ryanair flights (and all other flights to and from an airport) but these air control activities belonged to the respective airports and not to Ryanair.

Changes in any of the support, transformation, or control activities can improve service operations.

Planning of improvement of service operations is essential to service strategy.

SERVICE TRANSACTIONS

In addition to modeling service operations in formulating operations strategy, it is also important to describe the logic of the service transaction—how a customer uses the services delivered in the service operations model. Service providers make money in using service systems operations by transacting for a service with a customer. The logic of service transactions also needs to be strategically understood, as well as the operations systems model of the service.

For example, in the widespread enthusiasm of many of the new e-commerce business start-ups of the late 1990s ("dot.coms"), service transactions were often neglected in the enthusiasm of building operations. Then there were many cases where so much attention had been given by entrepreneurs to the technical system of the dot.com operations, without giving equally diligent attention to how to make money from the business transactions. We recall this happened in the case of Boo.com discussed in Chapter 2. And we recall that its example was then not unusual. After extraordinarily high stock market valuations of new dot.com businesses in 1998 and 1999, such valuations dropped dramatically in the spring of 2000 when the stock market began again to look at the business basics of generating revenue from services through the Internet.

In general, the transactional logic of service delivery consists of the sequence of decisions necessary to provide a service, including:

1. *Service referral.* A customer must arrange for a service by contacting the service deliverer, (e.g., selecting a doctor or opening a bank account).
2. *Service transaction.* Service delivery must be scheduled (e.g., making a

appointment and visiting the doctor's office or writing a check or making a deposit or withdrawal from the account).

3. *Selection of service application.* The appropriate application in the service must be selected (e.g., diagnosis by a doctor of the patient's illness or recording and accounting a banking fund transaction).

4. *Service application.* The selected application in the service must be provided (e.g., prescribing a drug or performing surgery or transferring of banked funds electronically or in cash).

5. *Payment for service.* Payment must be received by the service provider from the customer (e.g., billing a patient's insurance company or billing a client's bank account).

Information and other technologies are used in the different stages of the service delivery: such as devices (e.g., airplanes) and techniques (e.g., radar controlled air traffic) and information and communication technologies (e.g., reservation and scheduling tools) and professional knowledge (e.g., pilot flight skills). Improvements in any of these technologies can improve the transactions of service systems.

Changes in information and other technologies for service operations and service transactions are both essential to strategic improvement of service systems.

CASE STUDY: Apple Produces in 2000

Now we turn from the strategic issues of service operations to strategic issues in the operations of manufacturers (hard-good producers). We will look at the case of how Apple Computer strategically changed its manufacturing operations in the late 1990s after Steve Jobs took control again of Apple.

We recall from an earlier case study (Chapter 5) how Steve Jobs helped found Apple and then created the Mac model to save Apple from early IBM PC competition, using new applied knowledge developed at Xerox PARC. We also recall how Jobs left Apple after losing a power play to the CEO he had hired to replace him, and all the CEOs after Jobs failed to maintain Apple's competitive edge through innovation. Finally in 1997, Jobs was brought back to save Apple.

This case looks at his second rescue of Apple, but this time not with innovative products (as was the early Mac) but with improved operations: "Since returning three years ago to the company he founded, Jobs, 44, has worked the most unlikely comeback. . . . Left for dead in 1997 with mounting losses and shriveling market share, Apple is back to making the most stylish products in computerdom." (Burrows, 2000, p. 104).

In this description, we note that the new Apple products were only stylish and not innovative, as they were when Jobs first saved Apple with the Mac in 1985. Microsoft Windows 95 had finally caught up with the technical sophistication of the Mac operating system. Moreover, the CPU chips upon which both the MSDOS PCs and Macs ran were different but technically equivalent. After 20 years from their innovation, the personal computers had become commodity-type products. Jobs was distinguishing the Mac with fashion.

In addition to restyled products, what Jobs did to save Apple was to dramatically improve operations: "The company known for its incorrigible, free-spirited, free-spending ways has become a master of operating efficiencies. Jobs has slashed expenses from $8.1 billion in 1997 to $5.7 billion in 1999 by out sourcing manufacturing, trimming inventories, shifting 25% of sales to an online store, and slicing the number of distributors. (Burrows, 2000, p. 104)

When Jobs took over Apple again, he first reduced Apple's fifteen multiple product lines to a few that shared common components. Then he introduced new products in the iMac and iBook. These new products were primarily restyled rather than innovative in technology, but they did provide a replacement line for Mac owners. Jobs also addressed production, which then had 70 days of product inventory worth $500 million in warehouses at the end of each quarter—a substantial drag on profits. To improve Apple's operations, Jobs hired Timothy D. Cook, who had been a Compaq's procurement executive. Cook outsourced production of the printed circuit cards of Macs, which made the manufacturing job easier. He reduced the warehouses to nine regional sites, closing ten warehouses. Cooke reduced Apple's parts inventory down to only a day's supply. He persuaded key suppliers for Apple to establish production close to Apple facilities for just-in-time delivery. Overall, Cook succeeded in reducing the time for producing a Mac from four to two months.

Also Jobs refocused management as a team focused upon products, eliminating the chief administrative officer and making each executive responsible for everything related to the position's specialty across all products. Each Monday morning the executive committee met and made operating decisions for the week: "Says hardware chief Rubinstein: 'We don't sit around talking about how to drive up the stock or how to stick it to the competition. It's always about the products.' " (Burrows, 2000, p. 112)

Jobs' improvements of operations in product design and production capabilities increased sales and reduced costs—the right combination for profits. In the market niche of the consumer, education, and artistic professions personal computer market, Apple turned around its share, climbing from 3.8% in 1997 to 6% in the year 2000. Its gross margins had climbed to 30%; and its share price was back to $53 dollars, up eight-fold since Jobs returned.

Still there were many strategic challenges ahead. The first was to successfully replace the original Mac operating system with next-generation object-oriented software. The new operating system was to incorporate the features and structure of the NEXTStep program which had been purchased by Apple when it bought Job's NEXT Computer Systems in 1997

But even greater was the strategic challenge to innovate again in applications. Running the company on efficient production of stylish products that perform similarly to competitors will not enable the company to survive in the long term.

We recall there were two killer applications on the Apple that saved the company twice in the 1980s. The first killer application was the innovation of the first spreadsheet program, Visicalc, that ran on the Apple II and enticed the early adopters of PCs in the business market. The second killer application software was desktop publishing, which saved the Mac by finding it customers in the publishing departments of large corporations. Apple was looking for a third killer application. Apple was emphasizing movies. Its improved iMovie software enabled video editing. Jobs was hoping that the application of desktop movies would be as big for Apple earlier desktop publishing had been for the Apple Mac. Still in 2001, Apple held only a small percentage of the overall PC market and faced a long struggle toward continuing survival.

Case Analysis

In this case one can see the importance of efficient production operations when products are no longer very distinctive in performance. Jobs did have a substantial and loyal group of users to whom he turned for replacement of Macs, by introducing the new iMacs with faster speeds and by adding movie editing software and fashionable cases. This enabled Jobs to pull Apple from its death spiral in 1997. With improved management and production, profitability was regained in Apple.

Still one sees a long-term challenge when market share is beneath 10% of any market. Apple's strategy still needed innovation to survive in the long term.

Operational efficiency and innovation together provide right combination of both short-term and long-term strategy.

MODELING OPERATIONS OF A MANUFACTURING SYSTEM

It is important in a manufacturing business to model the whole system of enterprise operations. Operations in hard-good production businesses are complex because of the many processes and procedures and activities necessary to make the material transformations from natural resources into physical artifacts sold as physical products.

To construct a model, we can use the generic form of Figure 10.1 applied to an enterprise strategic model, which treats *resources and capital as inputs* and *sales and profits as outputs.*

Accordingly, in Figure 10.3, we indicate three planes: the middle transformation plane, supported by the support plane and controlled by the control plane.

> *As did the model of a service system, a model of manufacturing operations of a business requires three planes of description—support activities, transforming production activities, and control activities.*

Support Plane Activities

Projects in the support functions of a business are necessary to change operations for improvement. The kinds of improvements that are useful are in product, production, personnel, markets, finances, information technology, and other technologies. Accordingly it is useful to have and depict the explicit project activities attending to these areas for improvement as listed in Figure 10.3 on the **Support Plane**:

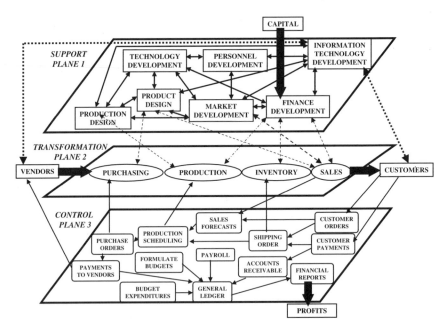

FIGURE 10.3 OPERATIONS MODEL OF A MANUFACTURING ENTERPRISE

- Product Design Program
- Production Design Program
- Personnel Development Program
- Market Development Program
- Financial Development Program
- Information Technology Development Program
- Technology Development Program

Strategic change in operations is planned and implemented in the form of specific projects—design projects, training projects (and programs), product-lines and brand projects, financial analysis projects, information technology projects, and technology research and development projects. Since all businesses need some kind of change annually to continue to adapt to the future, some projects in some of these areas will be occurring each year.

Long-term programs for change are usually organized within engineering departments, marketing departments, and research units and performed as discrete projects by multifunctional teams.

New product designs are performed by multifunctional design teams, led by engineers in the engineering department. Production improvement projects are performed by multi-functional teams led by manufacturing engineers in the engineering department. Product-line and brand-market analysis projects are performed by multifunctional teams led by sales personnel in the marketing department. Cost analysis projects for product-lines and new product-lines are performed by multifunctional teams led by financial personnel in the finance department. Personnel development projects are performed by multifunctional teams led by personnel people in human resources department. Technology development programs are performed in multi-disciplinary teams led by scientists and engineers in the research and development (R&D) laboratories. Information technology development improvements are performed in projects by multifunctional teams led by personnel in the information technology department.

Strategic change in operations is planned and implemented as discrete change projects performed by multifunctional teams led by personnel in the appropriate organizational department.

Accordingly, the management style for change in operations is a project management style as opposed to an entrepreneurial or professional management style. For a comparison of project management styles to the bureaucratic and professional styles see Betz (1997).

The double arrows within the support plane connecting these different areas of development indicate the kinds of informational interactions that occur between multifunctional projects in carrying out the projects for strategic change.

For example, new technology developed in R&D projects might be used in new product design, new production design, or in market and financial analyses and projections. Projects to improve information technology in the business can impact the processes of technology development, product design, production design, change in markets and financial performance of the business. Information technology development projects can also improve communication and interactions with customers and with vendors.

Financial analyses can look at the strategic implementations for operations change as to the requirements for capital. We recall that the enterprise strategic model does not seek to optimize capital but sales and profits so that capital in this model is treated as an enterprise input.

Transformation Plane Activities

The continual operations which produce and sell the products (services) of the enterprise are performed as a sequence of value-adding activities, as depicted in the **transformation plane** of Figure 10.3. Products and services are produced and delivered by acquiring appropriate supplies, materials, resources from vendors and then forming and assembling and using these to create the product sold to the customer. The activities in the transformation are organized in this sequence, and one can depict the direct value-adding activities in the enterprise value chain as:

- Purchasing
- Production
- Product inventory
- Product sales

Purchasing activities are usually organized in a purchasing department; production in a production department, and sales in a sales and marketing department. Product inventory is stored in product inventory warehouses until shipped to dealers and customers.

As an illustration of the complexity of production operations, let us recall the example of automobile production at Toyota in 1983, as illustrated in the earlier Figure 9.2. Toyota purchased both manufactured components and materials for Toyota's own manufacturing processes from various supplying vendors. Toyota purchased components such as electrical parts, bearings, glass, radiators, batteries, and so on. From other suppliers, Toyota also purchased processed materials, such as steel sheets and rolled steel, nonferrous metal products, oils, paint, and so on. Purchased components and materials were subjected to acceptance inspection.

Next, materials went through various production processes to be formed into parts (e.g., forging, casting, machining, stamping, plastic molding). In addition, some of the purchased components also went through further processing to be finished as components (e.g., heat treatment or additional machining).

Materials, components, and parts eventually were all used for three subassembly systems in fabricating the automobile:

1. Power subsystems. Engine, axles, transmission, steering assembly, etc.
2. Chassis subsystems. Frame, suspension, brakes, etc.
3. Body subsystem. Body, seating, windows, doors, etc.

Various plating and painting processes prepared the power and chassis subsystems for final assembly, and the body was painted for final assembly. Then finally the three major fabrication subsystems were attached together as an automobile. After adjustments and inspection, the product emerged as a completed automobile.

As we have indicated in Figure 10.3, strategic changes to transformation operations are implemented from strategic projects in the support plane. The dotted double-arrow lines between the two planes of transformation and support indicate the kinds of interactions

Control Plane Activities

We have noted how modeling operations as an enterprise systems uses both Porter's ideas of value-chain description of a business and also Forrester's ideas of dynamics of business operations. In a Forrester-style of systems dynamics model, one needs to relate the flow of information of business operations to the activities that produce products or services of the business.

In a hard-good business, the transformation plane describes materials flows as physical parts and materials are purchased from vendors and physically shaped and assembled into hardware products shipped to customers. The control plane describes the information flows which control the material flows of the transformation plane. For example, in the earlier case of Toyota, the automobile assembly line was the production system for assembling automobiles from parts supplied by vendors (internal parts division or external parts vendors). The financial accounting system instituted by Sloan controlled the information of costs of the materials flows in automobile assembly.

In a service business, the transformation plane describes the activities sequences of the service delivery (within which some activities may involve the application of hardware in the service). The control plane describes the flow of information in delivering the service. For example, in an passenger airline operation, passengers arrive at the airport, check-in luggage and board an airplane. The airplane has been provisioned and staffed. It takes off to fly to a destination, whereupon passengers and luggage are discharged from the airplane. The materials flow in this service of air transportation are the geographic transfers of passengers and baggage from a geographic origin to a geographic destination. The information flows controlling this service include scheduling and control of flights, making passenger reservations and payments, loading and deplaning of passengers and baggage, and scheduling crews, provisions, and maintenance of the airplane.

Within the control plane, information systems process information on the operations and report performance of the operations. These systems must include the ability to

- Receive customer orders
- Receive customer payments
- Create product shipping (or services) orders
- Forecast sales
- Schedule product production (or service delivery)
- Purchase supplies
- Pay vendors
- Formulate budget plans
- Control budge expenditures
- Pay personnel
- Control customer payables
- Account for finances
- Produce financial reports

The directional arrows in Figure 10.3 depict where the information system on the Control Plane directly impact the control of operational activities on the transformation plane. Customer orders control shipping orders and sales forecasts. Shipping orders control inventory depletion and co-control production scheduling. Actual sales control the sales forecasts, which co-control production scheduling, along with inventory depletion. Production scheduling controls the production kind and rate and build-up of inventory. Purchasing orders control the kinds and rates of purchases from vendors and payments to vendors.

The general ledger records customer accounts receipts and receivables, payments to vendors, payroll and budget expenditures. Financial reports (daily, quarterly, and yearly) summarize financial performance of operations and calculate profits.

Information systems record and control the performance of operations.

CASE STUDY: GE's Refrigerator War in the 1980s

We will look in detail at the activities on the transformation plane of a hardgoods producer—or manufacturing, as it is normally called. We examined the strategy for the design of products or services in Chapter 9, which belongs to the support plane activities. We will examine strategy for information, also on the support plane in Chapter 11. But now we will examine in more detail production strategy in operations, looking first at a case of strategic improvement in the operations of hard-good production—manufacturing. We will examine the theoretical ideas for thinking strategically about improved physical production.

Although the Internet and electronic commerce have emphasized the virtual aspects of information and communication, still much of what is ordered in electronic commerce (or traditional commerce) requires the manufacturing and delivery of hard goods—physical products. Therefore, operations strategy needs to begin with the examination of the production of the physical aspects of production—whenever the enterprise system deals in hard goods.

This case study occurred in the last quarter of the twentieth century when U.S. manufacturers had slipped behind Japanese manufacturers in both the design of physical products and in the manufacturing of physical products. It examines the strategic changes needed in one of General Electric's manufacturing businesses of the 1980s to survive the strong Asian manufacturing competition of the time. In this case we will see a historic example of a large firm redesigning one of its major businesses, product and production, in order to improve competitiveness and remain in the business—the business of large consumer appliances.

As one result of World War II in the twentieth century, United States manufacturers dominated world hardware markets for a period of twenty years after the war. Then beginning in the 1970s after European firms and Asian firms rebuilt capabilities, American manufacturers found themselves under competitive assault as one North American manufacturing business after another faced foreign competitors again, but ones who now often were producing lower cost and higher quality products. Then General Electric's consumer appliance business faced such competition. GE gave up its small appliance business, but choose to battle it out in the large appliance business.

Ira C. Magaziner and Mark Patinkin described the challenge GE then faced:

> Fifty miles south of Nashville . . . is one of the world's most automated factories. Had it not been built, U.S. households might soon have had yet another product, the refrigerator, stamped 'Made in Japan'. Instead here in the heartland, General Electric found a way to build products better and cheaper than those made by foreign workers paid one-tenth American wages.
> —(Magaziner and Patinkin, 1989, p. 114)

The GE manufacturing manager who helped manage the improvement was Tom Blunt:

> Tom Blunt still remembers the day in 1979 that he first stepped into Building 4, the plant in Louisville, Kentucky where compressors for GE's refrigerators were made. . . . The plant was a loud, dirty operation built with 1950s technology: old grinders, old furnaces, too many people. Finishing a single piston took 220 steps. Even the simplest functions had to be done by hand. Workers loaded machines, unloaded machines, carried parts from one machine to the next. The scrap rate was ten times higher than it should have been, 30% of everything the plant made was thrown out.
> —(Magaziner and Patinkin, 1989, p. 114)

In 1979, this sort of practice was common in American manufacturing. American industrial capacity had aged after the Second World War. In the 1970s, the OPEC cartel rise in energy prices had pointed out their obsolescence, while the accompanying inflation reduced incentives to make the investments necessary to improve manufacturing technology. But while the American manufacturers hesitated, the Japanese had not. For example, Japanese automobile manufacturing had continued to improve productivity during the 1970s so that by the end of the 1970s, the American manufacturers were only half as efficient. During the 1980s America became aware that its had lost manufacturing superiority and was rapidly losing manufacturing base in many industries.

Blunt, a manufacturing engineer, had only recently joined GE's Major Appliance Business Group (MABG) as chief manufacturing engineer for ranges, and he was worried about the manufacturing capability. He saw that manufacturing had to be improved at GE's Louisville plant, and he found that other managers in GE had begun to worry about the poor state of manufacturing. The major appliance group business's (MABG) profits and market share were declining. One competitor, Matsushita, was manufacturing better and cheaper compressors in Singapore and was selling them to the GE appliance subsidiary in Canada. Matsushita was also experimenting with new rotary compressor technology. And this was a technology GE had invented but only used in their air conditioners. Moreover, Whirlpool was moving its compressor manufacturing to Brazil to lower labor costs. In the fall of 1981 several manufacturers approached GE offering lower-priced compressors than GE could produce itself. GE's managers began talking about the strategy of sourcing rather than producing the compressor.

As a manufacturing engineer, Blunt didn't like that strategy. The compressor was the core technology of the refrigerator, important to performance and energy efficiency of the refrigerator. He was promoted to head of advanced manufacturing for refrigerators, MABG's biggest product. Since the compressor was the key component of the refrigerator, Blunt decided that GE had to innovate its compressors to keep its refrigerator business. For a product, some aspect may or may not provide a competitive advantage. If an aspect does provide a competitive advantage for differentiating the product, one should not outsource that aspect. In refrigerators, energy efficiency is one differentiating competitive factor; so that it would have made poor competitive strategy to outsource the compressor.

In 1981, Ira Magaziner was then a manufacturing consultant hired by GE to provide information on their competitor's refrigerator manufacturing costs. He visited all the compressor manufacturers in the world. When he returned to Louisville and gave his report to GE's MABG, he had documented a major cost problem for GE. It cost GE's MABG more than $48 dollars to make a compressor, while in comparison the costs at Necchi and Mitsubishi for the same compressor were $32 and $38 dollars respectively. In addition, other competitors were designing new plants that would further drive costs down, with Hitachi and Toshiba aiming at a cost of $30 per compressor and Matsushita and Embraco building new plants to produce compressors at $24 dollars apiece.

One reason was for their competitor's cost advantage was cheaper labor costs—$1.70 per hour in Singapore and $1.40 per hour labor in Brazil, compared with $17 an hour with benefits in Louisville. Another reason was productivity. It took GE 65 minutes of labor to make a compressor, whereas the cheaper labor was also faster: in Singapore it built one in 48 minutes, Brazil in 35 minutes, and Japan in 25 minutes. Their new plants and designs were more efficient.

It was obvious to Magaziner that GE's only hope was in a new design of a rotary compressor because it had fewer parts and could therefore be cheaper than a reciprocating compressor. GE had invented the rotary compressor for air conditioners but hadn't bothered to use it in refrigerators where it would have to stand up to harder use. Magaziner next went to see Tom Blunt and asked him if he could design a plan to produce the rotary. Blunt replied that

that's what he did for a living, if he got the chance. And Blunt would get his chance. Truscott, MABG's chief engineer, assembled a team of product design engineers and came up with a model they thought could be made cheaply. But the design required a precision in working parts to fifty millionths of an inch—more precise a tolerance then used by any mass-produced consumer equipment.

Next Truscott told Blunt to assemble a team to design a factory to produce it. Blunt assembled a team of about forty people, including engineering product-design people and manufacturing people (a concurrent approach). He put the product engineers and the manufacturing engineers in offices across the hall from one another. Together they refined the design of the rotary pump to the most automatable model, with less than twenty parts.

But there was a major technical bottleneck in the production system. Existing production equipment at that time normally could not meet the close tolerances required of the new product. Blunt and his engineers also had to improve the equipment. One of Blunt's engineers was Dave Heimedinger, who negotiated with suppliers of grinding and gauging machines for equipment with the needed processing accuracy. This was much higher that traditionally done, and Heimedinger and his team had to develop combinations of equipment that could produce the parts at the required low tolerances. Often existing production processes need to be improved to produce innovative products.

Finally, Truscott and Blunt had a design for an inexpensive rotary compressor and an automated factory to produce it. Should GE build it? The investment was high, at least $120 million dollars. Millions more would have to be spent on redesigning GE's refrigerators to fit around the new compressor. The GE engineers would have to sell the project to the finance people in GE's headquarters in Connecticut, who were wary of large capital investments in appliances, since recently they had lost a lot of money in a new washing machine plant that had failed.

To prepare for their visit to headquarters, Truscott and Blunt asked Magaziner (who was still consulting for them) to check the proposal for a new plant. Magaziner advised them that it was a close call, but he would recommend the investment. Then Truscott sent the proposal to Roger Schipke, then head of MABG.

Schipke, Truscott and Blunt flew from Kentucky to Connecticut to present the proposal to the CEO of GE, Jack Welch. Welch listened to their presentations. He agreed that it was strategically important to keep major appliances as a core business, and he understood that it would require a major investment to improve manufacturing. The new plant would be built in Columbia, 200 miles from Louisville.

This was a major business risk depending upon a technological risk. The compressor technology worked. Could the manufacturing technology produce it within the tolerances required? In addition to the technical risks, there were also people risks, since even the most automated production process is still a socio-technical system. GE would have to train its manufacturing people in new processes. Accordingly, management developed a training plan for its work force to staff the new factory and asked the workers to undergo extensive training. The workers were eager to get training, as they knew that jobs were lost when manufacturing companies became technically obsolete and hence uncompetitive.

The new plant was built, workers trained, and production of the new efficient rotary-compressor refrigerators began. Then one might have hoped for a happy fairy-tale kind of ending to this brave case of GE's strategic change, a kind of ending without any problems. But in real life and in business, there are always problems that have to be met and overcome. This story was no exception: "The celebration was short-lived. In January 1988, 22 months after the first compressor rolled out of the new factory, a problem surfaced. Some of the larger compressors- those in GE's bigger refrigerators- began to fail."

—(Magaziner and Patinkin, 1989, p. 121)

Although the failures constituted only a small percentage of the plant's total production, still a few failures could destroy GE's reputation of quality with customers. Management and engineers turned to the problem. Schipke formed a project team to analyze the problem. They worked incessantly, often through the night, for several weeks. It was a difficult problem because only a small portion of the compressors had yet failed in service. But finally they found the source of the problem: lubrication had not been properly designed and was allowing one of the compressor's small parts to wear more quickly than the designers had calculated. In engineering, defining the problem is halfway to the solution. Truscott and other GE engineers improved the compressor design to eliminate the lubrication problem.

Still, the management problem was not over. Schipke first approved a customer-service plan to go ahead and replace immediately at no cost to the customer any compressor that broke down in the refrigerators GE had already produced and sold. But it would still take several months to redesign the compressor for manufacturing. That meant production could continue during that time to make a product that would prematurely fail, or Schipke could stop production until the redesign was ready. Either way, it would cost a lot of money and lose customers.

Schipke cut through the Gordian knot of his manufacturing dilemma a hard, direct way. He didn't wish to lose GE's reputation for quality by shipping refrigerators that might develop problems. He decided to purchase older reciprocating compressors from abroad while engineering fixed the problem.

It was a tough decision, but it was the only way GE could keep major appliances as its core business. The final responsibility of the quality and cost of a product is manufacturing's responsibility. GE paid the price, and when the twenty-first century began, GE was still in the major appliance business.

Case Analysis

This case illustrates an important point about strategic innovation in manufacturing operations. To maintain competitiveness in hard good products, products must be periodically improved through innovation and new production processes must be periodically innovated. Strategies for product systems and production systems of an enterprise are interrelated issues.

STRATEGY FOR IMPROVING HARD-GOOD PRODUCTION SYSTEMS

Strategy for improving production systems of hard-good manufacturing enterprises needs to analyze the production system as an integrated system composed of unit production processes.

The hard-good production system of a firm is a series of process subsystems of unit-production-processes embedded within a socio-technical system of the organization of production. Improvement in a production system can occur in any aspect of the production system:

1. The boundary of the production system
2. Unit-production processes within the production system
3. Connections between unit-processes as materials handling technologies
4. Production system organization, communication, and control

Innovation may occur in any part of the production system, of its parts or whole:

1. New or improved unit processes of production
2. New or improved tools or equipment for unit processes
3. New or improved control of unit processes
4. New or improved materials-handling subsystem for moving workpieces from unit process to unit process
5. New or improved tools and procedures for production scheduling
6. New or improved tools and software for integrating information for design and manufacturing

Unit Process Innovation

Innovation can be implemented in the unit processes of a manufacturing systems as:

1. New unit processes
2. Improvements of existing unit processes
3. Improved understanding and modeling of the physics and chemistry of unit processes
4. Improved control of unit processes

Innovation of new unit processes for a production of a product may be required by:

1. A need to process new kinds of materials
2. A need to process with improved precision (improved purity or at greatly reduced tolerances)
3. A need to process with reduced energy consumption
4. A need to speed-up throughput of production or to scale-up the volume of production

New materials may need to be processed on new or existing equipment. Improvements in tolerances or quality or the reduction of energy or the scale-up in volume production will usually be helped by improved understanding of the physics and chemistry of the processes. This understanding can be used as models for sensing and process control.

Strategic innovations in unit processes should be introduced into the production system as discrete modules, previously debugged, into an existing production system.

New applied knowledge in innovation can improve unit process control in two ways:

1. Real-time control through intelligent sensing and control
2. Experimental design for processes one cannot presently model

The real-time control of unit processes is a complex physical and logical problem that requires:

1. Sensors that can observe the important physical variables in the unit manufacturing process
2. Physical models and decision algorithms that compare sensed data to desired physical performance and prescribe corrective action
3. Physical actuators that alter controllable variables in the physical processes to control the manufacturing process

CRITERIA OF STRATEGIC IMPROVEMENTS OF PRODUCTION

The measurement of strategic improvements in production is not simple but requires several measures of the concept of production quality. The strategic meaning of the quality of a product or service has three different meanings:

1. Quality of product/service performance. How well does it do the job for the customer?
2. Quality of product dependability. Does it do the job dependably?
3. Quality of product variability. Is the product/serviced produced or delivered in volume without defective copies/instances?

Performance is first a function of design. Poorly designed products or services can never provide a high enough quality of performance for the customer no matter how well produced or delivered it is.

However, it is also important to attend to the second and third meanings of quality (dependability and variability). In U.S. manufacturing until 1980, traditional manufacturing quality-control focused primarily upon the third notion of quality—product variability. Then the standard technique was to control quality after production, though sampling batches of produced products to determine if product variation was within acceptable specifications. Innovation in hard goods and materials strategically improves quality when it reduces deviation from target design specifications. When all parts of a product are on target specifications, the product's performance in field-use becomes more robust and independent of field conditions.

The quality of product variablity is important to customer satisfaction and to controling the costs of production. Strategy for production improvement needs to attend to the economic and competitive benefits in production operations, which include:

1. Production quality. Reducing rejects and improving production accuracy.
2. Production efficiency. Reducing material wastage and/or energy usage.
3. Production effectiveness. Increasing flexibility and variability of producible products.

4. Production capacity and throughput. Increasing the volume of production per unit time.

5. Production responsiveness. Decreasing the time of change-over for product variability.

6. Production cost. Decreasing the overall cost of a unit product.

CASE STUDY: Outsourcing Manufacturing to Guadalarjara

A basic strategic decision about production is whether or not to produce in-house or to outsource production. In production operations, competitive advantage can come from proprietary skills and tools in product design, production, or delivery. If a business outsources any of these activities, no proprietary advantage can be obtained, since the skills for design, production or delivery reside in the sources. Nevertheless, it can make strategic sense to outsource when the cost of investment in a production capability is large and cannot provide a competitive advantage. This occurred in electronics products in the late 1990s, as the components of the products were standardized (such as the central processing chips in personal computers) and assembly of the product from standardized components did not provide a competitive advantage.

This case looks at outsourcing of electronic products at the end of the twentieth century, resulting in a build-up of increased production capability in the part suppliers of the electronic industry, particularly in Guadalarjara, Mexico:

> Flextronics and other electronics-manufacturing service providers are the big beneficiaries of . . . corporate plans to outsource. . . . In a bid to boost return on capital and hone their core competencies, even the staid industrial giants of Germany and Japan are starting to sell off factories. They then award long-term contracts to outside suppliers—often the same companies that bought their plants.
>
> —(Engardio, 2000, p. 177)

Electronics manufacturing service (EMS) companies do not produce its own brand name products. A printed circuit board supplier, Flextronics, had grown by assembling printed circuit boards and other electronic components into completely assembled consumer electronic products. This outsourcing effort was large enough to begin building large firms of suppliers in EMS. Solectron Corp in 1997 was one of the first of these to have $3 billion sales. By 2001, five more EMS firms had attained $10 billion in sales, with Solectron growing to $20 billion. At that time, industrial sector was growing at a 20% rate.

The gross margins for these suppliers were running about 6% to 8% on sales, but because of economies of scale, they were generating about a 20%

return on equity. One of the ways of doing this was to use inexpensive labor in the final assemble of electronic products, as they are mostly snapped together. This provided business growth for the regions of the world where suppliers were building new production capacity. One of these locations was Guadalajara, Mexico:

> Four years ago (1996), cornfields filled the 125-acre industrial park on the dusty outskirts of Guadalajara, Mexico. Now glistening white factories and 4,000 workers turn out thousands of Ericsson cell phones, 3Com Palm Pilots, Compaq circuit boards, and Cisco routers each day.
>
> —(Engardio, 2000, p. 177)

Production expansion for the EMS firm of Flextronics had been placed here, and the manager, Brad Knight, had opened a one-million-square-foot facility. In 2001, Mr. Knight bought an additional 75 acres to triple the production capacity of the plant. This kind of production expansion in Mexico was then growing Guadalajara into one of the principal manufacturing centers for electronic products sold in the United States.

STRATEGIC OUTSOURCING OF PRODUCTION

A key decision in hard-good production strategy is what to make and what to have made by suppliers. As we saw in the concept of the industrial value chain, any company in an industry can vertically integrate down into the lower supplier areas of the chain or up into the upper device/systems assembler areas of the chain.

In the previous case, the circuit-board parts suppliers, such as Flextronics, integrated vertically into the device assembly sectors, as these were abandoned ("deverticalized") by electronic products firms in their outsourcing strategy.

Also as we saw, each sector in an industrial value-chain has different kinds of competitive factors, rates of innovation, proprietary knowledge, operating margins, capital investment requirements, economies of scale, and economies of scope. Therefore, a strategic decision to vertically integrate (up or down) or to outsource in the industrial value chain requires a strategic decision about competitive advantage and return on investment. Competitive advantage is lost in outsourcing but return on investment may be temporarily increased. There is no absolute answer as to whether is better to outsource or not. It depends upon the industry, times, and competitive factors in the industry at that time.

If quality in production is difficult to achieve and provides a distinct proprietary advantage, then production should not be outsourced. If quality is high from all suppliers, and all competitors use similar parts, the production can be outsourced. However, outsourcing production loses control over quality and cost.

Outsourced parts (or even assembled device) cannot differ in quality or cost from those of a competitor who purchases the same parts (assemblies).

Similarly, outsourcing parts loses control over the capacity of production and flexibility of production.

Businesses with outsourced parts are completely dependent upon their suppliers for capacity, quality, and costs.

For example, in the early 1990s, a major airplane manufacturer reduced its supplier list to tighten control over suppliers. But in the mid-1990s, the airplane industry had a large and unanticipated upturn in business. This airplane manufacturer could not meet the demand of its airline customers. It lost not only control over its costs but permanently lost significant market share.

In summary, the advantages to outsourcing production is a reduction of investment requirements in production facilities in the short term, but in the long term, the disadvantages are loss of control over quality, costs, and capacity.

COMPARING MANUFACTURING AND SERVICE OPERATIONS

Finally, let us compare operations strategy in both service and manufacturing businesses. We look at the generic knowledge competencies required for manufacturing or for services.

Manufacturing Knowledge Assets

For a manufacturing type of firm, knowledge assets can be classified for their use as

1. Knowledge competencies for a product development and design
2. Knowledge competencies for production design
3. Knowledge competencies for product marketing and distribution
4. Knowledge competencies for product maintenance and repair

5. Knowledge competencies for assisting the customer's applications of the product
6. Knowledge competencies for communicating and conducting transactions with customers and suppliers
7. Knowledge competencies for controlling the activities of the manufacturing firm

Service Knowledge Assets

For a service firm, knowledge assets can be classified as to their use as:

1. Knowledge competencies for using devices essential to service delivery
2. Knowledge competencies for supply and maintenance of devices used in service delivery
3. Knowledge competencies for services delivery
4. Knowledge competencies for services development
5. Knowledge competencies for assisting the customer's applications of services
6. Knowledge competencies for communicating and conducting transactions with customers and suppliers
7. Knowledge competencies for controlling the activities of the service firm

SUMMARY: USING THE STRATEGIC TECHNIQUE OF MODELING OPERATIONS

In formulating operations strategy, it is important to develop a detailed strategic model of the business as an operating system.

1. Model A Business In Detail As An Operating System
   - Describe and relate the planes of transformation to areas for strategic changes in operations and to information-flow control of operations.
3. Examine Wherein Changes In Operations Should Be Made
   - Forecast progress in technologies relevant to all the enterprise subsystems.
4. Formulate Strategic Projects For Operational Change
   - Identify needed projects for strategic change in operations in the different areas of the change plane of the enterprise model.

For Reflection

Select and compare two industries, one in hard good manufacturing and one in services, as to their process innovations during the twentieth century. (For example, automobiles or electronics and medicine or banking.) What were the underlying knowledge advances that paced the operating innovations?

CHAPTER 11

INFORMATION STRATEGY

PRINCIPLE

Information strategy has both strategic technical and business parts.

STRATEGIC TECHNIQUE

1. Form an information strategy team:
   - Divide the team into three strategy subteams:
     - E-commerce strategy subteam
     - Legacy system strategy subteam
     - Integration strategy subteam
2. Examine business process redesign
3. Develop information strategy

CASE STUDIES

Beauty Goods in Cyberspace
Kinko's Buys Liveprint.com
Dell Sells on the Internet
CDnow and Geffen Records

Changing IT in Allied Signal Technical Services

Reengineering Coats-Viella

SAP'S ERP Software in the 1990s

INTRODUCTION

Strategic Challenges of Information Strategy

By the beginning of the twenty-first century, information technology was pervasive in all businesses, so the strategic challenges of information strategy focused upon the following three issues:

1. Using the new electronic commerce medium
2. Upgrading information capability through replacing older legacy information systems
3. Improving the flow of information throughout a business ("enterprise integration")

Modern information technology in business consists of a system integrating the use of computers and communications in business activities. Change in information technology in a business/firm involves change in any part or all of the information system:

- Change in computer architecture, hardware, software, and peripherals
- Change in communication architecture, hardware, software, and peripherals
- Change in skills of using computers/communications systems

Information strategy is the part of a business plan to change the computer/ communications architectures, hardware, software, peripherals, and training. All such changes are aimed at improving business operations. Accordingly, information strategy consists of a technical part (the hardware/software architectures) and a business part (how improved information technology can be economically utilized).

Technical Part of Information Strategy

Information systems are complicated systems of technologies using software and hardware and connections. The overall functional scheme of what are the components of an information system and how these components are related and connected is called the "architecture" of the information system.

The technical part of information strategy consists of changes in the architecture of the information system.

Business Part of information Strategy

As we reviewed earlier, information systems facilitate and control business processes. Business processes include purchasing supplies, producing product, taking orders from customers, shipping product to customers, receiving payments, and so on (as sketched in Figure 10.3). Changes in information system architecture are guided by application of these changes to improvements in business processes.

Also we recall (from Chapter 3) that information strategy needs to be examined for *all* the potential impacts upon the different business functions. A cross-function information strategy team can use the strategic business policy matrix (as given in Figure 3.7) as a guide to reviewing the interaction between information system change and business model change. When examining the information strategy impacts, the planning team can use a list, such as in Figure 11.1, to summarize the areas of change needing planning.

Accordingly, in formulating information strategy, one must attend to the business component in the strategy, which focuses upon the applications of the information system to business purposes.

The business part of an information strategy consists of changes in business processes to use changes in information architecture.

Since businesses are ongoing repetitive operations, an changes in information systems architectures or in business processes can interrupt operations. Therefore, any change in operations, particularly changes in information systems and business processes, needs to be carefully planned and implemented to minimize operations disruptions and costs to the business.

INFORMATION STRATEGY'S IMPACT UPON:	*IMPACTS*
Product Strategy	**Changes in Product Design Processes**
Production Strategy	**Changes in Production Processes**
Marketing Strategy	**Changes in Marketing Processes**
Competitive Strategy	**Changes in Value-Added Processes**
Organization Strategy	**Changes in Business Processes**
Finance Strategy	**Changes in Financial Processes**
Innovation Strategy	**Changes in Knowledge Processes**
Diversification Strategy	**Changes in Planning Processes**

FIGURE 11.1   INFORMATION STRATEGY IMPACT MATRIX

Implementing an information strategy consists of plans and training to implement changes in information systems and business processes. It requires a multifunctional team consisting of both managers and technical personnel from the information technology unit and managers and technical personnel from the business functional units of the business.

CASE STUDY: Beauty Goods in Cyberspace

We begin by looking at a case of one type of a retail e-commerce business sector, cosmetic retailing. Many new dot.com companies were started in 1999 in this sector but few succeeded. Only a year later, business consolidation had already begun occurring:

> Last week's [January 17, 2000] purchase of Beauty.com by Drugstore.com for $42 million in stock marked the first of what could be a rash of consolidations among online retail categories cluttered with well-financed—but poorly patronized—Web stores.
>
> —(Tedeschi, 2000, p C1)

The retail category of beauty goods included perfumes and cosmetics, and more than 300 Web sites had been set up in this sector. In addition, established bricks and mortar businesses, such as WalMart and CVS, had begun selling the same items online through their Web sites. Obtaining a meaningful market share in the e-commerce retail of beauty products was a strategic challenge.

An additional challenge was obtaining the capability of selling the top-of-the-line products:

> But these purveyors of glamour have concerns more pressing than the sheer number of competitors. Analysts say that any online retailer hoping to make a dent in the estimated $15 billion beauty market cannot hope to succeed without the cooperation of the Big Three of the beauty industry—Lancome, Estee Lauder, and Clinique.
>
> —(Tedeschi, 2000, p. C1)

Clinique was owned by Estee Lauder. The prestige part of the market constituted about 25 percent of the total market (roughly $4 billion out of $15 billion); and the top three brands held 70% of the $4 billion. Clinique and Lancome were selling their goods on their own Web sites, and Estee Lauder was also expected to put up a retail Web site. The policies of all three companies was to forbid retailers to sell their brands online. Thus e-commerce retailers in the beauty business had could sell only more obscure prestige brands or mass market brands (e.g., L'Oreal's Maybelline and Procter and Gamble's Cover Girl lines).

The reason for this policy was a long-standing strategy to be careful about who sells their prestige brands:

> "Companies like Estee Lauder see their brands as the crown jewels," said Lisa Allen, an analyst with Forrester Research. "They pay close attention to things like how and where they're displayed, who they're next to."
>
> —(Tedeshi, 2000, p. C1)

Estee Lauder sold exclusively in the luxury retailers and did not allow its goods to be displayed in drugstores or next to mass market brands. As one cosmetic retailer commented: "In Saks, you buy the whole ambience." (Tedeschi, 2000, p. C1)

Case Analysis

This case illustrates the importance of the business part of information strategy. Simply putting up a Web site on the Internet (as the pioneers of e-commerce learned) was not the key to business success. Here the luxury (i.e., the large profit margin sector of the industry) intended to keep its luxury product sales for its own Web sites.

TYPES OF E-COMMERCE

When it emerged in the late 1990s, e-commerce developed in several kinds of business applications, including:

1. Internet
   - Internet portal services (e.g., AOL)
   - Communication services (e.g., AT&T)
2. Retail
   - Consumer products (e.g., Amaz0on, CDnow),
3. Markets
   - Commercial supply businesses (i.e., B2B),
   - Auctions (e.g., eBay),
   - Materials trading markets (i.e., commodity products)
4. Finance
   - Financial trading markets (e.g., stocks and bonds)
   - Financial services (e.g., banking, credit, mortgages)
5. Information
   - Reservations (e.g., travel, hotels)
   - Query and search (e.g., Ask Jeeves)

6. Entertainment
   * News, music, TV, etc.
7. Educational
   * Higher education (e.g., UMUC)
   * Industrial training

In each of these business applications, information strategy differed according to the customers and value-adding operations of the businesses.

CASE STUDY: Kinko's Buys Liveprint.com

Next we look at the technical part of retail e-commerce information strategy. The addition of Internet-based information strategy became important to the long-term prosperity of all businesses by the end of the twentieth century. Many existing businesses then decided to add e-commerce capabilities by buying a dot.com competitor. This case looks at one example of an existing dominant franchise business in copying and printing, Kinko's, which purchased a new competing dot.com start-up, Liveprint.com, in March of 2000:

> It's not particularly surprising that the king of copies is finally coming up with a Web vision. What is unusual, however, is that Kinko's strategy is to buy the majority stake in Liveprint.com, a little local Web-based new company in Alexandria (Virginia) and install Liveprint leader Rick Steele as chief executive of the new enterprise.
>
> —(Shannon, 2000, p. E1)

Kinko's Inc. was then a privately held company headquartered in Ventura, California. It franchised and/or owned local stores providing copy and printing services. In 1990, it had about 1,000 locations in nine countries and was opening about 100 new stores a year. With the growth of electronic commerce in the late 1990s, Kinko's tried to start a Web-based business:

> At first, says Kinko's chief executive Joe Hardin, the company tried to Web-ify on its own. But that didn't work, so Kinko's started looking for a company to help it figure out how to take services for small and home offices-such as corporate logo, stationery and business-card design—online."
>
> —(Shannon, 2000, p. E1)

In September of 1999 a mutual investor of both companies, Chase Capital Partners, introduced Kinko's CEO, Hardin, to Liveprint's founder, Seele: "We were looking for a way to tap the market online," says Hardin. But then he realized: "Internet was different from our speed. With them, we can move much quicker." (Shannon, 2000, p. E1)

From Liveprint's perspective being acquired by the larger and older Kinko's made sense: "Steele was ready to sell because he realized it would take boatloads of time and money to even try to build a recognizable brand in this market. "We weren't arrogant that we'd knock everybody's socks off with Liveprint," he says." (Shannon, 2000, p. E1)

The new company formed by Kinko's from the acquisition of Liveprint would be called Kinkos.com, capitalized at $40 million, and it would remain based in Alexandria, Virginia, where Liveprint was begun. Steele would be president of the new division of Kinko's, and Liveprint's sixty-five-person staff would form the nucleus. Kinko.com began its new marketing thrust with a multimillion dollar marketing deal with America Online. Kinko.com would let customers create business cards, newsletters and the like online rather than going to one of Kinko's real-world stores. As an overall strategy, Hardin and Steele hope the new company will also help change the image of Kinko's from that of a mere copy shop to one of a problem solver for businesses. From Steele's perspective, he saw the future of e-commerce as a mixture of bricks and mortar, physical stores and virtual Web sites: "Steele believes in what's known as the "click and mortar strategy" in which walk-in stores coexist with catalogue and Web presences." (Shannon, 2000, p. E9)

Case Analysis

One sees in this case that the innovation of the Internet had provided opportunities for a new retail market-channel for Kinko to reach its customers. But this was already being pioneered by a new start-up e-commerce company, Liveprint. Kinko's strategic choice was to start up its own Web site or to acquire Liveprint. Liveprint's strategic choice was to go it alone or be acquired (and Liveprint was less than one year old at the time of the case).

Kinko decided to buy Liveprint in order to quickly acquire the information technology and management to enter the e-commerce business. Liveprint decided to sell because it would have required much capital and time to grow a significant market share and survive.

In the earlier case of Amazon entering the book retail market over the Internet, it was ignored by other firms until it had demonstrated and grown a large market. However, Liveprint had entered later when everyone's attention was already on e-commerce. Liveprint decided it would not have the lead time to establish dominant market share over competitors.

TECHNICAL PART OF BUSINESS STRATEGY FOR E-COMMERCE

The typical architecture of the information systems of e-commerce retail businesses in the late 1990s (such as Liveprint and Amazon) were structured as in Figure 11.2. The starting point in the information architecture is the Web site on

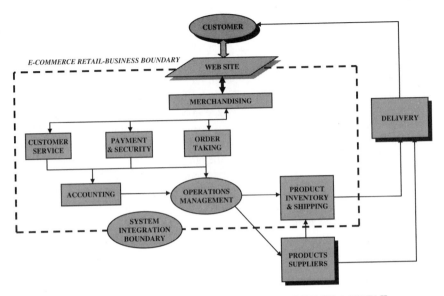

FIGURE 11.2 INFORMATION SYSTEM ARCHITECTURE OF A RETAIL E-COMMERCE BUSINESS

the Internet, which provides access to the e-commerce retail business by the customer. Within the business, the merchandising function supervises the appearance of the Web site, products presented, and interaction of the customer with the Web site. When the customer makes a purchase, the order-taking function records the order and informs the operations management function to direct product inventory and shipping to ship the ordered product. Operations management then reviews the product inventory and, as necessary, orders more products from product suppliers. When the customer purchases, the payment and security function bills the customer's charge card. Customer service records the customer's purchases, as information available for any subsequent interaction with the customer. The accounting function records the purchases and operations expenses. The whole of the business is tied together through a system integration function.

Normally, delivery is provided by external businesses, such as UPS or Federal Express. Product manufacturers provide products to the business or may even ship directly to the customer upon notification by the e-commerce business.

What is especially important to note in the business operations architecture is that the boundary of the business is not defined by a physical space but by a virtual space—the software system integration function.

This was the aspect of the new business strategy that was irritating the CEO of Barnes and Noble in the earlier case. It was not a bricks and mortar business, but a virtual business. Of course, there were physical spaces involved in the e-commerce business—space to house the personnel operating the business functions and space to house the physical products inventory. But what was missing was the space in which to interact with the customer. The physical retail bookstores of Barnes & Noble were absent from Amazon.com. In the case of Kinko's, its acquisition of Liveprint added electronic capabilities to physical capabilities.

Thus operations may change under information strategy, not only how one operates in a traditional business—but even how one "architects" the whole business. Strategic change in operations requires change in procedures and in the architecture of procedures—the business architecture—the set of policies that structure the business activities.

CASE STUDY: Dell Sells on the Internet

We next turn to the business part of information strategy. This case examines how one of the pioneers in the sales of personal computers used the Internet to reinvent its business—Dell computers. It illustrates the importance of the business part of an information strategy.

In the early days of the personal computer industry, one entrepreneur who first saw the personal computer as a standardized product was Michael Dell. He understood that the two product features that individualized the personal computer, its operating system and its central processing unit, had been standardized under IBM's brand name entry into the personal computer market in 1985—but without IBM proprietary operating systems or central processing (CPU) chips. These were produced by Microsoft and Intel. Therefore, all PCs (except Apple's Macs) would look exactly alike, and the competitive situation in PC manufacturers depended low-cost production and pricing. Dell focused his business strategy on direct sales by telephone to large organizations and on outsourcing PC assembly using the Japanese manufacturing strategy of just-in-time methods.

By the mid 1990s, only three new PC companies had succeeded as big, dominant companies: Microsoft in operating system software, Intel in CPU chips, and Dell in computers. In fact, Dell's direct sales by telephone and low-cost production, minimizing product inventory became a model that any business selling PCs had to emulate to be price competitive.

Then as the Internet emerged as a major business channel, Dell rapidly adopted its use to Dell's business model:

> Late in August (1999), 1,225 business men and women, ranging from information technology pros to buttoned-up CEOs, journeyed to Austin, Texas, like pilgrims flocking to Mecca. They filled half a dozen downtown hotels and endured three days of 101-degree heat, all for a chance to hear Michael Dell kick off Dell Computer's inaugural Direct-Connect Conference. Michael Dell, the

oracle of Austin, would reveal how he'd made the Internet part of his company's success and explain how they could too.

—(Roth, 1999, p 152)

Michael Dell had become famous in the business world in the 1990s, from his successful operations in direct phone sales and his control of product inventory with just-in-time assembly by suppliers. Dell's profit margins were large because Dell had avoided the investments of building production facilities when production gave no competitive edge. (PC's were assembled from standard purchased parts in part modules—cases, motherboards, disk drives, model cards, keyboards, monitors, etc.). The manufacturing challenge and competitive edge) lay in the computer part production, not in the computer assembly. Therefore, Dell had outsourced production to minimize production costs and avoid production investments. Because the PC was such a standardized product (with "Intel inside" and Microsoft Windows installed), this strategy of outsourced, just-in-time assembly and shipping worked very well—and made Dell investors rich and Michael Dell famous.

Now he was telling his new web business strategy to the world (and also selling new Dell web products). In 1999, Michael Dell was asked to speak at over 1,700 occasions, accepting 35 of these. Each of talk discussed how Dell was using the Internet to continue growing sales, as its online sales had grown to total 40% of total sales. What Michael Dell had done was to exploit this market channel of the Internet to continue to grow his direct-sales mode, moving from phone sales to web sales.

In 1994, Michael Dell had envisioned how the new Internet could be another channel for Dell's mode of direct sales to customers. He formed a strategic project team to launch Dell's Web site, Dell.com. The site provided technical support information for Dell computers and price guides to help customers select the appropriate components in buying a Dell computer.

Dell continued to develop the Website, and by 1996, Dell computers (PC desktops, laptops and servers) were selling on the Website at a rate of $1 million in daily sales. By 1999, the Website was receiving 2 million daily visits and selling $30 million worth of products each day. (To appreciate this high level of retail sales in 1999, then Amazon was receiving 11 million visits a day but only selling $3.5 million a day.)

Dell's successful high volume of sales on the Web generated intense business interest for hearing Michael Dell's talks. As Fred Buehler, then director of electronic business for Eastman Chemicals, commented: "Dell is clearly one of the top few companies that have really been successful on the Internet." (Roth, 1999, p. 154)

For example in 1997, Fred Buehler of Eastman Chemicals was so impressed with Dell's web success he had replaced all Eastman's PCs with 10,000 Dell

PCs. (Eastman Chemicals was a chemical producer in Tennessee, with about $4.5 billion sales in 1999.) Next, Buehler and other managers of Eastman began traveling back and forth between Tennessee and Austin to study the business part of Dell's information strategy. In July 1999, Eastman started its own Website, Eastman.com, using what they learned from Dell.

The business part of Dell's information strategy was a combination of instruction in business information strategy and sales of web products.

When the twenty-first century began, this combination was fueling Dell's new growth. There were then about 17,000 business Web-sites that Dell had helped corporate customers create, using Dell's model of its Website "Premier Pages.' " For example, Dell's own version of its Premier Pages shows each of its customers the configuration of every computer bought, the price paid, and a tmeline when a new version of the computer will be introduced. This kind of consulting to other businesses to help them set up their Websites facilitated Dell's sales. As companies turned to Dell to understand how to sell on the Internet, Dell had an opportunity to sell them its PowerEdge network server and PowerVault storage device.

The strategy of informing customers and then selling products to customers worked well for Dell. In sales of servers to about 300 corporate IT users (surveyed in 1999), 30 % bought their servers from Compaq but 15 % bought from Dell. When the surveyor asked these companies who they would buy from in the future, the number cited Dell grew. "Companies are confused; they don't know what the Internet means (says an analyst at one of Dell's biggest institutional holders). So when the CEOs get back (from Austin) to their offices, they call up their purchasing managers and say, 'Let's do business with these guys, they have their act together.' " (Roth, 1999, p 156)

Also Dell used the web for its supply chain management of its vendors. Dell's suppliers use a focused version of Premier Pages at the password-protected site Valuechain.Dell.com. They also used it to communicate with Dell about what orders they have shipped and about how Dell sees they are measuring up to Dell's quality standards.

Case Analysis

This case illustrates the importance of the business part of information strategy—as Dell used information both to create sales and to control the costs of production. As a pioneer in developing strategic business models for use of the Internet, Michael Dell was using his fame to provide information to other businesses, which in turn bought Dell's products.

BUSINESS PART OF INFORMATION STRATEGY FOR E-COMMERCE

The Internet is a medium of communication and a channel of marketing. An information strategy of a business needs to use the particular characteristics of this medium in its activities that add value to the businesses customers.

The business part of information strategy focuses how the information system contributes to the strategic business model. In the previous case of beauty products, we see that a key business ingredient for profitability was being able to sell in the luxury sector of the cosmetics market—a sector that the three brand leaders were holding for themselves. In the case of Kinko's, we saw that the business strategy was to integrate the bricks and mortar operations with electronic commerce. In the case of Dell, we saw that informing customers how to do business on the Web led to sales.

Marketing strategy requires identifying customers, their needs, products or services to satisfy their needs, and channels of reaching customers. E-commerce information strategy is one of the channels of communication between a business and its customers, as sketched in Figure 11.3.

From a marketing perspective, what is unique about the e-commerce channel is (1) that customers must find the business on the Internet and (2) the computer can assist in communications. We can understand the strategy requirements of these unique features by listing four strategic business challenges of information systems for e-commerce:

FIGURE 11.3 INFORMATION STRATEGY IN ELECTRONIC COMMERCE

1. *Attracting*
   - customers
   - needs
   - desires
   - finding site
2. *Informing*
   - interesting
   - clarity
   - products
   - matching
3. *Adding value*
   - inspecting
   - price
   - convenience
   - transacting
4. *Profiting*
   - revenue
   - costs
   - volume
   - economies

Attracting

As a marketing channel, what is particular to e-commerce is that it requires effort to attract a customer to the Web site of the business. A customer must take a positive action to find the business on the Internet. Channels of marketing vary in how much positive effort is required for a business to communicate with a customer.

For example, advertising in a broadcast mode (e.g., such as on radio, television, newspapers, magazines, etc.) requires no positive effort from the customer. The customer is bombarded with the advertisement via the communication medium over which it is broadcast. However, on the Internet such broadcasting has been minimal (so-called spamming via e-mail has been discouraged).

This is why a connection service such as AOL can be profitable as a means of providing advertisements when subscribers connect.

Except for Internet portal businesses, all other e-commerce businesses must be actively found by a customer (either by deliberately seeking or by stumbling onto a Web site).

This is the first of the four basic strategic challenges of e-commerce—attracting the customer and getting the customer to find the business. E-commerce information strategy must carefully identify the kinds of customers it hopes to *attract* and the *needs* of the customers.

However, only identifying the needs of a customer in terms of products or services as means to satisfy these needs is not sufficient in this medium of the Internet. Since the customer must actively want to use the Internet to find products or services to satisfy needs, the customer must also have the *desire* to use the Internet. So information strategy must also identify what are the desires of the Internet user, with respect to the customers product/service needs.

The customer may find the Web site through advertising or word-of-mouth or searches. Information strategy to assist the finding or searching or exploration activities of customers to the website are important for the e-commerce business channel. Also progress in information technology about the kind and nature of the search engine a customer uses to *find sites* is an important technical factor in e-commerce information strategy.

Informing

The second basic strategic issues is to get the customer to stay at a Web site—informing a customer. Once the customer has found the business's Web site, the next challenge is to have the customer perceive that the Web site is relevant to the customer's needs so as to stay there and use it. The first condition of a Web site is that it must appear instantly *interesting* to the viewer. One of the conditions of the Internet medium is the quick ability to click off to another site. Therefore, the appearance of the site has to grab the viewer's interest to keep the viewer there long enough to begin exploring the site. Grabbing interest is a combination of aesthetics and functional logic of the first appearance of the site. An important characteristic of the Internet medium is that it combine information, entertainment, and communication.

To explore a site, logical *clarity* is important. The scheme of organization and maneuver through the site must also be immediately apparent and clear to the viewer. This is where an understanding of the kinds of customers and their needs and desires is essential to a proper design of the organization and maneuvering paths through the site. How a site encourages a viewer to maneuver through it must be guided by having the viewer learn how its needs can be met by information at the site.

The site should show the *products* or *services* available at the site. They should be presented in a way to show an obvious way to satisfy the customers' needs—*matching* products or services to needs. One of the marketing lessons of successful catalog sales operations was that the presentation of products appealed to some entertaining desire and need of the catalogue viewer. Web sites must also be entertaining.

Adding Value

The third strategic issue in e-commerce information strategy is getting a customer to value the site, as in purchasing something through the site such as subscriptions to the site or goods and services purchased at the site or goods and services purchased from advertisers at the site.

In selling products on the Internet, after informing the customer that the business's products or services match the customer needs, the next step is to persuade the customer to purchase by showing how the product/service adds value above the price of the product or service. This can be accomplished by helping the customer benchmark competing products/services and figuring the cost or benefits of purchasing the product or service, so that an order is made and processed.

To do this, a customer must be able to adequately *inspect* the goods or services offered at the site. Inspection may be easy or difficult. For example, commodity type services or products with which the customer has previous experience may require little or no inspection—only pricing and timing. However, discriminating products, particularly of physical products, may require direct inspection, and the site may need to point to locations where the product may be directly inspected. Thus how much inspection and how to inspect is an important part of information strategy. Progress in information technology through adding multimedia perceptive experiences such as "immersive multimedia" (3-D sight and sound and touching) will be valuable to the inspecting needs of products sold over the Internet.

The *pricing* of products sold on the Internet is an important marketing decision. Are they priced the same, more, or less than bricks and mortar and why? Early on products sold on the Internet were often priced less than elsewhere to develop market share, but this led to problems with profits. A second issue is that of *convenience*. Is the Internet channel more or less convenient for transactions and why? Finally, a *transaction* of some kind is desirable to ensure that the customer values the service or products offered on the Web site.

Profiting

The fourth strategic challenge in formulating information strategy is to determine how profits are made from the site. For this one needs to determine the sources of income from the site. Viewers of the site may be *revenue* sources or advertisers on the site may be *revenue* sources or transactions occurring on the site may be *revenue* sources. Sales transaction income, advertisements income, membership fees, and other fees (e.g., action fees) are some of the ways e-commerce sites have generated revenue.

Costs must be determined for profitability. The e-commerce channel excels as scaling in volume so that strategies to increase volume of site usage are important. Finally, because of the importance of *volume* and *economies* of scale and of scope are important to address.

This combination of attracting, informing, value-adding, and profiting is together important for successful commercial use of the Internet. For example, David Kenny and John Marshall pointed out the amount of early disappointment for companies in using the Internet:

"Time for a painful admission: the Internet has been a letdown for most companies. Certainly, the Web is at the top of corporate America's priority list—the $10 billion that large U.S. Companies spent on Web site development in 1999 is evidence enough of that. Yet in any give month, only about half of the largest U.S. consumer businesses attract more than 400,000 site visitors—and a similar percentage of sites generate no commercial revenue at all."

—(Kenny and Marshall, 2000, p 119)

Kenny and Marshal go on to emphasize that a commercially successful Web site must attract repeat visits, with each visit adding incremental information:

"Does this mean the Internet is of no value to all but a handful of well-positioned companies? Not at all. What it does mean is that most companies need to discard the notion that a Web site equals an Internet strategy."

—(Kenny and Marshall, 2000, p 120)

An Internet strategy requires strategic consideration of all aspects of the Internet as an interactive marketing channel for attracting, informing, value-adding, and profiting from customer-company interactions.

CASE STUDY: CDnow and Geffen Records

Now we look at a case of e-commerce retail wherein strategy for *attracting* customers and for *adding value* to customers were separately performed in a strategic business alliance—between CDnow and Geffen Records.

Donna L. Hoffman and Thomas P. Novak recounted how a new e-commerce retailer, CDnow began from an early perceived need by a potential entrepreneur:

(In 1988) when Jason Olim was 19, a friend introduced him to Miles Davis's classic album *Kind of Blue*. Entranced, Olim went searching for more of Davis's recordings but was met with poor service and limited selection in traditional bricks-and-mortar retail stores. Out of that frustration was born a vision of a better way for music buyers to connect with music."

—(Hoffman and Novak, 2000, p. 180)

Six years later in 1994 after he learned about the new Internet, Jason Olim and his brother, Matthew Olim, successfully started a new e-business selling

CDs on the Web from a site at first hosted in the basement of their parents' home in Ambler, Pennsylvania. In their first month, they made a profit of $14 on sales of $387.

Of course, e-retailing was such a good idea that soon many other competitors began setting up competing sites, and like many e-commerce retailers of the time, they tried using banner ads on portals as a way to promote their site. We recall that service portals for Web access such as AOL made substantial revenues from selling banner ads to appear on their portal web sites. The practice was to price an ad according to the numbers of daily visits to a portal site. CDnow purchased such banner ads and found (as did other advertisers on the Web) that banner ads did not always translate to a visit to the advertiser's site and even of the visitor's to an advertiser's site, only a fraction actually purchased anything. Accordingly, when a Web-advertiser calculated the cost of attracting a purchaser to its site through advertising paying an advertising rate based on the number of visitors exposed to the banner ad, it would be very, very high.

Soon CDnow saw that the expensive banner ads were being a poor buy for them. For example, a Web publisher had demanded from CDnow a banner ad price of $70 dollars per thousand visitors exposed to the ad. Since only 1 percent of the customers seeing the banner ad ever clicked over to CDnow's website, CDnow ended up paying $7 dollars per visitor acquired through that link, and only a few of these visitors actually purchased something from CDnow. The cost to Cdnow of acquiring a new customer was $700 per customer. With an average profit of $10 per new customer, Cdnow was losing $690 on each new customer. That customer would have to return to CDnow over 69 more times, before Cdnow could make any profit. Olims began to think of better ways to attack customers to CDnow.

The better way occurred as a partnership with another business, Geffen Records. In 1994, Geffen had opened a website to promote their recording musicians. But when fans visited the site, Geffen wanted a way to sell records to these fans without their having to set up a direct sales operations themselves. Geffen Records approached CDnow to see if the latter might perform the sales function. The two companies reached an agreement. Geffen would put links on its site to carry fans directly to the Web pages devoted to Geffen artists at CDnow's site. This was a direct way for the two to sell more CDs, Geffen at wholesale and CDnow at retail.

This arrangement suited CDnow, and they paid a commission to Geffen for each visitor arriving from Geffen's site and purchasing records. This was a better deal for CDnow, since they paid only for real customers and not just exposure to potential customers, as in the banner ad payments. CDnow then began a strategy to find other recording company partners, and CDnow had a dozen more such partnerships by the end of 1994, growing to a few hundred by the end of 1995. CDnow gave a percentage of the revenue from sales back to the affiliate recording company, and thereby, from CDnow's perspective, developed an effective Web-based, virtual commissioned sales force.

CDnow continued to expand their affiliate program from professional record company sites, adding nonprofessional sites of music fans and also allowing unsigned musicians to put up a Web page at the CDnow site. CDnow pays commissions on such related record sales on a sliding scale based upon the volume of such purchases. Altogether, CDnow had added over 250,000 members to its affiliate program. By the year 2000, CDnow was acquiring its new customers about 40% through affiliate links and 60% directly.

Case Analysis

We see in the origin of CDnow that the entrepreneur Olim was motivated by a customer need for a much better inventory of records than was available in bricks-and-mortar stores. This was the same kind of need that Jeff Bezos had perceived when he started Amazon.com. The problem then was that such bricks-and-mortar retailers had to buy their inventory of books and records before selling them. This created a very expensive inventory cost and resulted in limited inventory selections for customers. Olim's and Bezos's value-adding idea was that the customer buying on a Website would expect a delay in receiving merchandise, in contrast to their expectation of immediate delivery of merchandise in a bricks and mortar store. This would give an e-commerce retailer an enormous inventory performance advantage over the older bricks and mortar stores—almost infinite inventory at zero cost.

Olim's and Bezos's ideas for profiting in e-retailing was very low inventory costs and their ideas for value-adding to customers in e-retailing was very high inventory selection—a winning combination. Then the problem was in *attracting* customers to their web sites. CDnow's solution of attracting customers to their Web sites moved from the early traditional advertising (e.g., banner ads) to an innovate affiliate site program of strategic alliances with recording companies and artists.

STRATEGIC ALLIANCES IN E-COMMERCE

Because of the ease in the Internet of linking to other sites, strategic alliances of businesses on the Internet provide opportunities to use the channel in creative ways to improve profitability. As the case of CDnow illustrated, it is likely that strategic business alliances will prove powerful on the Web for attracting the kinds of customers that create business revenue. Also as in the case of Dell, a strategic alliance may occur even within a firm between different parts of its businesses.

A strategic technique for considering such strategic alliances is illustrated in Figure 11.4. One can construct a strategic e-partners alliance matrix by listing:

- The basic e-channel functions across the top of the matrix as *attracting, informing, adding value, and profiting.*
- The potential strategic partners down the side of the matrix.
- Entering into the cells of the matrix the respective roles of the partners in the e-channel functions.

In the illustration of Figure 11.4, CDnow provides the adding value function and Geffen Records the attracting function. CDnow profits through gross margins on sales and Geffen by sales commissions. Both CDnow and Geffen provide informing roles in their sites. A strategic e-commerce alliance matrix can help synthesize new ways to create profitability for a business on the Internet.

Such strategic alliances are playing an increasingly important ways to do business on the Internet. For example, Hoffman and Novak observed:

> Impression-based advertising in the mass media will likely never completely disappear on the Web. But as the Internet continues to mature, advertisers will continue to seek out specific target segments of potential customers and the corresponding Web sites that can deliver those customers. That will contribute to the continued explosion in open revenue-sharing programs."
> —(Hoffman and Novak, 2000, p. 184)

In summary, we can say that three major issues have stood out in the evolution of information strategy for e-commerce stand out from this review:

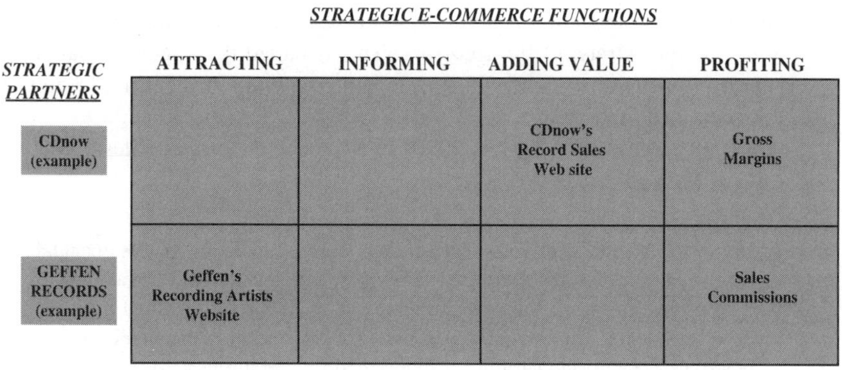

STRATEGIC E-COMMERCE FUNCTIONS

STRATEGIC PARTNERS	ATTRACTING	INFORMING	ADDING VALUE	PROFITING
CDnow (example)			CDnow's Record Sales Web site	Gross Margins
GEFFEN RECORDS (example)	Geffen's Recording Artists Website			Sales Commissions

FIGURE 11.4   STRATEGIC E-PARTNERS ALLIANCE MATRIX

1. There is both an information technology component to e-commerce information strategy and a business component.
   - What is critical to commercial success is integrating the information technology component with the business component.
2. An integrated e-commerce information strategy must attend to four critical and different functions:
   - Attracting customers
   - Informing customers
   - Adding value for customers
   - Profiting from serving customers.
   - What has been most difficult to achieve of these functions is relating the attracting to profiting.
3. What is evolving to work best and is unique to e-commerce is the capability of forming strategic alliances between different businesses on the Web to successfully relate attracting and profiting.
   - Strategic e-commerce business alliances will play major roles in e-commerce information strategy.

CASE STUDY: Changing It in Allied Signal Technical Services

We now turn from e-commerce to the strategic problem of replacing older information technology (IT) systems in traditional businesses. In addition to information strategy for e-commerce, information strategy for improved operations is another important and recurring strategy problem.

Progress in information technology has occurred most often within parts of the enterprise system and sometimes in the whole of the information system architecture (such as in e-commerce). In either case of progress in part or in whole, information technology progress over time has always made prior software and information architecture and computational platforms obsolete. Then replacement of a prior information system by a new up-to-date information system becomes strategically necessary. This is called the strategic challenge of legacy information systems. We now look at a case of a strategic replacement of a legacy information system.

This case occurred in the late 1990s in a service business, as described by Cathy-rae McNamara:

> In late 1995, AlliedSignal Technical Services (ATSC), realizing that its financial enterprise system was far behind the current technology, decided it was time to take advantage of one of the new commercially available off-the-shelf financial enterprise systems. The current, inflexible legacy business and financial systems were stifling growth, and varying customer reporting requirements were difficult

to fulfill. The customer reporting aspect was becoming a source of customer dissatisfaction, and the work reports were labor intensive, and therefore, costly.

—(McNamara, 2000, p. 2)

Allied Signal Technical Services was a division of AlliedSignal corporation that provided technical services to federal government customers. It used an accounting system software called Walker General Ledger, which when installed in the mid-1980s was state of the art.

In the information technology of the time, each functional operations area had its own stand alone software package, so that looking over the enterprise as a whole, information technology automation then was creating an enterprise system of isolated islands of technology. This isolation of data and information within functional islands of a business created the technical obsolescence in the next decade, after information systems were effectively able to communicate through network communications system (e.g., local area networks, wide area networks, Internet). Every department had a computer and database to handle the information needs of the department, and it was difficult to share information between departments. For example, payroll software could not exchange data with accounting software, so that every two weeks, employees' time sheets had to be entered by hand by payroll department staff. Moreover, payroll and human resources information were in separate software systems, so that any changes to employee status, salary, or other personnel data had to be submitted in paper form and entered manually.

In addition to the labor costs of hand entry of data, the aging equipment and changes to existing software had added to a maintenance cost of $2.2 million in the five years from 1990 to 1995. There was a large staff of programmers, administrators, and information technologists all together busy fixing and patching the old systems. Also the outputs of the systems, in terms of billing their customers, was unsatisfactory. As a service business, the company worked on cost-based billing to government customers and needed to report costs in formats tailored to individual customers. From data gathered from the various reporting systems, each month financial analysts spent many hours manually generating reports into other formats acceptable to the customers.

The overall impact on the enterprise of the outdated system were extra costs of multiple hand-entry of data, delay in the creation of hand-tailored reports, and high costs of continuing maintenance to an obsolete information system. Finally, top management decided it was time to replace the whole enterprise's information system. They formed an information strategy team to investigate the information systems problems and devise strategy. The information strategy team identified three major issues that needed to be addressed: systems were disconnected, data was inaccessible from one system to another, and the many of the functional classifications in databases were obsolete.

Next the team formulated the needs and specification for a replacement information system, compiling a needs list (McNamara, 2000, pp. 5–6):

Data

- Single-point data entry
- Validation of data at entry point
- Accurate employee absence and vacation usage registers
- Accurate and timely labor utilization statistics
- Detailed cost collection
- Ease of corrections to data with audit trail
- Electronic on-site data entry
- General ledger interface with cost ledger
- Integrated field location and Headquarters accounts payable and purchasing data
- Online viewing and corrections for all systems
- Provide detail to transaction level
- Responsive processing
- Revenue and profit and loss roll-ups to match reporting levels
- Systematic process for obtaining, maintaining and distributing cost data
- Timely labor corrections

Tracking

- Ability to track and monitor contract ceilings
- Ability to track purchase requisitions through the procurement cycle
- Accommodate tracking of subcontractor costs
- Inquiry function to determine if a specific invoice has been paid
- Online review of labor distribution
- Provide flags to prevent over-recording, double-billing, double-payment, etc.

Reporting

- All financial reports should be available for use online
- Downloadable reports to desktop applications
- Flexibility in closing charge numbers
- Flexible, customized reporting

- Integrated multi-user system
- Multiple contract reporting and cost collection
- Provide for automated billing
- Reduce/eliminate reversals and accruals
- Timely open commitment reports
- Weekly cost reporting

Maintenance

- Cost effective systems maintenance
- Whole system should be reengineered and not patched

One can see in this needs list that the strategy team was identifying the functional requirements of the replacement information system in terms of the management requirements of entering data on operational activities, tracking activities, reporting on activities, and maintaining the system. The strategy team summarized the intention of their needs list with a strategic vision statement for the new information system to implement an integrated financial software package and processes that could deliver valuable, accurate, and timely information to their customers.

To implement this vision with the detailed needs list, the strategy team recognized the need to change business processes. They discovered that old processes were as obsolete as the hardware and software. Improved information automation of old processes would simply result in disaster, and many processes needed to be reengineered. Accordingly, the strategy team provided specifications for the process engineering by identifying desired process outcomes to information solutions:

Desired Management Outcome	Information Solution
Validation at transaction level	Integrate all financial systems
Timely data	Weekly cost reporting
Eliminate non-value-added work	Streamlined time recording
Enable growth	Customizable project formats
Improve customer reporting	Standard account structure
Flexible financial architecture	Integrate information formats of Project/Account/Organization

This approach of reengineering all processes in anticipation of new information technology provided the basis for strategy. After a year of intense self-evaluation of needs and internal processes and customer needs, the information strategy team had a clear idea what they wanted in a new system that utilized off-the-shelf software.

Implementation of the new system began, with a target of January 1, 1997 to become operational. The computer network was extended from a local area network (LAN) to a wide area network (WAN) to link all field locations with headquarters. New hardware was obtained to host the new system and desktop computers were upgraded. Training in the new system was provided to all employees. The new system was begun on the first of the year:

> Finally, all users were trained and preparations were made for 'going live' . . . The (new) system was ready to start receiving input on January 1 . . . Of course, some problems did arise, but the IFS Team and all the other trained employees were available to assist end-users in resolving them.
> —(McNamara, 2000, p 13)

Process improvements were documented after the new system was operational, as summarized in Figure 11.5.

Case Analysis

The business model of this unit was contracted project work for government clients, billed as effort plus fee. Accordingly, work on each project contract had to be recorded and reported to the client for reimbursement of costs plus fee. The architecture of the information system focused upon tracking each contract and upon billing and reporting time and expenses on each contract.

One can see in this case that the replacement of a legacy information system is a big job and periodically needs to be addressed. In the pace of progress in information technology at the end of the twentieth century, probably new information strategy for replacing legacy systems was required every ten years. To prepare for replacement, top leadership commitment to a new information strategy was required, and a cross-functional strategy team was needed. It took about a

PROCESS	NEW SYSTEM	OLD SYSTEM
Labor Cross Charging	Automatic	24 hours/month
Labor Corrections	15 minutes online	30 minutes manually
Timesheets	Weekly online by end-user	Biweekly manually
No. of Journal Entries	95/month online	133/ month manually
Project Cost Inquiries	5 minutes online	30 minutes manually
Employee Expense Reports	Daily online	Weekly manually
Billings	Automatic	Manual
Purchase Orders	Realtime online	Daily batch manually
Project Cost Ledger	5 minutes online	4 hours runtime, manual input
Vendor Payments	Daily	Weekly batch processing

FIGURE 11.5   SUMMARY OF PROCESS IMPROVEMENTS

year to analyze the needs of the new system, reengineer process and train employees in the new system.

LEGACY SYSTEMS

Legacy information systems arise from a history in a company of installing earlier information systems in operations. But as progress in information technology continues, earlier systems become obsolete. Both the software and hardware running the software may become osolete. In replacing legacy information systems, information strategy should consider:

1. Anticipating progress in information technology
2. Selecting applications to business processes
3. Selecting desired kinds of operational improvements
4. Selecting the new structuring of control of processes

In the last part of the twentieth century, progress in information technologies was enormous and continuing after the invention of the computer and inventions in digital communications systems (particularly satellite communications, fiber-optic communications, computer networks). By the end of the century, many different kinds of software were found in a modern business. We can use the model of an enterprise system from Chapter 10 (Figure 10.3) to classify the kinds of software found in both the operations and authority structures of companies. In the three operations planes, one can identify where information tools for management were being used:

Support Plane 1

- Project management software
- Computer-aided design and engineering systems software
- Personnel records and evaluation software
- Training software
- Marketing analysis and forecasting software
- Financial analysis and spreadsheet software
- Investment management software

Transformation Plane 2

- Materials resource planning software
- Production scheduling software
- Computer integrated manufacturing software

- Inventory control and distribution software
- Point-of-sales software

Control Plane 3

- Relational database software
- Spreadsheet software
- Accounting systems software
- Planning and budgeting software
- Executive information and control software
- Vendor electronic ordering software

Not only do all organizations have operations structures (such as depicted in Figure 10.3), but all organizations also have authority structures. Authority structures order the relationships of power between divisions of the organization by defining scopes and levels of responsibility, accountability, and authority. We can use this concept to further identify where information technology assists management to manage with technology. Management software tools for authority structures are aimed facilitating planning, coordination and accountability functions. Examples included the following:

Management planning tools

- Relational database software
- Spreadsheet software
- Executive databases

Management coordination tools

- Project scheduling software
- Electronic mail software
- Phone systems software
- Teleconferencing systems software
- Groupware
- Internet browsers

Management accountability tools

- Accounting systems software
- Metrics monitoring systems software
- System activity monitoring systems software
- CEO database and monitoring systems software

So it was that when the twenty-first century began, a principal strategic problem with this range of business software was the inability of these different applications to share or communicate data with each other.

CASE STUDY: Reengineering Coats Viella

As we saw illustrated in the earlier case of Allied Signal Technical Services that information system redesign always needs some business process redesign. We next look at the procedure for re-designing functional business processes. Since legacy information system replacement depends upon continuing progress in information technology, periodically one should redesign the logic of business processes. Although one replaces a legacy system and/or reduces islands of automation as new software and platforms make this technically possible, yet the full business performance of new systems may not be achieved without also redesigning business processes. Thus information strategy always needs to be connected to operations strategy.

We next look at a case of strategic change in a business in the 1990s in both operations and in information systems in order to improve profitability and long-term competitiveness—a reengineering of production process in the British clothing firm. A. K. Bhattacharya and A. D. Walton reported upon a strategic change in the operations of Coats Viella Clothing Knitwear (CVCK); (Bhattacharya and Walton, 1998).

Coats Viyella was a strategic business unit of a larger, global company of the same name, and in 1993, the company had eight manufacturing plants in England and Scotland, with sales around £90 million and 4,000 employees. It supplied knitted wear to high-end retailers.

However at the time, management was not satisfied with profit margins, and there was a continuing problem of holding large inventories even in peak sales months. Also customers frequently canceled sales, complaining that they could not obtain the right stock, in terms of color or size. Management decided there was need for strategic change:

> It was clear to the company that they had to do something quickly. It was not that the company had suddenly become very bad in what they did. The market had changed slowly but surely, and the present market was very different from the one they supplied even three years previously, and the company was not restructured to respond to these changes."
> —(Bhattacharya and Walton, 1998, p. 712)

Called in as consultants, Bhattacharya and Walton visited manufacturing sites and reviewed the entire production process. They then suggested that the management board discuss three strategic issues:

- First, had the market changed? Was the customer demanding a heterogeneous product group now, compared to the more homogeneous group which used to satisfy customer needs?
- Second, what were the key factors for success in competition?
- Third, what where the company's present core competencies? Where they the kind of competencies needed for the factors for successful competition?

Organizationally, Coast Viella had a management board to discuss policy and do planning. This board discussed the above questions, and from these discussions a consensus emerged that the traditional strength of the company had been to manufacture limited number of basic styles of high quality garments at a very competitive price. But a consensus also emerged that the market was changing, with customers becoming more fashion conscious than before.

They agreed that management had been responding to this market trend by recently strengthening the company's design function; and this had doubled the number of styles sold to the customers, with concomitant sales growth. As a result, they believed that now the company had begun to develop a core design competency in addition to its traditional competency in production. So the conclusion in the strategic discussion of the management board was that design, as well as manufacturing, had begun to give the company its competitive strengths. But there was a problem now between design and production of several product lines. The manufacturing system had not been structured to simultaneously produce multiple product lines.

They decided there needed to be a change in production operations to produce three kinds of product lines:

1. *State-of-the-fashion products* (knitwear seasonal fashions). Products of high product complexity and high uncertainty in market demand (7 percent of present production).
2. *Current-trend products* (knitwear in fashion for a time). Products of medium product complexity and medium uncertainty in market demand (45 percent of present production).
3. *Traditional products* (classic knitwear always in fashion). Products of low product complexity and low uncertainty in market demand (48 percent of present production).

In designing and producing these three kinds of product lines, there were differences in knowledge about the yarn and in the levels of volume and sales demand times. From this strategic review of changes in market and impact upon design and production competency, the management board concluded that the company needed to change, to develop a new core competency in flexible manufacturing of different lines of knitwear produced in timely response to sales:

> New manufacturing capability which was flexible and responsive to Electronic Points of Sales information (EPOS) had become increasingly critical for business success, and this was a very different type of manufacturing capability from that which presently existed in the company.
> —(Bhattacharya and Walton, 1998, p. 716)

We can see that this kind of new manufacturing was in line with the general progress in production operations discussed in a previous chapter in using information technology to create a quick-response business.

To build this new competency, the management board decided to alter the organization and production processes of the business. Previously, the company had a traditional organizational structure, pyramidal and functional—sales, design, technical, yarn procurement, sundries procurement, yarn development, planning, quality, production, personnel, finance and accounts, industrial engineering, warehouse. Most functions were centralized at the head office, while production was divided among a number of process-focused factories (wherein four different plants did knitting and preassembly, shipping components to each of eight different assembly plants). The assembly plants were each subdivided into skill areas.

In this structure, the lead-time required for a new product introduction was about thirty weeks, and there was no single responsibility for the entire product introduction process. It was typical of manufacturing operations of the twentieth century with design an isolated effort. Designs were then (so to speak) tossed over the transom to manufacturing. This common practice usually left production plants with a frequent problem of late design implementations due to the need to go back frequently for redesign of a product that could be manufactured efficiently. This kind of delay universally lengthened product development times and raised production costs. (We recall that the case of using concurrent engineering design processes in the development of the Ford Taurus was specifically aimed at reducing the new model design and development time.)

The management of Coats Viella next changed its business processes for designing new products, producing new products, and handling the materials logistics in production. For this, three key business processes were reengineered :

1. *New product introduction process* (design). Management realized that they were now in a fashion-wear business, and their market share was a direct function of the number of their designs accepted by the customer.

The difference between what they were doing earlier and the new process was in viewing the various activities carried out by different functions in a process orientation. It was also not enough to come up with a large number of high quality designs. The new process needed to introduce new products that were manufacturable the first time.

2. *Garment supply process* (production control). Scheduling and controlling production units needed to be changed to permit flexible production of multiple product lines.

3. *Logistics process* (production system). The production system needed to be changed to automatically take a customer order and convert it into a feasible design and production plan—procuring all the parts to enable the production to take place at the right time, controlling the transfer between sites and outsourced dyeing, and being responsible for having the right stock at the right time in the warehouse for call-off by the customers.

Having identified the key business processes to be reengineered, the next step was to establish project teams to redesign and implement the new processes. The goals of the projects were to reduce lead times for fulfilling customer orders and to steady the production process by eliminating frequent instances of fire-fighting to rush jobs for customers and to correct production mistakes in customer orders.

After the new business processes were redesigned, these new process were implemented: Even during the first phase of implementation which concerned primarily the New Product Introduction and the Garment Supply process benefits were clearly visible. The chosen garments when piloted through the New Product Introduction Process caused far less turbulence as compared to the normal method of introducing new products . . . the lead time (was) reduced by four weeks. The average inventory levels (fell) from 8 to 10 weeks to 3 to 4 weeks. . . . The average labor efficiency in assembly plants has increased by 5 to 10 percent. Return To Lines [quality problems] . . . reduced by 30 to 40 percent . . . customer cancellations [decreased], which implies that there [was] an improvement in customer service.

—(Bhattacharya and Walton, 1998, p. 719)

Case Analysis

This case illustrates the importance of business process reengineering in conjunction with information technology improvement.

The product design process was reengineered as Coats Viella purchased new design software and computational platforms. The production scheduling process was redesigned with new materials purchasing software, new machine control systems, and new sales projection information from point-of-sale data.

Both the authority structure as well as the operations structure needed to be changed for improved performance of business processes.

In the previous case of Allied Signal, the need to upgrade the information system (legacy system) drove the need to re-engineer business processes. In this case, the need to reengineer businesses processes for improved profitability drove the need to upgrade information systems.

Sometimes the technical part of information strategy drives the business component and sometimes the business component drives the technical component.

REDESIGNING BUSINESS PROCESSES

As we saw in the cases of Allied Signal and Coats Villela, improving operations for profitability or replacing legacy systems both required changes in business operations as well as in information systems. Improvement of business practices through deliberate reexamination and improvement of specific procedures was strategically needed to improve operations. The continuing pace of computational and communications technologies means that periodically businesses need to look at their operational and authority structures and reengineer business practices for improved efficiency and effectiveness. Also as we saw earlier, authority structures, operations, structure, and informal cultures might all need to be changed in substantial redesign of business processes.

The implementation of redesign of business processes requires extensive employee participation in redesign teams. Many companies in the early 1990s used a specific training process to educate its employees to participate in work-process redesign. This was called Total Quality Management (TQM). At Allied Signal, employee participation in redesigning business processes was facilitated by training them in TQM, which emphasized the importance of quality in all operations of a business and the importance of measuring progress in improving operations (Hammer and Stanton, 1993; Bounds et al., 1994; Elzinga et al. 1995).

TQM guided the efforts of multifunctional business teams to think systematically about improving business operations by

1. Describing, modeling and measuring performance of current business operations
2. Defining the purpose and function of the business operation

3. Defining goals and targets for improving the processes of the business operations
4. Reviewing progress in new information technology capabilities
5. Specifying requirements for redesigned processes
6. Redesigning business operations with new information technologies
7. Planning implementation of new business processes and procurement of new information technology products and services
8. Implementing new business processes and new information technologies and providing training for utilization of new processes and technologies

The focus of reengineering business processes to use new information technology can be to improve operational productivity, quality, capacity, responsiveness, and flexibility.

Productivity is the ratio of outputs of a process compared to the inputs. Quality is the ratio of outputs meeting a minimum standard of functionality (e.g., six sigma, zero defects, customer satisfaction, percent repeat sales, etc.). Capacity is the quantity of outputs a process can produce in a given period. Responsiveness is how long it takes for an output to be produced from an input. Flexibility is range of outputs that a process can produce from inputs.

CASE STUDY: SAP'S ERP Software in the 1990s

In the 1990s new software packages were developed specifically to integrate the enterprise—then called enterprise resource software (ERP). The architecture of ERP software used a centralized database in which software put and used information.

One pioneer of this approach was the German-based firm of SAP AG Corporation (*Business Week*, November 3, 1997). Its software package DSP R/3 was composed of four application categories: accounting, manufacturing, sales, and human resources. Altogether, there were seventy modules in the whole package, and they were interconnected through a centralized database.

With installed SAPR/3, a customer could electronically order products from a manufacturer, and then the ordering module checked the customer's credit and communicates price, approving the order. Next the R/3 inventory module checks the stock of the product and fills as much as the order as possible, issuing shipping orders. Then the inventory module informs manufacturing of the need for producing more product. The R/3 manufacturing schedules production in the appropriate factories; and the R/3 human resources module calculates labor requirements for production. R/3 material planning module then issues purchase orders for production materials. The customer is informed by the R/3 software of the schedule of delivery of the completed order and is billed for the sale. R/3's forecasting and financial modules inform management of the ongoing sales, expenses, and profitability of operations.

Historically, a software application central to the function of a department was bought and installed in that department. For example, accounting software was bought and installed by the finance department, personnel software by the personnel department, purchasing software (MRP) by purchasing and production, computer-aided-design software (CAD) by the R&D/engineering department, and so on. None of this software shared data nor was usable in any but the installed context.

The SAP ERP software provided one approach to solving isolated information problems, using a massive central database storage information architecture. Yet this was only one approach, and as the twentieth century began other approaches were being innovated. The difficulty of installing ERP and the rigidity of the centralized database made the ERP solution to information integration costly and inflexible.

The rapid use of the Internet in the late 1990s provided a new path to information integration by having all software communicate through the Internet. Whatever information architecture approach taken to enterprise integration, a centralized database or a decentralized Internet communicating architecture, the need to integrate information throughout the enterprise was a basic strategic need.

Case Analysis

Because of the problem of islands of automation which emerged in the 1980s, ERP was the most rapidly growing new software application in the early 1990s. However, after the expansion of the Internet, ERP began to become obsolete as a sole approach to enterprise integration. Newer technologies using the Internet and its standards made it easier to construct appropriate interfaces between the information systems in the different business functions.

ENTERPRISE INTEGRATION

Historically, the evolution of progress in pieces of information systems did result in organizations putting together their information systems piecemeal and thus created what was called "islands of automation". The basic reason this occurred was not only the piecemeal progress in IT, but also the basic nature of organizations.

All organizations have both authority structures and operations structures. Authority structures divide the organization by business function and assign responsibility and power to manage these functions. Operations structures are the procedures for carrying out the transformation processes across the authority structures.

We recall that organizations are established to perform repetitive activities of value-adding in transforming resources to sales, thereby creating profits through economies of scale and scope. Business processes are designed within the organization to order these repetitive activities as operations, controlled by procedures. The design (or redesign) of the types and sequence of activities and the procedures to control these activities forms the "operations structure" of an organization.

The operations structure provides the procedural means of achieving the mission of the organization.

We recall that modeling operations requires three planes of description (as shown in Figure 10.1). A model of this kind describes the operations structure of a business organization.

In addition to an operations structure, all organizations have an authority structure, which is established from the need to assign power and accountability in an organization to operate divisions-of-labor of the operations structure. Also we recall (from Chapter 1) that in a diversified firm, there are usually at least four levels of management hierarchy:

1. Firm level. Board, CEO, and firm executive team
2. Business level. President and business executive team
3. Department level. Department head and staff
4. Office level. Office manager and assistant

Authority structures are usually expressed by the organization chart of a business. Operations structures are seldom explicitly described, and the procedures of activities in the operations structures are usually maintained as separate volumes of procedural instructions (on paper or digitally). Accordingly, when business process redesign is undertaken, flow diagrams of the operations structure usually need to be created.

All organizations have both operations structures and authority structures, both of which may be needed to be changed in redesigning business processes and replacing legacy systems.

Now it is this very essence of organizational nature that creates the problem of islands of automation:

1. Pieces of information technology are implemented as a functional business unit's IT system within the scope of the business unit's division of labor and under the authority of the business unit and specifically to serve the rationality of business processes within the unit.

2. Other division-of-labor units with their authority structure have difficulty interfacing with information tailored to another unit's IT system.

3. Islands of automation arise from the *lack of proper interfaces* between the different functional units' IT systems.

Harold Kerzner emphasized that this kind of lack of horizontal communications across functional units of a business creates the islands of automation (Kerzner, 1984, p. 5). In Figure 11.6, we picture this as showing the authority structure of a multi-business firm and of a business within the firm, with lines delineating boundaries of functional authority and islands of automation.

Typical authority structures within a business are organized by business function with a manager for each function of

- Production
- Marketing
- Finance

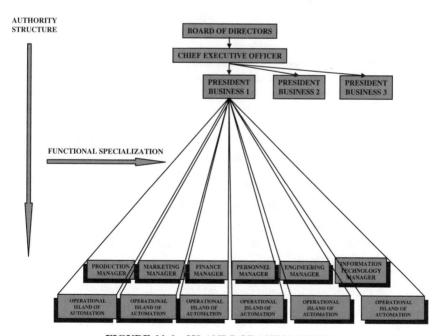

FIGURE 11.6   ISLANDS OF AUTOMATION

- Personnel
- Engineering
- Information technology

It had been under the authority of each functional manager to determine the kinds of software and business processes in the divisional unit. Accordingly, within each business function unit, a separate set of software (and/or platforms) might have been applied that did not necessarily automatically interface with software (and/or platforms) in other business functions.

Islands of automation within a company arise from differential progress of information technologies in the different functional areas of the company without concomitant attention to developing the proper communication interfaces between business functions.

To solve this problem, legacy information systems need to be periodically improved or replaced with improved communication between information systems in the different functional areas of a business. And this kind of legacy change will require re-examination and possible redesign of the business processes in the different functional areas.

To eliminate islands of automation, proper information interfaces need to be constructed between the different areas of business functions to ensure the flow of information throughout business processes.

Also information strategy needs to construct communication between the business and its suppliers and customers.

Integration is difficult because business processes are controlled by software applications specific to the business function involved in a process. Each functional application requires specific data in specific formats and outputs analyzes and information specialized to the function. Communication between business applications requires

- Selecting data from one application to be given to another application
- Reformatting data from one application to another
- Inputting information specific to the an application and interlacing this with data from other applications.

The selection of data from other applications to be used in a specific application is a nontrivial task because the categories of data in one application seldom exactly match up with the categories of data needed in another application. Therefore data often needs to be recategorized and reaggregated between applications. Moreover, data for one application is often incomplete for another application, and new data for the second application needs to be added into the data from the first.

In addition to the different categorizations of similar data used in different applications, the interpretation of the meaning of these categories frequently shifts between applications. What a category of data means in one application may be very different in another application, and related categories of data may need to be combined to provide the right kind of interpretation. Moreover, the accuracy needed in data may vary between applications.

Accordingly, the exchange of data between applications is seldom a simple problem. Integrating information across an enterprise system requires translation of data from one application to another and not mere transmission of data.

In addition, the integration of information system also requires attention to the quality and timing of information. The quality of information depends upon the accuracy of data entry and maintenance of accurate data. Which business unit is responsible for inputting and maintaining data is important to control for quality.

And also the required timeliness of data may vary from business function to business function. For example, machine control in manufacturing requires instantaneous data, whereas financial control of operations requires daily, weekly, monthly summaries and projections.

In summary, data integration across a business enterprise is a complex problem of information, involving meaning, quality, and timeliness—all of which may vary by business process application.

Information strategy for enterprise integration requires the re-design of information systems and business processes to improve the coordination of business processes.

SUMMARY: USING THE STRATEGIC TECHNIQUE OF INFORMATION STRATEGY

Information strategy has two components, technical and business, that need to be integrated. Three strategic challenges need to be addressed: e-commerce, legacy systems, and enterprise integration. All information strategies for these may also require business process redesign.

1. Form an information strategy team
   - Divide the team into three strategy subteams:
   - E-commerce strategy subteam:
        Have one subteam examine e-commerce information strategy
        Construct an e-commerce strategy matrix, which examines the busi-

nesses within the company as to type and strategies for attracting, informing, value-adding, and profiting.

- Legacy System Strategy Subteam

  Review progress in information technology that would incrementally improve and/or dramatically improve the information processing capabilities of the existing system.

  Decide whether or not these improvements can be easily added to the existing system.

  When improvements can easily be made, identify, and specify strategy projects to implement change.

  When improvements require replacement of the existing system (in part or in whole), identify and specify a legacy replacement strategy project.

- Enterprise Integration Strategy Sub-team

  Examine the information integration of current system.

  Identify business processes and functions where information communication is not working well.

  Examine how communication can be improved.

2. Examine business process redesign

   - Schedule meetings of the whole information strategy team, including the subteams on e-commerce, legacy, and integration to explore commonalities and synergy in creating new information strategy.
   - Explore what business processes require redesign to:

     Improve operations

     Improve e-commerce channels

     Exploit improvement or replacement of legacy systems

     Improve enterprise integration of operations
   - Identify the new information technologies that would accompany these redesigns.

3. Develop information strategy

   - Identify desired strategic changes in information technologies and business applications.
   - Estimate order-of-magnitude costs of change and benefits of change.
   - Identify necessary strategy projects for implementing change.

For Reflection

Examine the stories that appeared from 1995 to 2001 in business magazines (such as Fortune, Business Week, and the Economist) about the new dot.com businesses

Select some of the businesses using the Internet differently (e.g., portal services, communication companies, retail e-com businesses, auctions, etc.). Why did some of these fail and some survive? Trace the stock price history of each. In retrospect, can you think now of strategies that might have made them profitable, given their substantial initial investment funding?

CHAPTER 12

DIVERSIFICATION STRATEGY

PRINCIPLE

Successful strategic management of a diversified corporation requires proper interactions of competency and trust between firm-level executives and portfolio business executives.

STRATEGIC TECHNIQUE

1. Identify the reasons for diversification
2. Establish core competency strategies
3. Analyze the corporate industrial/business portfolio
4. Review interactions between firm-level and business-level staff
5. Properly manage strategic acquisitions
6. Properly manage strategic innovation

CASE STUDIES

Cisco's Acquisitions Strategy
RCA Dies in 1985
Chase Grows by Acquisitions

AT&T's Strategic Thinking in 2000
Perils of Sunbeam
3M Diversifies Through Innovation

INTRODUCTION

We next focus upon the problem of strategy in a multibusiness company—diversification strategy. Previously, we have been examining the issues for formulating strategy in single business firms. But how does strategy look like at the top of a multi-business firm, a diversified firm?

We recall (from the Chapter 1) that the strategy focus for single-business and multi-business firms differ greatly. In a single-business company, the competitive situation is principally in the marketplace, wherein it provides value directly to customers and against competing products and services. In contrast, the competitive situation for a multiple-business company is principally in the capital markets, wherein it provides value directly to investors. Whereas competitive strategy for the single-business company must focus primarily upon its products and services, competitive strategy for a multiple-business company must focus primarily upon its rate of return and value accumulation of the firm.

Accordingly from the perspective of the diversified corporation, the businesses owned by the corporation can be conceived of as a kind of "business portfolio" of the corporation. In the last part of the twentieth century, this concept of "portfolio" became a popular way to view the strategic management of a diversified firm, as expressed by Lowell Steele:

> Multi-business strategy focuses first and foremost on portfolio optimization—what mix of sources of revenue is desired and what allocation of resources will best bring about this preferred mix. [Yet] multi-business strategy must include other components, such as corporate organization and culture, management style, the conventions that guide behavior, and the acquisition or development of new resources that will be required to support a different business portfolio.
>
> —(Steele, 1989, p. 179)

We will use the strategic idea of a business portfolio to characterize diversified firms, but as Steele suggested, we will look at four particularly important factors that affect the successful management of a diversified business portfolio:

1. Market positions of its businesses
2. Industrial dynamics of businesses contexts
3. Leadership of its businesses
4. Core competencies of the whole corporation

CASE STUDY: Cisco's Acquisition Strategy

One of the reasons for diversification is innovation, and we now return to the earlier case of Cisco to see how it managed to stay ahead of the rapid pace of innovation technology through strategic acquisitions of other high-technology new businesses.

We recall that upon becoming CEO, Chambers launched an aggressive continuation of business acquisitions so Cisco could continue to be at the forefront of advancing information technologies and expand its product lines. Also we recall that the stock market grew through the 1990s, and Cisco's stock soared with very high price/earnings (P/E) ratios. Chambers' was able to use Cisco's highly valued stock to acquire other companies.

Ethernet technology was connecting computers into LANs. Asynchronous Transfer Mode (ATM) technology was connecting LANs. Routers were connecting into the Internet. Cisco's information technology challenge was in tying LANs into WANs. In 1995, Ethernet technology was still the preferred LAN technology. For WAN networking, ATM was preferred.

ATM was a hardware-based switching technology that transmitted data faster than routers and could be used to connect a finite number of LANS together, with resulting high-speed communication between LANs. ATM allowed a digital emulations of traditional switch-based phone networks and could bridge between data communications and telephone communications.

Chambers' first major acquisition as CEO was to acquire StrataCom, one of the leaders in ATM, with 22 percent of the market. Based on Cisco's stock price of $47.75, the deal was worth $4.5 billion to StrataCom:

> Chambers took StrataCom CEO Dick Moley out for dinner in a private room at the Westin Hotel in Santa Clara, where he placed a semiformal offer on the table. Less than two weeks later, Chambers and Moley made a joint announcement [of] a stock-swap agreement in which shares of Cisco stock would be exchanged . . . [for] Stratcom stock on a one-to-one basis.
>
> —(Bunnell, 2000, p. 81)

Next Chambers implemented systematic procedures for such acquisitions, with tasks to select, acquire, and integrate new businesses. First a Cisco business-development team scouted for new companies with a technology that would be needed by Cisco to maintain its technology progress and fill out its product-lines (so it could be a one-stop network supplier). Next, the team looked to see if there could be a shared strategic vision:

> The prospective company had to be moving in the same direction as Cisco. Chambers looked to see if their visions were the same—about where the industry was going, what role each company wanted to play in the industry. . . . The

product strategies of the two companies had to complement each other rather than compete.

—(Bunnell, 2000, p. 66)

Then the team looked for a compatibility of company cultures. Since the acquisition would become part of Cisco, personnel would have to adapt to Cisco policies and conventions. Chambers did not want too large a cultural shock to prevent newly acquired employees from quickly becoming happy with their new home. A third criteria was financial, that the product line of the acquired company would produce sufficient profits through rapid growth to soon justify the acquisition price.

Next, the Cisco team would have to persuade a new company to allow itself to be bought. Many departments of Cisco would be involved in negotiations, so that all aspects of the acquisition would be apparent to both potential partners, such as issues about human resources, business development, engineering, and financial and legal points.

Valuation of the Cisco stock in the acquisition was the critical issue—the buying price. For example, when Cisco acquired Grand Junction in 1995, the company had been planning to go public at a Goldman Sachs estimate of a $200 million IPO. Cisco offered 5 million shares of Cisco, then worth $346 million, and Grand Junction accepted. In 1996, Cisco acquired Nashoba Networks to produce and sell token-ring switches. Cisco also used pricing strategy against competitors not acquired. For example, in 1996, Cisco cut by half the price of its networking switches, then undercutting a rival's switches by 50 percent.

Integrating the new company was the final step. The long-term value of the new acquisition to Cisco was the capability of its employees to manage, design, and produce new cutting edge products in its product line. Accordingly, retention of the acquisition's management, technical, and sales people was important to the real value of the purchase. Cisco moved quickly to orient and integrate its new employees to its policies, and Cisco stock options were a big attraction. Cisco retained the top executives by letting them continue to run the company within the Cisco fold and play a major role within Cisco.

Case Analysis

Two important strategic lessons stand out in this illustrative case. First, Cisco's business acquisitions were carefully selected to support its continuing technological advancement and fill out its product lines. Second, Cisco management paid careful attention to retaining and integrating employees of its new acquisitions. When one acquires a business, one of the major assets are the skills and knowledge of the acquired employees. When one loses key acquired employees or fails to inspire their dedication to their new firm, one loses major value in buying a business.

REASONS FOR CORPORATE DIVERSIFICATION

Cisco was able to pursue its acquisition strategy because of its then very high stock valuation due to its rapid growth. We recall that in terms of determining the share price, modern stock market investors in the second half of the century valued growth and potential future earnings. For this reason, corporate diversification began in the second half of the twentieth century for

- *Growth by innovation.* Launching new product lines and new businesses, financed by cash flows from existing businesses.
- *Growth by acquisitions.* Growth by buying businesses with lower P/E ratios, as encouraged high valuation by financial markets of growing firms.
- *Surviving economic cycles.* Economic recessions affected different industries differently, and businesses in different industries can soften the losses.
- *Improving coverage of markets.* Extending coverage of niches in a market might improve overall market share.

Growth by Innovation

As we saw in the Cisco case, an important reason for diversification is to expand or create new markets for the firm through innovation.

Another example of a firm that used innovation to strategically diversify was the Ethyl corporation. Ethyl began as joint-venture of General Motors and Exxon to produce the lead antiknock gasoline additive, which had been invented at General Motors for high-compression gasoline engines (but which General Motors did not wish itself to produce). Ethyl first sold antiknock gasoline in 1923, and for nearly forty years it was a one-product company. In early 1960s, Ethyl was acquired by the Albemarle Paper Manufacturing Co., and pressure was there to diversity to survive (Gottwald, 1987).

At the time of the 1962 merger of Ethyl with Albemarle Paper, Floyd D. Gottwald, Jr. was elected executive vice president of the newly merged Ethyl Corporation. He had originally joined Albemarle Paper Manufacturing Company in 1943 as a chemist, and advanced to president of Albemarle. Later he commented on the merger that set Ethyl to diversifying:

> When Albemarle purchased Ethyl in 1962, Ethyl was clearly a one-product company with a wealth of pent up talent restless to exert itself. Under the previous joint

owners, GM and Exxon, there had been virtually no opportunity for commercialization of the many possibilities that had emerged from forty years of research on improving or finding a better antiknock. For good reasons of their own, the previous owners had preferred to keep Ethyl a one-product company. Our change of perspective in 1962, as we sought to diversify, could not have been more dramatic. . . .
—(Gottwald, 1987, p 27)

But which businesses to acquire? Ethyl chose to diversify toward areas in which it had a strong underlying knowledge base. The Ethyl research program had grown out of their original focus on lead antiknock compounds, whose original rights they had acquired from General Motors. Ethyl developed chemical skills and innovations by branching out from its original chemistry of lead antiknock compounds and also from other research on aluminum alkls chemicals. This kind of research branching focused Ethyl's original business acquisitions, and those acquisitions stimulated further innovation through research branching.

As Ethyl began acquiring businesses, it was originally focused on its existing research base in chemicals and paper. In 1963, Ethyl acquired Visqueen film for plastics. In 1967, it acquired Oxford paper, and in 1968, IMCO Container.

Ethyl had an interesting research program in aluminum, which didn't work out but motivated it to acquire the William L. Bonnel Company in 1966 and Capital Product Corporation in 1970. Ethyl continued acquiring businesses throughout the 1970s:

As we reached the mid-1970s and Ethyl's success seemed assured, our acquisition program entered phase two—broadening the base. In 1975, we acquired the Edwin Cooper Division of Burmah Oil to give added strength to our existing lube additive lines. Harwicke Chemical, purchased in 1978, expanded our insecticide business to include synthetic pyrethroids. In 1980, we acquired Saytech, Inc to extend Ethyl's basic bromine position into flame retardants.
—(Gottwald, 1987, p. 27)

The result of this diversification program over twenty years dramatically altered the structure of Ethyl's businesses. The diversification program at Ethyl had also looked at many other businesses that it chose not to acquire. In these cases, the businesses had no relation to Ethyl's research strengths. An important factor in Ethyl's acquisition program was that it not only had a financial strategy but a plan for innovation. Moreover, the diversification was fortunate in timing, since later in the 1970s the U.S. government began legislating lead additives out of gasoline as a health hazard.

Growth by Acquisitions

Another reason for corporate diversification was that it enabled a firm to escape from the confines of a low-growth or low-return industry.

For example, the first U. S. conglomerate, Textron, was created by Royal Little for this reason. In 1923, Little founded a company called Special Yarns Corp. in Boston, Massachusetts. The 1930s depression was hard, and the company struggled to stay alive. After the war, the textile business turned out to be highly cyclic, with a low return on capital. One of the reasons for this was that the industry expanded production capacity by reinvesting profits, reluctant to pay out high dividends or taxes (Little, 1984). On June 30, 1952, Roy Little held a special stockholders' meeting to change the articles of association to buy businesses outside of textiles. His first acquisition was the Burkart Manufacturing Co., which had begun by making horse blankets in St. Louis and then turned to making auto seat stuffing. Little then bought two more companies in 1954, Dalmo Victor and MB Manufacturing. In 1955, Little bought Homelite Corp.

High stock valuations have always facilitated the building of growth by corporate acquisitions. For example, in the United States, a decade of widespread corporate conglomeration occurred in the late 1950s and early 1960s when the robust stock market provided growing companies with high P/E ratio valuations that enabled such companies to buy other companies with no growth and much lower P/E stock valuations.

For example, in 1954, Litton Industries began from a company called Electro-Dynamics Corp, which was taken over by Roy Ash and Tex Thornton. Ash and Thornton changed the name to Litton Industries and acquired at least twenty different businesses, using high P/E ratios (as high as 47 at one time). In the early 1960s, they acquired two companies with problems, Ingalls Shipbuilding and Royal McBee. In 1968, Litton's quarterly earnings declined for the first time. For a long time thereafter, the shares were down, and finally the company returned to being a strong performer (Little, 1984).

In the 1990s, high valuations of new Internet businesses allowed a similar strategy for corporate growth, which we saw examples of in the earlier cases of AOL acquiring Time Warner and of Cisco Systems acquiring new technology companies.

Surviving Economic Cycles

Another reason for diversification is to counter the financial impacts of common business cycles. Different businesses are affected differently in a recession.

One example of a merger put together specifically to counter the effect of business cycles was the formation of the Martin-Marietta Company in the 1950s. Martin was an aircraft firm and Marietta a construction materials firm, selling cement and crushed concrete. The merger followed a belief that defense and domestic economies are often on opposite cycles. In the 1970s when defense was booming and construction was down, a hostile raid was made on the firm. To avoid being bought up, the company bought out its own stock and compiled an enormous debt. Then later to reduce that debt, Marietta was sold off. After the end of the Cold War in the late 1980s, the U.S. defense industry began strategic changes, and Martin merged with Lockheed to form Lockheed Martin.

Improved Coverage of Markets

Another reason for diversification was to cover a market. Examples we have seen of this were the cases of Durant's assemblage in the early 1920s of many different auto model business into General Motors and of Cisco's acquisition of networking component companies in the 1990s.

STOCK MARKET VALUATION OF BUSINESSES

Since for a diversified firm, the financial market provides its immediate performance context, it is useful to review the criteria for market valuation of stocks.

The traditional criteria for valuing share price is called its price to earnings ratio (P/E). The meaning of this lies in its inverse ratio E/P. This is a measure of the present return on an investment in a stock at a price P as the fraction of annual earnings E at that price, or E/P.

For example, consider a company with a *P/E* ratio of 10. Suppose the company's share price is $P = \$200$ dollars and the company earned $E = \$20$ dollars that year. The stock valuation would be *P/E* = 10. This can be interpreted as the present return of the investment is 10 percent percent return based upon inverting the *P/E* ratio as $E/P = 10/100 = 10$ percent return.

The inverse of the P/E ratio calculates the present rate-of-return of the company's performance.

This measure is fine for a constant rate of earnings in a company, but it undervalues a company if its earnings are continually growing. One needs to value the company not at a present rate of return but at a future rate of return. Accordingly, a growing company is valued at a higher P/E ratio than a constant-rate sales company. Just what P/E ratio for a growing company is reasonable depends upon the rate of growth and the enthusiasm of a stock market.

For example, a P/E ratio of 20 means that the present rate of return of a company's share is E/P = 1/20 = 5 percent. So the company must double present earnings in the future to gain a future 10 percent return. A P/E ratio of 40 would mean that the present rate of return of a company's share was E/P = 1/40 = 2.5 percent. So the company must quadruple present earnings in the future to gain a future 10 percent return.

Figure 12.1 illustrates this general criteria of growth as a factor in stock valuation by influencing P/E ratios for four general patterns in businesses—growing businesses, steady businesses, cyclic businesses, and declining businesses:

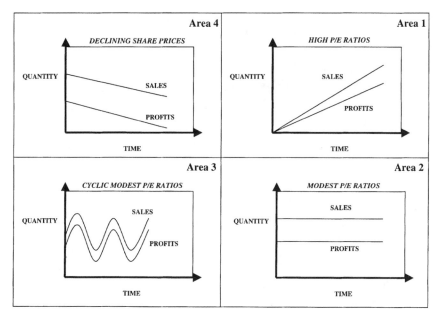

FIGURE 12.1   GROWTH AS A GENERAL FACTOR IN MARKET VALUATION OF BUSINESSES

Area 1: Businesses growing in both sales and profits. Shares are valued with high P/E ratios. Depending upon the state of the market, high P/E ratios of growing firms have varied in the range from 20 to 200 in the U.S. market in the second half of the twentieth century.

Area 2: Businesses constant in sales and profits. Shares are valued with modest price/earning ratios. Depending upon the state of the market, modest P/E ratios of growing firms have varied in the range from 7 to 17 in the U.S. market in the second half of the twentieth century.

Area 3: Businesses with cyclic sales in economic cycles. Shares are cyclicly valued with modest price/earning ratios. Depending upon the state of the market and the economy and the dividend policies of a business, modest P/E ratios of cyclic firms have varied in the range from 7 to 17 in the U.S. market in the second half of the twentieth century.

Area 4: Businesses with declining sales. Shares steadily decline in price. End games of declining businesses have frequently ended with acquisition by another company or in bankruptcy.

In summary, the critical variable for the valuation of a company's share price has been the P/E ratio assigned by the market in a given state of the market. And this P/E ratio has had a great range of variation, particularly for newly growing businesses during times of a stock market boom.

This is the reason the growth became the single most important strategic performance variable for diversified firms in the late twentieth century.

CASE STUDY: RCA Dies in 1985

We next turn to the factors for successful diversification. Often CEOs have found that it was easier to put together a diversified business portfolio of a company than to successfully manage such a portfolio. We recall that the General Electric CEO Jack Welch had become famous because he had been one of the very few CEOs at the end of the twentieth century to have successfully managed a very large conglomerate. In the mid-1980s, many large conglomerated firms were taken over and restructured. This case of the demise of RCA is a particularly poignant story.

RCA began as a high-technology firm in the 1920s, assembled by David Sarnoff to consolidate patents in radio technology. RCA became the premier radio producer in the United States in the first half of the twentieth century and a major innovator in electronics, innovating after the second World War both black & white and color television.

But the CEOs who succeeded Sarnoff began a corporate strategy of growth through conglomeration. When Griffiths became CEO in the 1970s, RCA's financial position was poor. Griffith instituted stronger financial controls, increased factory automation, decreased the payroll, and began a process of divestiture. He sold Random House (a book publisher), an Alaskan telephone system, two food companies, and manufacturers of x-ray equipment, aircraft radar, and mobile radios. He turned again toward the electronics business, increasing the research budget from $112 million to $197 million (Nulty, 1981).

During the 1960s, RCA acquired the National Broadcasting Corporation (NBC) and Hertz, the leading car-rental company, both major businesses. In addition, many other businesses were acquired. Over the course of the 1970s, RCA had taken several heavy write-offs. In 1971, RCA wrote off $490 million in withdrawing from the computer business. In 1981, RCA wrote off $230 million: $130 million in TV-picture-tube operations; $59 million in truck leasing, which Hertz had begun in 1978; and $34 million on TV shows in NBC's inventory.

In 1980 Griffiths purchased a financial firm, CIT Financial, for $1.4 billion in cash and stock. This doubled RCA's debt and left the company highly leveraged. RCA's per share earnings had increased from a little over $2 to about $3.50 from 1976 to 1979. But from 1979 to 1981, the per share earnings dropped to minus 19 cents (Ehrbar, 1982).

In 1981 Thorton Bradshaw became chairman and chief executive of RCA Corp., and in 1984, Bradshaw sold off CIT Financial (for nearly what RCA had paid for it).

One of the unfortunate aspects of RCA's management of their diversified acquisitions was that the acquired companies often performed better both before RCA acquired them and after RCA sold them. For example, Hertz had prospered under Robert Stone, initially after being acquired by Robert Sarnoff. From 1971 to 1977, Hertz profits increased fivefold, to $131 million. But RCA replaced Stone in 1977, and Hertz profits declined from 1979 through 1981. In 1985, RCA sold Hertz to United Airlines.

Other companies acquired by RCA and then sold off also prospered afterwards. For example, Banquet Foods was acquired by RCA and later sold. It did well before and after RCA. Bradshaw commented: "We didn't know how to run it, and we should not have had it" (Ehrbar, 1982, p. 67).

In December 1985, GE acquired RCA and took it apart. Today the high-tech firm RCA no longer exists. RCA was killed by a succession of CEO strategic blunders from bad diversification investments. Historically, one can see that it takes a series of poor leaders to kill a really big company, but it can and has been done.

Case Analysis

Not all conglomerate acquisitions have proved valuable. Some acquisitions redefined the core businesses of a company. Others could not be integrated into the company or perceived as valuable enough for management attention or resources. In the 1970s and early 1980s a large number of corporate divestitures occurred as sales of divisions to other companies or leveraged buyouts.

For example, Royal Little, the founder of Textron, commented about that time in the United States:

> The spinoff trend started in 1972. The conglomerates had bought too many small companies, and they began selling the ones with the least growth potential to put more capital into the most promising divisions. . . . By 1972, of course, I was out of Textron, but I was running Narragansett Capital Corp. . . and we were able to pick up dozens of divested cash cows that weren't growing. We used leveraged buyouts, which enabled us to give the managers a piece of the action. So in a way I had the best of both conglomerate cycles-when they were diversifying in the 1950s and 1960s and which they were selling off in the 1970s.
>
> —(Little, 1984, p. 60)

Are conglomerates good or bad? Neither. They are simply another form of business organization, relying primarily on financial mechanisms for control of performance. Like all companies, when well managed they do well. But they, too, have their problem times when the value of the individual businesses of a diversified firm adds up to more value than the stock price of the conglomerated firm, and this happens when not all businesses of a firm are being well managed. The poignant comment by Bradshaw expresses the key to any diversification strategy: Don't acquire or get into a business that you don't know how to manage.

Managing any business includes understanding the market of a business and the operations of value-adding for that market. George White and Margaret Graham nicely summarized the essential questions about operating a business (White and Graham, 1978):

- Are we doing the right job?
- Are we doing the job right?

These are critical problems for conglomerate diversification—first buying the right businesses and then successfully managing acquired businesses.

MARKET-POSITION ANALYSIS OF BUSINESS PORTFOLIO

One of the problems RCA leadership had with managing diversification was in using a particular strategy then popular for a multi-business firm, called business portfolio analysis. In the 1970s and 1980s, strategy for a company's portfolio of businesses often only looked at the market position of a portfolio business. For example, in 1975, George Day summarized this kind of simple portfolio analysis in terms of a business's market share and the growth potential of its market, as shown in Figure 12.2.

Looked at only from these dimensions, the most obviously desirable businesses to own in a corporate portfolio would be those with high market share in a high growth market (Area 1 of the chart). For this reason, business strategy consultants in the 1970s named these kinds of businesses "corporate stars." Also, they named businesses with large market share in low-growth markets as corporate "cash cows" (Area 2). Businesses with low market share in low growth markets (Area 3) they named corporate "dogs" (a name that offended those of us who love dogs). Finally businesses with low market share in a high growing market were called "problem children" (Area 4). And these names suggested an obvious set of strategies:

- One should invest in the stars of the businesses in the portfolio.
- One should only milk revenue from the cash cows businesses without further investments in them.

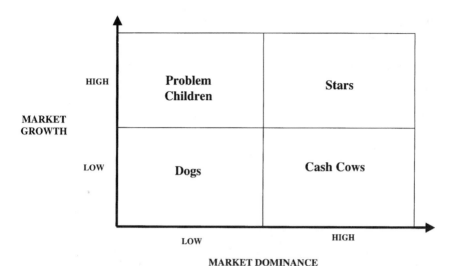

FIGURE 12.2   MARKET-POSITION ANALYSIS OF A BUSINESS PORTFOLIO

- One should decide whether or not to keep the problem children businesses.
- One should gets rid of the dogs of the businesses in the portfolio.

However as RCA experienced, the use of this simplistic portfolio approach betrayed two hidden and fatal strategic flaws:

- First, such simple strategy ignored any corporate strength as a whole which could benefit individual businesses of its portfolio.
- Second, such simple strategy ignored the industrial contexts of business and treated all businesses as existing within the very same industrial structure and dynamics.

RCA became more valuable in parts than as a whole, so GE acquired it and took it apart.

In Area 1 of the chart, we know now from the concept of the industrial life cycle that star businesses are only possible in the early phases of an industrial life stage. We also know that as all industries mature, any successful business in that industry must over time come to reside in Area 2; thus all large businesses must become eventually cash cows. Calling them cash cows is simplistic because one should never starve the cows in any large dairy operation, upon which one depends for cash flow. Moreover, for businesses in Area 3 such so-called dogs may still be profitable if they dominate in a good market niche. For example, a luxury market leader will often exist in small areas of a large market with no growth and still be profitable. Finally for businesses in Area 4, the only businesses that can long exist as problem children are specialty businesses in the early phases of an industrial life cycle.

In summary, a market-position basis of portfolio analysis is a useful but only partial way to strategically look at a diversified company's business portfolio and needs to supplemented by other views, particularly:

• Overall corporate synergism
• Industrial life-cycle contexts of businesses

As happened often in the 1980s, using only a partial market-position analysis of a business portfolio resulted in a poor corporate performance. Companies often destroyed their cash cows through under investment in maintaining their competitiveness. They often wasted money on stars that never even made it big through lacking proper competitive advantages. They often didn't know how to handle and solve the challenges their problem children. In short (as in the case of RCA) leadership of many conglomerate companies who used only a portfolio analysis strategy didn't know how to properly run the businesses of their portfolios.

After the widespread popularity in the 1970s and 1980s of this simplistic approach to corporate strategy, the astute business observers began criticizing the approach. For example, James Brian Quinn suggested the need for more complexity in analyzing businesses:

> Perhaps the most difficult task for top managers is to balance the needs of existing lines against the needs of potential lines. This problem requires a portfolio strategy much more complex than the popular four-box Boston Consulting Group matrix found in most strategy texts.
> —(Quinn, 1985, p. 81)

Allan Kantrow emphasized several factors should go into a company's diversification strategy:

> It is of great importance to identify and assess the nature of the relationship among a company's distinctive technological competence, its organizational structure, and its overall strategic orientation.
> —(Kantrow 1980, p. 12)

Richard Hamermesh and Roderick White argued that a typical flaw in simple business portfolio analysis was to ignore effects of the organizational relationships within a corporation. The aspects of interaction that are important are autonomy of the business unit managers, line responsibility for direct and complete control of key functions, and incentive compensation of business unit managers linked to unit performance. They argued that

> The nature of a business unit's relationship to headquarters can have as much effect on its performance as its competitive position and the industry's environment.
> —(Hamermesh and White, 1984, p. 103)

In summary, while the market position of a business is an important factor in analyzing a business portfolio, other factors are just as important:

1. Corporate strategic management should carefully select the businesses of the corporation with regard to their industries.
2. Business strategy in handling a portfolio of diversified companies needs not only to optimize short-term gains but also provide for long-term growth and strengthen integrative competencies that make the corporation more than a sum of its parts.
3. Corporate strategic management needs to pay close attention to the relationships between company's distinctive technological competencies, its organizational structure, and its overall strategic orientation in constructing, maintaining, and expanding its business portfolio.

CASE STUDY: Chase Grows by Acquisitions

Next we will examine how to properly perform a business portfolio analysis that includes the industrial context. First we look at a case wherein growth by acquisition did pay attention to the industrial context, Chase Bank's acquisitions of J. P. Morgan and of Hambrecht and Quest.

The financial services market was divided into banking services for individuals and businesses. For individuals, financial services was divided into large consumer markets and a small but profitable market of very wealthy individuals. In the business market, it was divided into financial services for large businesses and small businesses. And for the life stage of businesses, it was divided into venture capital for new businesses and banking services for mature companies.

Chase Acquires J. P. Morgan
On September 13, 2000, Chase Manhattan announced its merger with J. P. Morgan & Co: They stood on the dim stage in midtown Manhattan this morning, a study in patience as they waited to announce the banking industry's latest blockbuster deal. Chase Manhattan's Chairman William B. Harrison Jr. crossed his arms and bounced gently on his toes. J. P. Morgan Chairman Douglas A. Warner sat at a simple table, his hands folded, calmly looking at the packed auditorium. . . . They soon made it clear that in their own deliberate ways, they will be turning

two venerable U.S. banks into a single financial services powerhouse with the resources to compete on a global scale.

—(O'Harrow and Day, 2000, p. E1)

This latest merger of Chase was only another step in its strategy for the last fifty years of growth by acquisitions. Back in 1955 Chase Manhattan Bank was formed by a merger of Chase National Bank and the Bank of the Manhattan Company. In 1995, Chase Manhattan merged with Chemical Bank.

Chase had been not the only New York bank growing through acquisitions. Chemical Bank had grown by acquiring the Corn Exchange Bank in 1954, the New York Trust in 1959, and Manufacturers Hanover in 1991. (Manufacturers Hanover had been formed in 1961 by a merger of Manufacturers Trust and Central Hanover Bank.)

In the second half of the twentieth century, the U.S. financial services industry underwent a steady consolidation of firms for growth and market position:

The deal . . . marks the latest in a series of huge transactions in the rapidly consolidating financial-services industry. . . . Many of today's global financial giants are U.S. companies that have leapfrogged over their European counterparts. In effect, the globalization of finance has become the Americanization of finance. . . .

—(Lipin et al., 2000, p. A1)

But even this new game was only another historical phase in a world of continuing strategic change for banking in the United States over the last 200 years. For example, the Manhattan part of Chase Manhattan traced its business history back to 1799 when it was formed as the Bank of the Manhattan Co. (then with the support of Alexander Hamilton and Aaron Burr). Manhattan was important in financing the early economic development of New York City in the early 1800s,

The Chase part was established in 1877 by John Thompson and four partners as the Chase National Bank (taking its name after Salmon P. Chase, the treasury secretary under the Abraham Lincoln).

J.P. Morgan & Co. also had a venerable and even more famous business history. Morgan traced its origins back to 1838 when American businessman George Peabody began a merchant bank in London and took on a partner in 1854 named Junius S. Morgan. After Peabody's retirement, Junius Morgan ran the firm and changed its name to J.S. Morgan & Co.

In 1861, Junius' son, J. Pierpont Morgan set up a New York sales office to sell European securities underwritten by the London office, and Pierpont named the office, J.P. Morgan & Co. J.P. Morgan & Co became a key financial firm in financing the United States' building of its enormous railroads and mining, steel, and utilities industries. In 1907 when a major financial meltdown threatened to halt the U.S. economy, and J. Pierpont Morgan infused money into banks and propped up the stock market to stop the U.S. Panic of 1907.

Later in the great depression of the 1930s, the U.S. Congress passed the Glass-Seagall Act, forcing the separation of commercial and investment banking in the United States. The investment bank part of Morgan was split off as Morgan Stanley, and the commercial bank part remained as J.P. Morgan & Co.

Over time, U.S. banking has undergone continual change, and the new world of global financial services was just the latest phase in a long business history.

In the year 2000, another reason for the merger of Chase and Morgan was that the different sectors of the banking world had different markets. Chase was a big commercial bank with a very large credit-card business and a very large consumer lender for house mortgages and car loans. J.P. Morgan raised capital for large corporations and advised them on mergers and acquisitions:

> J.P. Morgan had been criticized in recent years for not being big enough and for not expanding its retail-customer base more aggressively. . . . Morgan's asset-management arm has long coddled the super-rich, writing their wills, managing their estates, selling their family heirlooms and baseball teams. . . . [In contrast] Chase had been said to need a more beefed-up investment banking division. This marriage answers those complaints. . . .
>
> —(O'Harrow and Day, 2000, p. E3)

In the merger, Chase exchanged 3.7 shares for each J.P. Morgan share, at that time a premium for the J.P. Morgan stock. The merged company would be called J.P. Morgan Chase & Company. It would have $662 billion in assets, about 90,000 employees and a stock market value then of $95 billion. Then only two banks were larger: Citigroup with $792 billion in assets and Bank of America with $679 billion in assets.

Chase Acquires Hambrecht Quist In the second part of this case, we look in 1999 at an earlier acquisition by Chase of a venture capital firm, Hambrecht & Quist. At the time of the merger, Jimmy Lee was CEO of Manhattan Chase and Dan Case was CEO of Hambrecht & Quist. Hambrecht & Quist was a small but high-quality investment bank for new high technology businesses: "The deal brings together two worlds, the Fortune 500 clientele of Chase and the red-hot business of technology IPOs (Initial Public Offerings)." (Serwer, 2000, p 121)

In the life stages of new high-tech businesses, start-up capital usually comes from individuals—entrepreneurs and "angels." The entrepreneurs who start new high-tech firms often use personal savings, mortgages on homes, and credit cards to begin a new business. Sometimes wealthy individuals, so-called angels, also invest in start-ups. Venture capital funds usually provide expansion capital after first sales are made by a new start-up.

An exception to this usual pattern occurred in the late 1990s when many e-commerce businesses received initial start-up capital from venture firms. For a brief period in 1998 and 1999, the Internet frenzy encouraged several venture capitalists to provide start-up funding for concepts of new e-commerce businesses, even at lavish scales and sometimes without even good business plans (as was the earlier case of Boo.com).

When the electronics and computer innovations occurred in the United States in California's Silicon Valley of the 1960s and 1970s, a new sector of banks grew to specialize in the IPO offerings of new high-tech firms. These were Hambrecht & Quist, Alex, Brown, Robertson Stephens and Montgomery Securities (then together called the HARMs of the San Francisco Bay Area: "The HARMS—a group of smallish, technology-oriented investment banks—were vibrant businesses, but they had come under pressure over the years (in the 1990s) as big New York investment banks tried to invade their turf. One by one, they sold out." Serwer, 2000, p. 122)

The head of H&Q was Dan Case (who happened to be the older brother of AOL's Steve Case). H&Q was the last of the HARMS, not yet acquired by New York banks. From Case's perspective, he saw that H&Q also would need a partner, but he also thought that many of experiences of other HARM mergers had been "pretty negative" for the acquired partners. However, he hoped that partnering with Chase might be better because Chase had been built upon a series of mergers that were properly put together.

In a merger, two strategic issues are always upon the mind of leaders in the merger: the business viability of the new merged firm and the place in the new firm of the top executives of the old firms. By 'pretty negative' in the perspective of H&Q's CEO, Dan Case, was the knowledge that the former CEOs of the acquired HARMS did not always do well in the corporate cultures of the acquiring New York investment banks. Still, Case could see H&Q's need for an investment bank partner. For example when Amazon.com issued new stock in 1998 (recall the case of Amazon versus Barnes & Noble from Chapter 1), neither H&Q nor Manhattan Chase benefitted from banking fees generated in the offering. Of this particular offering, Chase's Lee commented: "You know we run one of the biggest high-yield businesses on Wall Street, and I'm sitting at my desk one day in New York, doing some work for Barnes & Noble, and all of a sudden I hear that Amazon.com is going to do a $500 million high-yield deal and I didn't even know about it. That's when I knew that I had better do something." (Serwer, 2000, p. 122)

It had happened that this same incident had bothered Case, as the head of H&Q: "Case nods his head. The Amazon.com deal was troubling to him too, but for a different reason. H&Q, you see, helped take Amazon public back in 1997. And while Case knew all about the coming transaction, his firm was unable to pitch the business, never mind win it, because H&Q had no junk bond desk." (Serwer, 2000, p. 122)

Because it had no junk bond desk, H&Q had no clientele who would buy large amounts of junk bonds and so could not offer the service to companies. In contrast, Chase had such clients for junk bonds, but it was not networked into new high-tech firms (such as Amazon) so was unable to pitch for and gain their banking business. The bank that served Amazon had been Morgan Stanley which had a "full service" investment bank for both IPOs and bond marketing.

The merger of H&Q for $1.35 billion dollars into Chase then represented only about 2% of Chase's market capitalization and was not big in affecting Chase's bottom-line. It would provide Chase with a fuller-service product line for serving the growing high-tech firms: "'We bought H&Q because we needed a footprint in the technology space,' says Lee." (Serwer, 2000, p. 124) From H&Q's, perspective, the merger seemed hopeful, as Case commented: "We fit well together. There is no overlap of businesses." (Serwer, 2000, p 126)

In investment banking, large deals generate profits; and large deals are intensely personal contact and trust—which is why the people part of the deal mattered. In melding H&Q personnel into Chase, a major strategic issue was personnel compensation, since the two sectors had different compensation policies. Chase reserved a $200 million compensation pool, from which H&Q bankers were given generous packages at nearly twice as much as they had made, and the packages would only pay out over time, providing an incentive for the H&Q bankers to stay.

Case Analysis

Long-term growth strategy for Chase had been through mergers and acquisitions. Chase had grown through previous mergers of banks and had become a dominant bank in the mid-level but large commercial markets of credit cards and consumer loans.

Chase merged with J.P. Morgan to add investment banking for large companies and wealthy individuals and to grow assets to survive as a major global bank. J.P. Morgan's interests in merging was also to survive in the global market. Also, Chase had added a market niche by acquiring Hambricht & Quist to enter the high-tech IPO banking market.

INDUSTRY/BUSINESS PORTFOLIO ANALYSIS

The Chase case illustrated how the industrial contexts of acquisitions can serve as a guide for acquisition strategy. When a diversified firm acquires a new business, it also is entering the industry in which the business operates. The competitive conditions and life stages of the particular industry of a business will constrain the opportunities and profitability of the business.

We recall that an industrial life cycle is the pattern shown by all new industries that originate upon innovation of a new core technology. Markets of the new industry grow and mature as the rates of innovation in industry slow down. In a new industry, rapid business growth is possible. In a mature industry, the market size is stable, and growth must come from the market share of a competitor. Moreover, in a mature industry with excess capacity for market demand, profit margins will be limited by strong price competition.

To properly analyze business portfolios, one needs to add a kind of industrial-context analysis to the previous technique of analyzing a business portfolio in terms of market position. We can do this by embedding the market-position analysis within a larger space of industrial-context analysis, as depicted in Figure 12.3:

- The larger space of industrial context can be characterized by the two dimensions of stage in industrial life cycle and size of industry.
- Any business may be located in context as to whether or not its industrial context is in a new industry or in a mature industry and as to whether or not that industry is small or large.

Multiple-space analysis of a business portfolio allows businesses in a company's diversification portfolio to be compared not only along dimensions of market position but also along dimensions of industrial context.

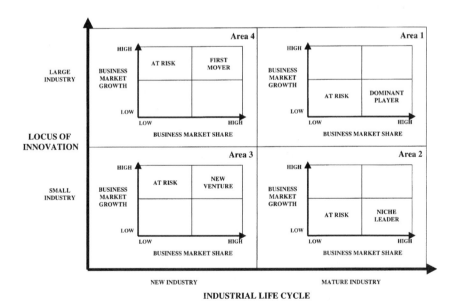

FIGURE 12.3 INDUSTRY-CONTEXT ANALYSIS OF A BUSINESS PORTFOLIO

Area 1: Businesses in Large Mature Industries

Consider Area 1 of Figure 12.3, in which a business exists in a mature industry of large market size. Therein no business can experience large market growth except at the expense of competitors. Accordingly, most businesses surviving in a large, mature industry will either have large market share and be a dominant player or have small market share and be at risk.

We recall that a dominant player will necessarily be a low-cost and high-quality leader in its industry in order to maintain or gain market share. For example, we recall that in the last part of the twentieth century, General Motors was no longer a low-cost or high-quality leader in automobile manufacturing and lost market share from above 50 percent to 29 percent. We also recall that then Japanese automobile leaders in cost and quality were Toyota and Honda, who gained market share from GM and became dominant players.

Other firms with small market shares in large, mature industries are continually at risk and are eventually bought up or go bankrupt. For example, Chrysler's market share declined to less 10 percent of the U.S. auto market in the 1960s, nearly went bankrupt in 1981, and was acquired by the German auto firm Daimler in 1998.

Strategically, it makes good sense for a large diversified firm to maintain businesses in its portfolio who are dominant players in large industries because the cash flow and return on assets can be very large.

For example, we recall how Jack Welch managed the diversified GE through the strategic policies of owning only businesses in large markets and having their managers be number one or number two in the industry. Upon becoming CEO, Welch sold off businesses that were in small, mature industries or were not a dominant player in a large, mature industry.

Area 2: Businesses in Small Mature Industries

Consider Area 2 of Figure 12.3, in which a business exists in a mature industry of small market size. Businesses in mature industries of relatively small size still face competition for occupying positions in the niche. For long-term survival in a niche, a business also must be a low-cost, high-quality leader. Businesses with small market share even in a small market are at risk for bankruptcy.

However, it can make good strategic sense for a smaller firm to own a portfolio of niche leaders in mature industries of small markets. These can provide substantial cash flow and return-on-assets investments. It does not make good strategic sense for a very large firm to own a portfolio of niche leaders because the contribution to revenue of even a niche leader cannot match the attention required to monitor or invest in the business. It is a truism that running a business in a niche industry requires just as much attention as running a business in a large industry.

Management attention demands in any business, small or medium or large, are equally great.

Area 3: New Businesses in New Industries

Consider Area 3 of Figure 12.3, in which a business exists in a new industry of small market size. New business ventures are of two kinds, high-tech new businesses and low-tech new businesses. High-tech new businesses occur in new industries founded upon basic innovations and initially have only small market size (and later the market grows larger). (Examples in the late twentieth century were information technology businesses and biotechnology businesses.) Low-tech new businesses occur in existing industries, and often are regionally localized kinds of businesses. (Examples in the late twentieth century were regional service businesses and franchised service businesses, such as fast food or auto service businesses.)

In a new industry, all new high-tech ventures when they began are at risk unless their market share rapidly grows to make them profitable early. For example, many e-commerce retail businesses failed when profitability was not established early. Early sales and early establishment of a significant market share is important to the survival of a new high-tech venture.

For a diversified firm to enter a new high-tech industry, it can make strategic sense to acquire a newly established business. But two strategic criteria should guide such acquisitions:

1. The new high-tech venture should be an innovative technology leader.
2. The management of the new high-tech venture should have shown managerial capabilities to make significant sales and to establish profitability.

The acquisition of a new high-tech venture by an existing large firm should pay strategic attention not only to acquiring a new technology but also a management team capable of successfully commercializing the new technology.

Area 4: New Businesses in New Large Industries

Consider Area 4 of Figure 12.3, in which a business exists in a new industry of large and growing market size. We recall that as the industrial life cycle of a new industry enters its rapid market growth phase, many competitors enter and later die. We also recall that the first mover in a new industry is the business that first makes appropriate investments in:

1. Continuing to advance the new technologies
2. Developing large-scale production capacity

3. Developing a national distribution capability
4. Developing the management talent to grow the new firm

Acquisition of a first mover in a new industry for an existing large business is usually difficult because of the high stock evaluation such companies usually obtain. Usually, such firms go on to become very big firms in their own right or acquire older large businesses. Cisco Systems was an example of a first mover in the computer network equipment business. It became a large firm, acquiring many other new high-tech ventures along the way. An example of a first mover acquiring an older large firm was AOL's merger with Time Warner.

OPPORTUNITY COSTS OF STAYING IN A BUSINESS

An area of major strategic decisions in diversified firms, is whether or not to make investments in improving the exiting businesses in the firm. Over time all businesses require periodic investment in order to continue being competitive. Investments need to be made in improving products and services, improving production and delivery, improving communications and control. Accordingly, the diversified firm must yearly make decisions about investments in its existing businesses. These investments are kinds of 'opportunity costs' of continuing to keep a given business competitive.

One standard way of judging opportunity costs is to use a discounted cash-flow approach. A discounted cash-flow calculation judges the value in the present of a future return. It does this by comparing an expected future return of a particular investment to one alternatively invested in a financial market with a known rate of return (such as a savings account). For example, suppose one invests $1000 dollars in a business improvement expected to return !0% in one year ($1000 + 0.1 \times 1000 = 1000 + 100 = 1100$). One can compare this to the alternative of despositing instead that same $1000 dollars into a bank's savings account for one year at 5% ($1000 + 0.05 \times 1000 = 1000 + 50 = 1050$). The discounted cash value of the investment in the business improvement is the difference of the return from the investment compared to the return from the alternate 'safe' market investment, which in this example would be $1100 - $1050 = 50.

Now this way of looking at the value of an investment is valid as long as *one does not care which business produces the return on the investment*. This is a purely 'financial' perspective, a perspective entirely appropriate to being in a banking business (since banks care not in what business they invest, only the risk and return of the investment). However, in any particular business, a decision not to invest in improving this business may in the future eventually result in the loss of the entire business. Accordingly, one should not evaluate a proposed investment in improving an existing business *only on the basis of a discounted cash flow calculation.* One also should look at the potential impact of the business improvement investment (or lack of it) upon the business survivability. What will the investment contribute to keeping the business competitive in the future?

We recall from the concept of the industrial life cycle, that as an industry matures technologically, its market saturates, becoming a replacement market and a market growing only through demographics of the market. As this is happening, the companies in the industry consolidate (or fail) until only a handful of companies survive in the industry. The 'magic rule' (which is to say a rule-of-thumb-through-experience) is that only about 5 companies will ultimately survive in any mature industry. This rule suggests that the minimum market share a business needs to keep going for long term survival is about 20%.

For example, we saw that the U.S. automobile manufacturers declined in number during the twentieth century, until in the 1990s only three major U.S. automakers had survived—GM, Ford, and Chrysler. Beginning from about 1970 through 1990, GM lost its once dominant 50% of the market of the U.S. Auto market down to 34% by the end of the century (as foreign auto markets in the last quarter of the twentieth century acquired about a third of the U.S. Auto market). During that same period, Ford hung onto nearly 20% of the U.S. market; while Chrysler struggled around the 9–10% level. According to the 'market-share-survival rule of thumb over the long run, only GM and Ford would remain independent, while Chrysler was at risk. And it did happen that in the 1999, Daimler Benz acquired Chrysler.

Therefore, in evaluating financial investments for improvements in the businesses of a diversified firm, management should strategically look at the opportunity costs of the investments in a way which evaluates impact of the investment upon the business's ability to maintain a survivable 20% market share in the industry.

CASE STUDY: Perils of Sunbeam

We can see how a strategic decision to allow a major business of a firm to decline can be made inadvertently (and ineptly) by top managment of a diversified firm by looking at the case of Sunbeam. As we saw in the case of RCA, diversifying a firm may be sometimes a lot easier to do than later successfully managing a diversified firm. Not only should the analysis of the business portfolio be multispacial (as above), but top management still needs to know how to manage the businesses of the portfolio. We next look at this kind of problem about management that often diversified firms, such as at RCA, have displayed about not properly managing their businesses.

Sunbeam is a case of a once very good company, but terribly mismanaged in its last two decades:

> Sunbeam was the company that grew up with modern America. Its ingenious inventors devised the first automatic coffee-maker, the first pop-up electric toaster, and the first mixmaster. . . .
>
> —(Byrne, 1999, p. 38)

Yet in the 1980s and 1990s, the once great Sunbeam bankrupted not just once but twice by really poor top corporate management. In 1987 it was forced into bankruptcy and again in 2001:

> (From 1998 to 2000), Jerry W. Levin has managed to bring some order to the chaos that once prevailed at the Sunbeam Corporation, the household appliance maker that installed him as chief executive after the notorious ouster of Albert J. Dunlap in 1998. But it has not been easy.
>
> —(Tanner, 2000, p. B6)

Sunbeam from 1897 to 1987   The story of Sunbeam goes back a hundred years. Chicago Flexible Shaft Company was founded in 1897 by John Stewart and Thomas Clark and made agricultural tools with flexible shafts (e.g., sheep-shearing machines). In the early 1900s, it introduced the brand name of "Sunbeam" for its new lines of electrical appliances for the home (e.g., toasters, irons, and mixers). In 1946, the Chicago company officially changed its name to Sunbeam.

Another source of Sunbeam's main appliance product lines, its electric blankets, is traced to the 1920s to an inventor named Pop Russell. He had the idea of creating electrical heating pads to replace hot water bottles that were extensively used around 1900 to warm cold beds (when no one then had central heating in their homes). He started a business called Northern Electric Co. and sold these innovative heating pads in drug stores. A customer who ran a tuberculosis sanitarium in New York State asked him if he could make a heating pad as large as a blanket. The customer had patients who slept outside in the cold (sleeping in the cold was thought to be helpful in tuberculosis treatment). In the 1930s, Russell invented the electric blanket, consisting of stiff wires coated in asbestos. A more flexible heating wire was developed during World War II to heat flying suits of pilots for flight at high, cold altitudes.

In the 1950s, Oster Company bought Northern Electric, and Sunbeam bought Oster in 1980. Sunbeam engineers invented new electric wires for electric blankets that adjusted heat levels by sensing body temperature and eliminated the lumpy thermostats previously used in electric blankets. With this innovation, Sunbeam captured the market, wiping out the competition.

Thus in the early 1980s, Sunbeam was a healthy company with excellent home products. But then a terrible thing happened to Sunbeam. It was purchased by a corporate conglomerate, Allegheny International.

Allegheny International (AI) had been put together in 1982 by an experienced business person named Robert Buckley, but he soon managed it into bankruptcy:

> To understand AI, you must first know its boss. The company is undeniably the creation of the ambitious 62-year-old Buckley. He grew up in New York, earned a law degree at Cornell University and gained his first extensive business experience at General Electric Co. In nine years he rose to manager of union relations of the Schenectady New York plant.
>
> —(Symonds, 1986, p. 57)

From there, Buckley moved to running Standard Steel and then became president of Ingersoll Milling Machine Co. In 1972 he moved to Allegheny Ludlum Industries, which then was a specialty steel producer located in Pittsburgh. In 1975, Buckley decided to sell off the steel business and bought Sunbeam Corp and Wilkinson Sword Group. In 1982, he changed the name of Allegheny Ludlum Industries to Allegheny International, with sales of $2.6 billion dollars. The purchases of Sunbeam and Wilkinson tripled AI's revenue over the earlier steel business but also added substantial debt.

Not content to learn how to manage AI's new acquisitions, Buckley went on expanding his business portfolio by investing in real estate and in energy. He set up a new realty unit and bought the Dover Hotel in midtown Manhattan in New York City and an office building in Houston called the Phoenix Tower. He built an elaborate corporate headquarters in Pittsburgh. He invested in energy, but only ended up drilling a dry well in Texas. Added to the recent Sunbeam and Wilkinson acquisitions, these other real estate and energy investments created substantial debt for AI, and in contrast to Sunbeam, they generated no profits.

Meanwhile, Buckley also greatly increased corporate overhead by providing extensive and expensive executive perks to himself and his corporate team. AI bought a Tudor home in Pittsburgh for Buckley's use and a condominium in Ligonier, Pennsylvania, that backed onto an exclusive golf course. AI also purchased a controlling interest in a condominium project in Florida in which Buckley himself bought three units. Buckley and his executives flew around the country and the world in a fleet of five business jets. AI was generous to its corporate executives by providing $30 million in personal loans to them at a below market interest rate of 2 percent:

> Even while AI's executives were sowing the seeds of their financial problems, their level of compensation was stirring discussion in Pittsburgh corporate cir-

cles. In 1984 when it earned a paltry $14.9 million, AI paid Buckley over $1 million in cash, more than the chiefs of the two larger Pittsburgh companies.

—(Symonds, 1986, p. 59)

In 1981, AI's net income had been $80 million but declined to $45 million in 1982 and to $30 million in 1983 and to $15 million in 1984. AI's executives were living well but were not properly taking care of the business. Buckley's bad real estate and gas investments created substantial losses. In 1985, AI's real estate unit lost $63 million, and the energy unit lost $30 million. In 1985, AI went from the previous year's modest profit of $15 million down to a very steep loss of $110 million. AI's stock had dropped from $54 dollars a share to $17.

Why were not AI's board of directors paying attention? "Where was the AI board while all this was happening? One possible explanation for its passivity and relative generosity is that a number of outside directors have financial ties to the company." (Symonds, 1986, p. 60)

One example of an apparent conflict of interest was that one of the directors, upon joining AI's board in 1983, received a consulting arrangement from Buckley worth $50,000 dollars (for services not to exceed five days a year). Also the director's consulting firm received $163,000 for work done in 1985 (Symonds, 1986).

While AI's board was not properly watching over the company's performance, poor Sunbeam suffered:

Under the portly Buckley, the appliance company with some of America's beloved brand names became a victim of neglect and abuse. Sunbeam was compelled to return nearly all of its profits to Allegheny's corporate center in Pittsburgh, and was starved of capital to update its factories and refresh its product lines.

—(Byrne, 1999, p 39)

In 1986, Buckley was fired. In 1987, Allegheny filed for bankruptcy. For the next two years, Sunbeam management had to struggle on under the cloud of a bankrupt corporation—uncertain of ownership, starved for resources, and not even knowing whether there was a long-term future for the company. Would there be a happy ending to the story of Sunbeam? Unfortunately, no.

STRATEGIC MODELS IN A DIVERSIFIED FIRM

We pause in the case to review how strategic models of a diversified firm differ from the corporate level to the business level. These kinds of strategic models assist the analysis of cases like that of Sunbeam, wherein there are severe strategic conflicts between the level of the firm as a whole and the level of a business in the firm's portfolio.

We recall (from Chapter 3) that an appropriate strategic model for a the diversified firm level is a strategic firm model and for a business in its portfolio is a strategic enterprise model. We illustrate this in Figure 12.4, wherein there are shown the two levels of a diversified firm and its business units. On the upper firm level, the firm as a whole is depicted as a strategic firm model, with inputs as sales and profits from its businesses and outputs as resources and capital. On the lower business-unit level, two of its businesses in its portfolio are depicted as strategic enterprise models, with inputs as resources and capital and outputs as sales and profits.

Lines connecting the two levels are shown as to how the profits from the outputs of the business units feed the input profits to the firm and how sales from the outputs of the business units feed input sales to the firm. Furthermore, some of the resources as outputs from the firm should feed needed capital as inputs to the business units.

The strategic business models of a diversified firm emphasize the flows of:

1. Revenues from profits on sales from portfolio businesses up to the firm
2. The needed flow of resources as capital from the firm down to the portfolio businesses

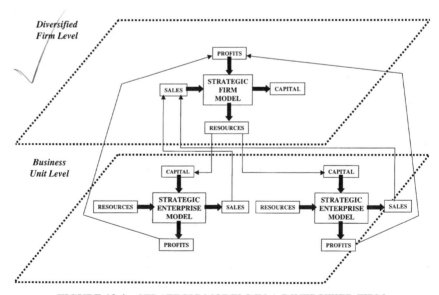

FIGURE 12.4   STRATEGIC MODELS IN A DIVERSIFIED FIRM

Using these strategic models, we can see how Buckley's mismanagement of his diversified firm of Allegheny impacted Sunbeam as a business unit by not properly feeding any resources of capital back into Sunbeam. Sunbeam was staved for investment in new products and improved production for fourteen years, from 1975 to 1986 under Buckley and from 1986 to 1989 in bankruptcy.

Accordingly, this long period of neglect caused Sunbeam's product lines and manufacturing capabilities to really fall behind global industrial standards and that was really serious. (We recall in the case of GE's appliance business, how GE management had to make an enormous investment in manufacturing capabilities and factories in the mid-1980s to survive growing foreign competition with modernized manufacturing techniques.) The last major product innovation Sunbeam was able to achieve was the innovative electric blanket in 1981, due to earlier product research and development funded before the effects of capital deprivation by Allegheny began to take its toll on Sunbeam's innovation capabilities. The strategic starvation of capital for nearly fourteen years by corporate headquarters and bankruptcy had left Sunbeam a company with antiquated manufacturing facilities and dated product lines.

CASE STUDY: Perils of Sunbeam, Continued

We now return to the saga of Sunbeam in 1989. Then two investment fund managers, Michael Price and Michael Steinhardt, seeing an inherent good business deal in the unfortunate situation of Sunbeam, bought up AI's debt cheaply and took control of the bankrupt company. With the assistance of investment bankers, Paul Kazarian and Michael Lederman, they restructured the company as Sunbeam-Oster.

Then Kazarian managed the company back to profitability. He consolidated the eleven divisions of the company into four. He coordinated the selling of all the company's products to major retailers and coordinated purchasing operations. Under Kazarian's first year of leadership, Sunbeam's operating earnings became $96 million, turned around from the previous year's loss of $95 million (Byrne, 1999, p. 81).

In August 1992, Kazarian took the company public again, raising $250 million by selling 24% of its common stock. Price and Steinhardt had essentially acquired Allegheny for $1.50 a share, and the public offering went at $12.50 a share. Their initial investment of $130 million was then worth $1.1 billion, but they still had 76% of the stock to sell to realize it all. They wanted Sunbeam had to be built up for an even larger sale—and quickly—but this would take time. Kazarian had a bitter falling out with Price and Steinhardt over how long it might take, and Kazarian left Sunbeam.

Next Price and Steinhardt hired Roger Schipke from General Electric to run Sunbeam. We recall that Schipke ran GE's appliance business when it reengineered its refrigerator line in the late 1970s. Schipke had spent twenty-nine years at General Electric, running the large appliance division and had grown it from $2.5 billion to $5.6 billion during the eight years he led it. Price offered Schipke a base salary of 1 million and stock options to run Sunbeam. When he arrived at Sunbeam in 1993, he found things still troubled. He saw Sunbeam had out-dated manufacturing processes, poor financial controls, no marketing, and aging product lines. No investments were being made in Sunbeam's future, and it was being run as if it were a leveraged buyout, focused only on paying down debt. Schipke thought it would take time to rebuild Sunbeam. He began improving the company's operations, such as building a new $80 million manufacturing facility.

Yet he found that the major shareholders, Price and Steinhardt, were reluctant to invest much in Sunbeam's long term future, as their interests were short term. They wanted to sell out of Sunbeam, but it was difficult to sell the company. In 1995 neither sales nor profits were substantially up at Sunbeam, as the Schipke turnaround was taking time. Also Sunbeam's market shares were still declining (e.g., blenders were fell from 41% market share in 1993 to 38% in 1995, food mixers from 25% to 19%, toasters from 4% to 3%, gas grills from 49% to 44%, folding outdoor furniture from 58% to 44%). The long-term problems at Sunbeam from its long days with AI and in bankruptcy and desperate rescue were continuing to show up in its decline.

Price and Steinhardt became impatient with Schipke, and within two years of his arrival, they decided to replace him. Schipke resigned. In mid-1996, Price and Steinhardt hired Al Dunlap, not to run Sunbeam but to prepare it for sale. At the time, this was Al Dunlap's reputation—an executive who could get companies ready for sale. His most recent position then had been as head of Scott Paper Co. In eighteen months he had been CEO of Scott, he had driven its stock price up by 225%. To do this, he cut costs. He fired 11,000 employees. He cut investments in production improvement and in research. He then sold Scott to its rival Kimberly-Clark.

But Kimberly-Clark then was surprised by what they had bought. Going into the merger, Dunlap's financial figures for Scott had projected a fourth quarter 1995 income of $100 million. Yet Kimberly-Clark found Scott really lost $60 million on that quarter (a difference of $160 million from the projection). In the first three months, Kimberly-Clark needed to spend $30 million on immediately necessary plant and equipment maintenance that Dunlap had earlier canceled. Also the market share of Scott products began declining. The short-term changes Dunlap had made in Scott to boost its share price was having bad longer term consequences on Scott's competitiveness and profitability.

But this was the kind of short term thing one can do to sell companies—pretty-up their bottom-lines by short-term cost cutting before the sale. Dunlap appeared to be good at that. Earlier he had begun fixing up companies for sale at Lily-Tulip, from which he gained $8 million. Next he had worked for Sir

James Goldsmith for three years also earning more millions. In Australia, he left Kerry Packer, also several million dollars richer. Then his brief running of Scott brought him $100 million. Dunlap had a track record of getting companies sold. Price and Steinhardt thought that Dunlap was just the kind of CEO they wanted! They were enthused and offered him incentives. Price offered Dunlap an annual salary of 1 million, 2.5 million in stock options, and 12 million of restricted stock. On the day of July 19, 1996, when Dunlap's appointment as CEO of Sunbeam was announced, its stock which had been at $12.50 a share jumped to $18.63 a share, so powerful was Dunlap's reputation with the Wall Street crowd as a turn-around artist.

Dunlap then set about 'fixing up' Sunbeam for sale. On November 12, he announced to Sunbeam's board that he would eliminate half of Sunbeam's employees, cutting 6,000 jobs. He would reduce manufacturing facilities from fifty-three to thirty-nine. He also reduced sales by selling some lines of Sunbeam's businesses (including Sunbeam's outdoor furniture products, clocks, scales, and decorative bedding). By these cuts, he claimed he would save $225 million annually. But cutting costs alone would not prepare Sunbeam for sale. How did Dunlap plan to increase sales? He promised the board to introduce at least thirty new products a year and double the company's revenue from $1 to $2 billion a year.

But four days later on November 16, P. Newton White (who Dunlap had just brought in to run Sunbeam for him) quit, and White's abrupt departure would be seen in retrospect as a clear signal of the enormous difficulties to fix Sunbeam. A top leader always implements through his or her operating executives. One can strategize battles, but the general on the ground leads the battle and implements the plans. This lack of a good operating executive like White would turn out to be a major reason for Dunlap's failure at Sunbeam. Dunlap could not keep capable executives working for him or allow capable people to work effectively for him. And within a month of becoming CEO, Dunlap lost or fired three of the operating executives at Sunbeam who had been responsible for ninety percent of its sales. Later Newton White commented on why he had quit after only two months at Sunbeam. He saw that it would take two to three years to begin fixing up Sunbeam and enormous amount of change to turn the company around, and he didn't want to stay that long.

After cutting many jobs, closing plants, and losing executives, those managers who did stay on with Dunlap were under tremendous and impossible pressure to increase sales. Dunlap couldn't sell the company merely by dramatically cutting costs, he also had to dramatically increase revenue. But Dunlap's plan was simply unrealistic. In order to double revenue, Sunbeam would have had to perform five times better than it had and better than its industrial competitors. In one year, Sunbeam would have had to increase its operating margins to 20%, compared to its margins of only $2.5%.

To meet Dunlap's unreasonable demand for impossible sales numbers, managers began resorting to devious schemes. Their jobs depended upon meeting Dunlap's impossible numbers. They couldn't tell Dunlap about problems: "In a meeting with Al, you are not there to tell him anything," said Bill Kirkpatrick, who worked with him at both Scott and Sunbeam. 'You are there to listen. If you didn't hit your numbers, he would tear all over you.' " (Byrne, 1999, p 154)

Dunlap's remaining subordinates created a sales fiction to boost apparent sales, by using an unusual accounting technique called "bill-and-hold sales." In November 1996, the gas grills division asked its major retailers to purchase merchandise they would not yet need for six months. The deal Sunbeam's managers offered was a major discount to the retailer along with the nice conditions that the retailers neither had to pay for nor receive merchandise for six months after the billing. Sunbeam's managers rented warehouse spaces and stored the merchandise there at Sunbeam's expense. So they billed the sale in November and shipped the merchandise, but not to the customer nor had the customer to pay until the next year. A Sunbeam manager later admitted: "More and more it became impossible to make the kinds of numbers that Al (Dunlap) thought should be made," conceded a top executive. "Al was so concerned about revenue that we sacrificed margin." (Byrne, 1999, p. 163)

Also in this way, they sacrificed future sales. There could be no more sales to that retailer that spring, because their 'bill-and-hold' merchandise was already languishing in a rented warehouse.

Yet the stock market was still impressed with Dunlap's reputation, and Sunbeam's stock climbed into the $30 range. Dunlap had succeeded with the stock market, but too well! That price was too high to sell the company to another company. Desperately, Dunlap began looking for another company to buy as an alternative way to boost Sunbeam's sales and revenues. He eventually focused on the Coleman company.

Coleman had been acquired by another conglomerate builder, Ronald Perleman, who owned a large number of companies, such as Revlon Inc and Marvel Entertainment Group. When Coleman's management had attempted to take Coleman private through a leveraged buyout in 1989, Perleman stepped in and bought Coleman. Perleman owned 82 percent of the stock, and he would be willing to sell Coleman to Sunbeam, but not cheaply. Coleman's stock was then selling at a low $16 dollars a share. Coleman had lost $42 million in 1996 and was likely to have a loss of $2.5 million in 1997 (Byrne, 1999, p. 195). Perleman had installed Jerry W. Levin, a former Revlon chairman to turn around Coleman.

When Dunlap approached Perleman to sell Coleman, Perleman asked for $30 dollars a share. Dunlap at first was angry and then desperate. He agreed to the price. In March 1997, Perleman then sold Coleman for a 14 percent stake in Sunbeam, which then translated into $1.6 billion value (because of Sunbeam's then highly valued stock of $45 dollars a share). Also, Dunlap had

Sunbeam make cash offers for Signature Brands and First Alert. To pay for all these acquisitions, Sunbeam's board authorized the sale of $500 million convertible bonds by Sunbeam.

But then public perception of Sunbeam began to see the reality past Dunlap's reputation. A team of Morgan Stanley officials did a due diligence analysis of Sunbeam to prepare for selling the bonds. They spoke with Sunbeam's managers, retailer customers, and auditors and found Sunbeam's sales were below projections. The information they learned would have to be made public to prospective buyers of the bonds. They learned that sales had slowed because Sunbeam's major customers, Wal-Mart and K-Mart, already had large inventories on hand from Sunbeam's previous fall's bill-and-hold sales routine. They learned that Sunbeam could not sell any more products to them for at least another half year.

After the Morgan Stanley team discussed this with Sunbeam's executives, one of them phoned Dunlap and told him that Sunbeam would have to issue a disclosure of the lower projected sales: "Dunlap kept screaming into the phone. He knew the announcement would tank Sunbeam stock and could even jeopardize the success of the debenture offering. It also wouldn't help his reputation as a turnaround master." (Byrne, 2000, p 213)

An ambiguous press release was drafted, merely indicating that sales the next quarter would be lower than earlier estimates of about $290 million. But the next morning, even this news caused the stock of Sunbeam to decline by 9 percent to $45 a share. Worse yet, stock analysts began asking questions about Sunbeam. The whole story would soon get out.

Matthew Schifrin, a reporter for *Forbes* magazine obtained Sunbeam's 10K statement filed March 6 with the Securities & Exchange Commission and wrote the first revealing story about Sunbeam's peril on May 4, 1998: "Seven months ago, 'Chainsaw Al' Dunlap declared victory in turning around . . . Sunbeam Corp. . . . But since the middle of March its stock has fallen nearly 50% from $52 to a recent $28. This turnaround hasn't turned and isn't likely to." (Schifrin, 1998, p. 44)

Schifrin summarized the story to then. After Dunlap was hired in July 1996, he had fired 6000 of Sunbeam's employees (half of the then 12,000). Dunlap cut back on Sunbeam's product offerings, focusing Sunbeam's products of grills, humidifiers and kitchen appliances. Dunlap had Sunbeam take a massive writeoff in 1996 of $338 million (including $100 million of Sunbeam's inventory). Schifrin further wrote: "Wall Street sat back and waited for a miracle. In 1997 Dunlap announced one. Sunbeam reported record results, with sales up 22% and earnings per share of $1.41—up from a $2.37 loss in 1996. As is their wont in these heady days, the analysts didn't look beyond the reported figures. If they had, they might have seen that Sunbeam was coming apart." (Schifrin, 1998, p. 44)

Finally Schifrin was able to look behind the reported figures. He found that

Sunbeam had boosted the sales figures for the last quarter of 1997 by selling grills at a discount and as 'bill and hold'. He learned that the discounts and bill-and-hold sales had accounted for most of Sunbeam's apparent revenue gains. But these reduced margin and prevented future sales. They were the reason for the very poor first quarter of 1998. Sunbeam's customers had simply stopped ordering new grills.

Schifrin's article then began accelerating this story of Sunbeam toward its denouement. Sunbeam's major share holder, Michael Price, read Shifrin's article and phoned Schifrin, telling him: "If you're right, it looks like fraud.' " (Byrne, 1999, p. 261)

On June 4, Sunbeam's executives gathered to review the second quarter prospects of 1998. They saw that the company would fail to meet their second quarter objectives. Then on June 6, a second article about Sunbeam appeared in Barron's financial weekly newspaper by Jonathan Laing. He pointed out that Wall Street was shocked by Sunbeam's sudden reporting of a loss of $44.6 million with a sales decline of 3.6% for the first quarter of 1998: "In a trice, the Sunbeam cost-cutting story was dead, along with 'Chainsaw Al' Dunlap's image as the supreme maximizer of shareholder value." (Laing, 1998, p. 17)

Sunbeam's stock had fallen 50% from its recent peak, and Laing further detailed the results of Dunlap's brutal management style at Sunbeam: "Many of the new products have bombed in the marketplace or run into serious quality problems. Moreover, Sunbeam has run into all manner of production, quality and delivery problems. . . . Dozens of key executives, members of what Dunlap just months ago called his Dream Team, are bailing out." (Laing, 1998, p. 17)

Laing expanded upon Schifrin's report that the earlier 1997 'earnings' really were not correct: "Sad to say, the earnings from Sunbeam's supposed break-through year appear to be largely manufactured. . . . Sunbeam jammed as many sales as it could into 1997 to pump both the top and bottom lines. . . . The company also pumped millions of dollars of goods into several national small-appliance distributors on such easy payment terms as to call into question whether a sale ever took place." (Laing, 1998, p. 18)

After Laing's article about Sunbeam appeared, Sunbeam's board of directors met early the next week. They finally began asking questions about the real problems at Sunbeam. But Dunlap reassured them that there was no truth to the article and that bill-and-hold as an accounting technique was a standard industry practice. Next the board asked about the second quarter results for Sunbeam, but instead of telling them the dismal truth about the major losses for the present quarter, Dunlap began to complain about attacks upon him and offered to quit. This surprised the board, and Dunlap abruptly left the meeting: "When Dunlap stormed out of the board meeting, the four outside directors sat incredulous and quiet . . . Howard Kristol broke the silence. Of all of them, he had known Dunlap the longest . . . 'That is complete bullshit,' he blurted out . . . ' " (Byrne, 1999, p. 296)

Finally, the four directors knew that something was really wrong, and they had a big problem. They decided to talk again later in the week. On Saturday, they met again but without Al Dunlap. Instead, they invited David Fanin, who worked as Sunbeam's attorney, to brief them. Fanin had loyally worked for Dunlap for the last two years; but he had to tell the board the truth. He told them that operations at the company had seriously deteriorated, and sales were far below projected targets. He added that Dunlap was not in real contact with the business and talked to no one.

The directors decided that Dunlap needed to be replaced. At 2:20 p.m., the outside directors met again in New York, and placed a conference call to the inside directors of Sunbeam. Dunlap was in Florida. Fanin began the board meeting, with Dunlap on the phone. Peter Langerman, one of the outside directors of the board, spoke first: "Al, the outside directors have considered the options you presented to us last Tuesday and have decided that your departure from the company is necessary." (Byrne, 1999, p. 324)

Next another outside board member, Elson, moved to a adopt a resolution to remove Dunlap from all positions with the company, and the motion was adopted with all votes in the room affirming the motion. Dunlap was fired.

Ronald Perelman, who had acquired a large share of Sunbeam through its acquisition of Coleman, then appointed Jerry W. Levin as CEO to run Sunbeam. Sunbeam lost $1.2 billion dollars between 1998 and 2000. Total debt was $2.4 billion, and Sunbeam's stock dropped to $1.75 in the year 2000.

Levin tried hard to keep the company going, repairing damage. But the debt load of Dunlap's improvident acquisitions proved too heavy, and on February 6, 2001: "The Sunbeam Corporation filed . . . for Chapter 11 bankruptcy protection, the latest step in a nearly three-year effort to recover from an accounting scandal and a series of acquisitions that left it with $2.6 billion debt." (Atlas and Tanner, 2001, p. C2)

Sunbeam still had some strong core products, such as Mr. Coffee and Oster blenders but needed to develop new small-appliance product lines to survive. Its annual $200 million interest payments left it with no funds to develop new products:

> As CEO, Levin won back the confidence of retailers such as Wal-Mart Stores.
> . . . But with those huge interest costs, it just wasn't enough. Chapter 11 gives
> Sunbeam some breathing room. 'We'll get rid of a couple billion dollars in debt,'
> said Levin. 'And we'll do it relatively painlessly.' Tell that to Perelman, whose
> 14% stake in Sunbeam—originally worth $588 million—has been wiped out.
> —(Haddad, 2001, p. 62)

When Sunbeam next emerged from bankruptcy, it would be a private company owned by the former lenders of $1.6 billion of its loans: Bank of America, Morgan Stanley Dean Witter and First Union.

Case Analysis

The case of Sunbeam illustrates the kinds of conflicts that can occur between the short-term financial interests of owners of a diversified firm and the long-term survival and competitiveness of businesses within the firm. Sunbeam's perils resulted from successive mismanagement by two CEOs. First its acquisition by Allegheny placed it in peril when Allegheny went into bankruptcy through profligate executive spending and foolish investments. Later rescued from bankruptcy by two investors, Sunbeam was returned temporarily to modest health and was successfully taken public. But the impatience of the investors to sell the rest of their shares in Sunbeam led them to finally hire a CEO with similar short-term interests. He was expected to quickly boost the company's apparent profits and sales in order to sell the company. But his drastic cuts, unreasonable demands, and overpriced acquisitions finally resulted in a second bankruptcy of the once proud and prosperous company.

STRATEGIC MANAGEMENT IN A DIVERSIFIED FIRM

From this cautionary tale of Sunbeam, we can see that strategic management of a diversified firm is a complex challenge. In particular, there are six critical factors for successful strategic management that need to be recognized:

1. The need for relationships of trust between levels of management
2. The impact of unequal power relationships between a holding company and the businesses of its portfolio
3. The effect of long-term and short-term differences of control over finances between the firm and its portfolio businesses
4. The possible results of differing incentives and rewards for levels of management
5. The inherent conflicts of interest of different levels of management
6. The influence of external forces on business valuation

Relationships of Trust

In a diversified firm, firm-level executives must depend upon the business-level executives reporting upon them. They must depend upon them for both their commitment to success and integrity in operations.

As we see starkly illustrated in an extreme form in the Sunbeam case, relationships of proper trust and respect between the CEO and his management were simply missing. Sunbeam's operating mangers were not allowed to perform with integrity and resorted to tricks to temporarily satisfy the CEO's insistence to just meet the numbers. We recall from an earlier chapter, that GE's CEO Jack Welsh had emphasized the strategic importance of management integrity (values) as well as "meeting" the numbers.

Management integrity cannot be maintained when a CEO's "numbers" are arbitrarily high targets for sales and profits and are not realistically achievable because of:

- A low-margin and mature nature of the industry (such as the consumer appliances in which Sunbeam operated)
- Any savage cutting of the core capabilities of production and products without regard to a company's long term survival.

In a diversified firm, a constructive and positive interdependence of reasonability and commitment to success between the firm-level executives and business-operating executives is critical to long-term corporate success.

Unequal Power Relationships

This kind of truism about integrity and feasibility should be obvious and therefore practiced by all management. Why is it not always practiced? Because there are unequal power relationships between the top-level of the firm and its operating levels of portfolio businesses:

1. A firm can buy or sell any of its portfolio businesses.
2. A portfolio business cannot buy or sell its owner firm.
3. Yet the long term success of a firm is dependent upon the successes of its portfolio businesses.

And as in any unequal power relationship, trust can be abused. It is important for the leadership of a diversified firm not to abuse the trust of its operating business managers in the short-term, because the success of the firm in the long-term depends upon the continuing short-term successes of the portfolio businesses.

For example, because of this unequal power relationship, it was possible for Sunbeam to be abused, not by one CEO of its holding company but by two. In the early 1980s, the executives of Allegheny abused their privilege of owning Sunbeam through not returning sufficient funds to Sunbeam to maintain its competitiveness and by extravagant spending and unwise investments, which forced the firm into bankruptcy. A decade later, it was possible for another CEO to again abuse the privilege of owning Sunbeam's businesses by forcing extreme cuts and unreal performance targets and adding extravagant investments that brought Sunbeam again to bankruptcy.

The leadership of a diversified firm has both the responsibility and necessity for being a good caretaker of the businesses that a firm has the privilege of owning—both for the short-term good of each of its businesses and for the long-term good of the firm.

Long-Term and Short-Term Differences of Control

That statement of responsibility may at first appear to look suspiciously like a goody-goody kind of ethical imperative. But it isn't. It is really only a very practical and basic imperative for running any diversified firm. It emphasizes the basic differences of perspective and capability in any diversified firm.

In a diversified firm, short-term financial control is always in the operations its portfolio businesses and long-term financial control is in the strategic plan of the firm.

For example, both the CEOs of Sunbeam who wanted greatly increased revenue (to handle debt of their acquisitions) could not control the real growth and margins of their acquired companies. Both CEOs had corporate financial abilities to acquire businesses for long-term strategy but no short-term operational abilities to foster sales and revenue growth in their portfolio businesses.

In contrast, the operating managers for Sunbeam's businesses had ability to efficiently operate these businesses in the short term (provided they had been allowed to do so by their top leaders). But in the long-term, they had to depend upon the firm's leadership and ownership for their investment needs and long-term control.

Accordingly, top-leadership of all diversified firms have limited control their short-term future and more control of their long-term future. In contrast, the portfolio businesses have limited abilities to control their long-term futures but more control over their short-term futures.

Differing Incentives and Rewards

In the 1980s, practices of executive pay began to turn from large executive salaries and perks and bonuses to substantial salaries and perks and very large stock options.

It has been estimated that in 1987, only 2 percent of the total value of the U.S. stock market was held as employee-owned stock or stock options, but by 1994, this portion had grown to 5 percent and by 1999 to 9 percent (Rosenberg, 2000, p. C1). This was nearly a five-fold increase to almost one tenth of the total stock market value as being owned by employees.

This way to reward employees of publicly owned businesses provided a major change in the incentive to the executives of firms on the short-term versus long-term perspective on business performance. It increased the importance of growing and maintaining the share value of a firm in the short term.

Accordingly, the incentives and rewards can differ for different levels of management in a diversified firm. The executives at the firm level may be rewarded with stock options that motivate short-term share-price value; while the operating managers of the businesses of the firms may be rewarded by salaries and bonuses for the operating profits of the firms, which can be structured to motivate for both short- and long-term perspectives.

Inherent Conflicts of Interest

Such differences in rewards for the firm-level executives and the business-level managers can create conflicts of interest in that the firm is trying to optimize short-term capital value, whereas the business-level managers are trying to optimize longer-term competitiveness and profitability.

As we saw in the case of Sunbeam, extreme conflict can occur when the firm does not put proper investment capital back into its businesses to help maintain their long-term competitiveness.

External Forces on Business Valuation

Stocks create a value of a return from investment either through dividends paid out annually by the company or by any increase in the price of the stock. Accordingly, earnings of a corporation can be used for paying out dividends, investments for improving businesses, or acquisitions. Earnings used for dividends provide an immediate return to the shareholder, while earnings used for improvement or acquisitions may provide future capital accumulation to raise share value. There is an important trade-off in optimizing shareholder value of a company, between how earnings are used for immediate return-on-investment or future return on investment. Economically, this trade-off should be made to balance appropriately short- and long-term shareholder value and short- and long-term competitiveness of the company. However, external forces can make an important influence on this balancing, particularly when a government's tax policies bias this balance. And in the world of the twentieth century, U.S. federal government tax policy biased corporate strategy strongly against dividends and toward stock appreciation.

We recall that government policies can often make an important impact upon the environments of a business. In the United States at the close of the twentieth century, federal government income tax policy had a major external impact upon business policies. The federal government taxed returns on investment from stocks very unequally compared to dividends and stock appreciation. Wealthy individuals who owned stock would have any dividends from their stocks taxed at a top income tax rate of 36 percent by the year 2010. In contrast, gains on sales of their stocks held at least one year would be taxed at a lower capital gains tax rate of 28 percent. This tax rate difference of $36 - 28 = 8$ percent had in effect created a 22 percent tax penalty on returns by dividends rather than by appreciation. One can see how government policy of the United States in the last part of the twentieth century encouraged corporations to pursue strategies that aimed at continually increasing stock prices, as opposed to a traditional business practice of sharing earnings with investors through dividends. Thus earnings were often used to buy growth through acquisitions, even when a company could not properly manage acquired businesses (as we saw in the case of RCA).

Moreover, as in the case of Sunbeam, one saw the kind havoc wreaked upon Sunbeam by a crude short-term gain strategy of trying to force corporate growth, as opposed to a traditional strategy of regaining market share of a company in a mature industry and generating substantial earnings and dividends over the long term. The tax policies of the United States biased twentieth-century corporate strategy away from traditional dividend strategies toward quick capital-gain strategies, making it difficult to properly run companies in mature industries with little growth but steady earnings (as all successful companies eventually become).

COMPLEXITY IN STRATEGIC PLANNING IN A DIVERSIFIED FIRM

Complexity in strategic planning arises from these kinds of differences in relationships, power, control ability, incentives, and interests. Taken together, they create very different perspectives on the nature of the firm as viewed from the top-down of the holding firm and the bottom-up of its portfolio of operating businesses. This makes the strategic management of a diversified firm a complex problem.

We can illustrate this complexity in Figure 12.5, which depicts the two levels of strategic planning at the diversified firm level and at the business unit level.

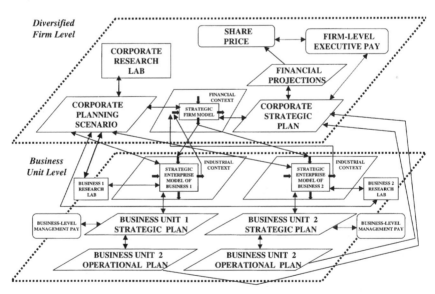

FIGURE 12.5 PLANNING IN A DIVERSIFIED FIRM

This figure is complicated, but it will illuminate some of the sources of problems about complexity in strategic planning.

Diversified Firm-Level Strategy Interactions

Begin looking at the firm level, wherein a corporate planning scenario summarizes the anticipated changes in the future for both the firm as a whole and all of its portfolio businesses. As indicated by the connecting lines, the formulation of the planning scenario influences the formulation of the strategic business models—strategic firm model and strategic enterprise models of each portfolio business.

As indicated by the connected line, the projected revenues from each of the strategic enterprise business models feeds the revenue input of the strategic firm model. In turn, the resource output of the strategic firm model provides capital inputs to the strategic enterprise models of the portfolio businesses.

Still looking at the firm level and as indicated by the connected arrows, the corporate planning scenario stimulates changes in the strategic firm model; and from this the corporate strategic plan is formulated. Financial projections for the firm's future are then derived from the corporate strategic plan, which then are used to inform the stock market and influence the share price of the firm.

Now to the extent that executives of the firm have substantial stock bonuses, then the corporate strategic plan will likely be strongly influenced by desires to see rising share prices in the short-term. This is one of the important complexities in strategic planning. How executives are rewarded will influence the corporate strategic plan in terms of goals and time frames of the plan.

In formulating the planning scenario, inputs are required from corporate research and from the business research labs of each portfolio business. It is the responsibility of research labs at both the corporate and business levels to be anticipating and preparing new technologies for the futures of the businesses of the firm. If a firm does not have a strong corporate research laboratory and strong research laboratories for each portfolio business, the firm as a whole cannot derive a competitive advantage from innovation. It will lack the ability to create future knowledge assets.

Business-Level Strategy Interactions

Next look at the business-unit level of strategic planning. There each portfolio business will formulate a strategic enterprise model of its business in each particular industrial context. As indicated by the connecting arrows, both the business models and the industrial context will influence the formulation of the corporate planning scenario.

After the formulation of each strategic enterprise model, each portfolio business will create a long-term strategic plan and a short-term operational plan for its business future.

Note that at the firm level, a short-term operational plan is not needed since all business operations are performed not at the firm level but at the portfolio business levels. What is needed at the firm level is the strategic plan to review and approve of the investment needs of the portfolio businesses as these needs arise in the future.

At the business level, the rewards for the portfolio business executives and mangers may differ from the rewards for firm level executives in that stock options in the firm may or may not be substantial portions of the business managers' remuneration. To the extent that business managers are rewarded principally with salaries and bonuses, they will likely have a balanced interest in both the short-term and long-term futures of the business. But since they cannot control the ownership feature of the businesses' long-term futures, they endure uncertainty about how long their job tenure may last, particularly if and when the business may be sold by the firm. This is one of the sources of conflict in interests and perspectives that we earlier discussed between firm-level executives and business-level executives.

Thus a strategic problem of a diversified corporation is that firm-level executives are removed by at least one level from direct experience in any of its acquired businesses. Firm-level executives must depend upon the operating executives the portfolio businesses to know a business is doing the right job and doing it right. Successful strategic management of a diversified firm lies directly in the skills and dedication of the operating executives of the firm's businesses and indirectly in the strategic policies and investments of the firm-level executives.

Cooperation and mutual trust between the two levels of executives are essential; yet there may be inherent conflicts in the long-term and short-term interests of the two sets of executives.

In summary, strategic planning is complex not only because the many conceptual units in the process but also due to the differing interests and perspectives between the firm and portfolio business executives.

CASE STUDY: 3M Diversifies Through Innovation

Note that in Figure 12.5, we indicated a particular kind of business functional unit in the strategy process model, and these were the research units of the corporate research lab and the business research labs. The reason for this is that one of the important ways that a diversified firm can show synergy as a whole is through its corporate capability of being an innovation leader. We next turn to the concept of corporate synergies of diversified firms as creating a firm that is more than a conglomerate structure. We will look at a case of a very large firm that was successfully built not upon acquisitions but upon diversification through innovation, the 3M company.

In 1980, 3M made about 45,000 products (including product variations in size and shape). They were diverse products from sandpaper and photocopiers to tape, skin lotions, and electrical connectors. The source of this product diversity was 3M's innovations, with 95 percent of the company's sales ($5.4 million in 1979) having come from products related to coating and bonding technology. This was 3M's secret for diversification success—it knew its businesses, for it had invented most of them (Smith, 1980).

3M prized innovation, even giving annual prizes to its best innovators. The Carlton Award was named after Richard Carlton, who was president of 3M from 1949 to 1953. It was given annually to a few scientists who made major contributions to 3M's technology. For example, Paul Hansen developed a self-adhesive elastic bandage sold by 3M under the Coban label. Dennis Enright developed telephone-cable splicing connectors. Arthur Kotz developed an electronic system for microfilming directly from computers. W. H. Pearlson's research in fluorine chemistry led to the development of agrichemical products for 3M and to their Scotchgard fabric-protection spray (Smith, 1980).

By 1980, forty business divisions had been created from these kinds of products developed from 3M's research. Earnings for 3M had risen each year from 1951 through 1980 (except in the 1972 oil-crunch year). Growth through innovation had been 3M's strategy since its early days.

Minnesota Mining and Manufacturing Co. began in 1902 at Two Harbors, Minnesota, when local investors purchased a mine. The mine was supposed to have contained high-grade corundum, a hard mineral used in abrasives. Instead, the corundum was low grade, useful only for ordinary sandpaper. Sandpaper even then was a commodity business with low profit margins. The disappointed investors decided to look for products with higher value.

The new company sent its sales personnel searching for innovative ideas. They went onto the shop floor of their customers to look for needs for which no one was providing a product. In automobile factories, they saw workers choking on dust from the dry sandpaper. They reported this to 3M, and researchers created a sandpaper that could be used when wet. This was the first step in starting 3M's technological capability—adhesives and coatings. It was also the first step in 3M's success formula—communication between sales people and researchers for innovation.

The next product also came from observations by the salespeople. They also noticed in automobile plants that workers had a difficult time keeping paint from running on two-tone cars. Richard Drew, a young technician in 3M's lab, invented masking tape. Another famous 3M product also came from Drew's inventiveness. In 1930, Drew conceived of how to put adhesive on cellophane. Cellophane itself had been invented it DuPont in 1924. Then a sales manager at 3M, John Borden, created a tape dispenser with a blade on it, and Scotch Tape was born.

About half of 3M's innovative products have come from 3M salespeople looking for needs in their customers shops and offices—market pull. The other half have come from bright research ideas looking for applications—technology push. For example, 3M's research lab came up with a synthetic fabric from pressing rayon or nylon fibers together. It was unique in that it had no weave (sort of like felt material). They first thought of using it for disposable diapers, but it was too expensive. Then they thought of using it for seamless cups for brassieres, but again it was too expensive. The health care division came up with the right application—surgical masks, which would be more comfortable than woven masks because they could be pressed into the right shape—and hospitals could afford it.

Case Analysis

In 3M, the organization was structured around product lines created from 3M's innovations. The fact that 3M's divisions were created from innovations contributed to an organizational culture in which innovation was seen to be of high priority to the corporation. Furthermore, each product-line division had an associated product-development department to maintain and improve the technology and applications of product lines.

Above the divisions, the 3M board and chief operating officer ran the corporation with a vice president of R&D. Reporting to the VP of R&D were two central corporate units—a corporate research laboratory and a new business development unit. Thus 3M was organizationally structured for business diversification by innovation—divisions created from innovations and a central corporate research laboratory creating new innovations, which were nurtured into businesses by a new business development unit.

Ed Roberts pointed out that in 3M's culture, the top management commitment to innovation was clearly articulated policy: "From top to bottom 3M's management provides active, spirited encouragement for new venture generation. Many at the company even speak of a special eleventh commandment: 'Thou shalt not kill a new product idea.' " (Roberts, 1980, p.139).

The company also provided multiple sources of funding support within the company for new ventures. Any of different groups could provide funding. If an idea is taken to one group and turned down, the proposer is free to try his idea with any other 3M group.

The 3M company uses product teams, which it calls "business development units" (which we have called "venture teams"). Early in the development of a new product, a complete product team is recruited from research, marketing, finance, and manufacturing. These teams are voluntary, in order to build in commitment and initiative. The incentives to join a new venture for 3M employees are the opportunities for advancement and promotion that the sales growth a new venture might provide. 3M also emphasizes direct financial measures of performance for each new venture: return on investment, profit margin, and sales growth rate.

Edward Roberts summarized some of the lessons applicable to creating corporate growth through new innovative business ventures. He emphasized the importance of having proper organization, top management committed to innovation, appropriate funding for innovation, innovative product teams, proper reward systems for internal entrepreneurial activity, and proper performance measures for gauging the contribution of innovation to the corporate bottom line (Roberts, 1980).

CORE COMPETENCIES IN A DIVERSIFIED FIRM

Innovations are an important kind of core competency in a diversified firm. For example, 3M's core technical competency was in surfaces and adhesives and was the base for 3M's long term growth. 3M was diverse in its products and businesses, but all were centered around a core technical competency. In the case of Cisco, its core technical competency was in network connections, and Cisco expanded its technical competency through strategic acquisitions. Some diversified corporations, such as 3M, create their businesses from resources of the firm as a whole, providing a competitive core competency. Other firms such as Cisco, add to their core competencies through strategic acquisitions.

The idea of a corporate core competency is the idea the firm level of a diversified company can provide resources to its portfolio business in some-other assets other than financial investments.

Core corporate compepencies can be in shared technology, in shared marketing strengths,in shared financial strengths, or in shared managerial strengths.

For example in innovation core competencies, Jon Didrichsen distinguished between corporations that had a strong kind of technological branching competence, such as 3M in adhesives and coatings, and those that have a broad technological competence in a scientific area, such as DuPont in chemistry (Didrichsen, 1972).

Later, C. K. Prahalad and Gary Hamel argued that a strategic core competency was necessary for diversified corporations in the conditions of global competition:

> During the 1980s, the top executives were judged on their ability to restructure, declutter, and delayer their corporations. In the 1990s, they'll be judged on their ability to identify, cultivate, and exploit the core competencies that make growth possible—indeed, they'll have to rethink the concept of the corporation itself.
> —(Prahalad and Hamel, 1990, p. 79)

Formally Reviewing Core Competencies

Core competencies can be formalized through a procedure for reviewing and judging them. One way to do this is through a specific corporate-level group to identify core corporate products, which some companies have called a "corporate core competencies and products" (CCC&P) committee.

This kind of committee should be composed of the senior executives of each business group of the diversified firm along with senior scientists from each business group and from the corporate laboratory. The committee should retreat for a session in reviewing the industrial value-chains of the different business groups and forecasting technical directions in these value chains. From this review, the committee should identify technological trends that would lead to restructuring of any or all of these industrial value-chains.

The committee should next organize task forces of first-line researchers, sales managers, and production management from the different business groups to further detail and evaluate the trends. The task-forces should identify core products for the corporation that vertically and horizontally would provide competitive edges for many businesses of the firm.

The executive CCC&P committee should use the material from the task forces to identify a small set of corporate competencies and core products and then formulate a strategic corporate business plan to acquire and implement these competencies.

The CCC&P committee should next identify any restructuring of industrial sectors in which these competencies may be of advantage and provide this vision to strategic business units to be used in their planning processes. In the strategic scenario process, the allocation of resources should be reexamined in light of the identified core competencies and core products and in light of and envisioned restructuring of industrial value chains.

Core products are key components (or materials or subsystems or services) produced in some of the businesses of the firm that provide competitive advantages in the products of other businesses of the firm. Core competencies are the knowledge, skills, and facilities necessary to design and produce core products.

Core technological competencies are corporate assets; and as assets, they facilitate corporate access to a variety of markets and businesses. For competitive advantage, a core technological competence should be difficult for a competitor to imitate.

Also the identification of core technological competencies requires management to look upstream in the economic value chain and to decide where and in what to make or buy for one's business products. Management must also look downstream in the economic value chain toward the final customer applications to determine which technologies most visible to the customer affect application performance.

Failing to identify core competencies is a kind of opportunity loss for a company. That failure is due to the inability of management to conceive of a company as other than a mere collection of discrete businesses. If management cannot conceive of strategic totalities—other than financial control—the concept of the corporation becomes merely that of a holding company. Management then will fail to use other competitive factors, such as technology, as corporate competitive weapons.

SUMMARY: USING THE TECHNIQUE OF DIVERSIFICATION STRATEGY

Now we summarize the ideas in this chapter as a strategy technique:

1. Identify The Reasons For Diversification
   - Different reasons for diversification will determine different strategies of diversification.
2. Establish Core Competency Strategies
   - Different kinds of core competencies will influence different strategies of diversification.
3. Analyze The Corporate Industrial/business Portfolio
   - The current business portfolio should be reviewed as to the proper market positions of the businesses within their industrial context.
4. Review Interactions Between Firm-level And Business-level Staff
   - Properly integrate corporate strategic performance measures with staff reward packages. Smart people usually do what they are really paid to do.
5. Properly Manage Strategic Acquisitions
   - Management of the successful integration of acquisitions into corporate structure is essential for the success of the acquisition.

6. Properly Manage Strategic Innovation
   - Continual innovation within portfolio businesses of their product, production, operations efficiency is necessary for their long-term competitiveness and profitability.

For Reflection

Select a sample of large businesses in the twentieth century of two kinds, those with diversification within an industry (such as the food industry) and those with diversification across industries (such as RCA or GE). How did they fare and why? From the second half of the twentieth century, select and compare some large diversified companies in the different countries United States, Europe, Japan, and Korea. How did they fare and why?

CHAPTER 13

KNOWLEDGE ASSETS

PRINCIPLE

Knowledge assets are created from research and development activities.

STRATEGIC TECHNIQUE

1. Create a knowledge pathway map for each business
2. Develop a research capacity for each path
3. Formulate a research plan

CASE STUDIES

Iridium and Teledesic
Napster
The First Commercial Computers
IBM's System 360 Computers

INTRODUCTION

In developing a strategic business model, we should understand the basics of knowledge strategy as an input to strategic thinking. We conclude our survey of modern strategy theory by examining a strategic issue about which information technology has created a new emphasis, knowledge assets. In the second half of the twentieth century after the innovation of the computer; strategic emphasis was upon the handling of information in a business enterprise, information systems. When the twenty-first century began and as information technologies continue to advance in both computers and communication, the new strategic emphasis is on knowledge in a business enterprise, knowledge assets.

Earlier in accounting for the value of operations, accounting systems traditionally valued the physical means of production, such as equipment, buildings and facilities. But when investments in information technology exceeded investments in physical equipment (as began to happen in the 1990s), the accounting problem of valuing the virtual means of production (such as software, information databases, value-adding knowledge) became a major strategic issue. We turn to examining the strategic concept of knowledge assets.

We have seen that strategic vision needs to identify strategic preparation. What new knowledge and skills are needed to implement strategic vision? The generation and acquisition of new knowledge improves the knowledge assets of a company. Knowledge assets consist of information, skills, understanding, intellectual processes that enable a business to add and deliver value to customers. Creating and implementing changes in knowledge assets to improve competitiveness is the focus of knowledge strategy.

Knowledge strategy consists of the creation of new knowledge and its transformation into economic utility. In the modern world, this is a complex process, having its ground in science & technology and its transformation in business innovation. We need to understand these contexts of knowledge creation and transformation to effectively formulate knowledge strategy for a strategic business model.

CASE STUDY: Iridium and Teledesic

We begin by examining a case of knowledge strategy within a communications businesses strategy—the case of Iridium and Teledesic. Both were planned to a create global communications systems using satellites. The historical setting was in the year 2000, when satellite technology was continuing to progress and the communications capability of satellite systems needed to handle multimedia information technology. This case illustrates how new knowledge can be applied to new kinds of services. However, not all new commercial ventures in new high tech products or services succeed. It is precisely how knowledge is applied and at what price that determines commercial success or failure.

Iridium.   Iridium was an early venture to provide wireless phone service on a global basis. It was a venture planned and partly financed by Motorola. Other investors included Sprint and BCE of Canada. Motorola built and operated the communication satellite system and designed and produced the handheld phones. In 1997, Iridium launched its first satellites into low orbits about the earth. But by 2000, the company was bankrupt and planned to destroy their satellites: "In the coming months, Iridium, L.L.C., the bankrupt global satellite telephone company, will begin sending 88 giant satellites spiralling in toward Earth, ending one of the colossal corporate failures in recent memory." (Barboza, 2000a, p. C1)

The financial loss to Motorola was large, $2.5 billion. Iridium never even came close to getting its planned 1.6 million subscribers by the year 2000: "After spending more than $5 billion on a system that promised to communicate 'with anyone, anytime, virtually anywhere in the world,' Iridium could muster only about 55,000 subscribers, not enough even to pay interest on its start-up costs." (Barboza, 20001, p. C1)

Iridium phones didn't work well because the phones were too large to hold comfortably in the hand and didn't work inside buildings (they had to directly "see" the satellites without obstruction). As one observer of the time, James Grant, commented: "It was a technology that didn't live up to its hype or its billing." (Barboza, 2000a, p. C2)

Moreover the customer subscription prices had been set too high at $7 dollars a minute, with an additional cost of $3000 for an Iridium phone. This price was in contrast to cellular phones at only a few dollars for the phone and pennies per minute for connect time. Cellular phones did have limited range but connected to long distance services. The numbers of corporate customers really needing access from lonely global places and willing to pay that price were few.

After Iridium's bankruptcy, a former president of Pan American World Airways, Daniel Colussy, acquired the assets of Iridium for $25 million. In December 2000, The United States Defense Department contracted with Colussy's new Iridium Satellite company for $36 million a year to obtain unlimited use of the global phone network: "Iridium Satellite is the only company that can offer encrypted wireless service worldwide, allowing Defense Department officials to discuss classified information anywhere in the world over the company's satellite phones." (Jaffe, 2000, p. A18)

This contract provided one third of the revenue needed for the new Iridium company; and it needed another 60,000 customers to pay 80 cents a minute to break even. The billions of investment in Iridium had been lost to Motorola, but Colussy might operate and eventually build a more modest system, profiting from that failed investment.

Teledesic. Still others continued to have a vision of a global phone service like Iridium. Another satellite service was being planned by Craig McCaw was called Teledesic.;

> Mr. McCaw's investments include a maze of advanced wired and wireless services. But none presents a more formidable puzzle than Teledesic, his futuristic

scheme to dot the heavens with a . . . satellite network for voice and data communications . . . Mr. McCaw first announced (it) in 1994 and the start of the service has been delayed from 2001 to 2005. Still, the venture Mr. McCaw has referred to as 'an Internet in the sky' remains by far the most ambitious commercial project in space.

—(Feder, 2000, p. 1)

In the year 2000, Hughes Electronics, PanAmSat, Intelsat, Lockhead Martin were providing telecommunications systems primarily for broadcasting applications from geostationary-orbited satellite systems. Low-orbit communications systems were being used for paging with low bandwidth capacity (provided by Orbcomm and Vita), cellular phone service with medium bandwidth capacity (ICO Global, Iridium, Globalstar). Teledesic was to be a low-orbit system with high capacity for full Internet service.

When Iridium had gone bankrupt and its 88 satellites in orbit saved by Colussy's investment and reorganization, Teledesic was still on the drawing board. Teledesic was to be a competing global satellite communications system, envisioned by Craig McCaw. McCaw had obtained both regulatory approvals and interested major investors, such as Boeing ($100 million), Motorola ($750 million), and Prince Alwaleed bin Talal of Saudi Arabia ($200 million).

McCaw had begun in the communications business as an inherited family business. His parents were J. Elroy and Marion McCaw, whose financial assets included the ownership of a small cable television business in Seattle, Washington. McCaw attended Stanford University and had worked in the cable business during summers. Suddenly his father died of a stroke, and McCaw assumed primary management of the family businesses.

Subsequently, McCaw expanded the family's cable business investments, adding paging businesses and then going into the then new cellular phone systems. McCaw borrowed large sums and bid aggressively for new communication properties,. He was always confident that the acquisitions would grow rapidly enough to cover debts. He built a nationwide network of cellular phone licenses as McCaw Cellular, and in 1996, he sold it to AT&T for $11.5 billion.

In 2000, McCaw was building and held controlling interests in Nextel, Nextel Partners, Nextlink, Nextband, and Nextext—provided businesses high-capacity telecommunications and Internet service. Nextel provided wireless phone, paging, data and teleconferencing services, with revenue or $3.3 billion and a net loss of $1.3 billion in 1999. Nextel Partners provided digital wireless services in small markets with revenue of $33 million and a net loss of $112 million in 1999. Nextlink provided local phone service to business customers, with revenue of $274 and a net loss of $559 million in 1999. Nextband and Internext offered high-speed wireless connections to offices and a 15, 000 mile fiberoptic network.

Early in 1986, McCaw had looked at the challenge of providing communications service to Asia by placing satellites in orbit above the Indian and western Pacific oceans. In 1990, one of his employees suggested that satellites in low-earth orbits for a global phone system would avoid communications delays that occurred in systems using geo-stationary satellites in high orbits. But low-earth orbits required more satellites. Teledesic was being planned as a low-orbit system, but would require at least 800 satellites and control of this complex system.

In May, McCaw and his group invested $1.1 billion to take control of then bankrupt ICO Global Communications, a satellite venture: "Before the ICO deal, (still McCaw's) apparent prosperity could not quell reports that Teledisic's progress had stalled. Indeed, if anyone but Mr. McCaw had been involved, Teledesic would probably have been dismissed long ago. 'But McCaw is someone you never bet against.'" (Feder, 2000, p. 7)

Case Analysis

Iridium was planned and launched with high prices and marginal technology by Motorola, a company not in the service business but in the communications equipment business. Since Motorola had no direct experience in the communications services business, its engineers did not quite get the system design right (e.g. its handheld phones were too large for customer's comfort, they wouldn't work inside buildings, and the service was too expensive).

From this one can see that the building of a commercially successful system requires some major up-front design decisions about equipment and operating systems. Iridium's potential advantage by rushing these design decisions could have been a very desirable position of first mover. Perhaps if Iridium's initial big investment funds could have been used less on start-up costs and more to cover the longer term costs necessary to build the business and correct first-design limitations then Iridium might have succeeded. (We recall from the earlier case of Osborne that new ventures frequently need sufficient capital to survive initial problems and competitive challenges, particularly to get the product just right the second time and to build a customer base for competitive pricing.)

Teledisic had Iridium's case as an illustration of what not to do and was beginning slower with more careful planning. Also it had an entrepreneurial investor, McCaw, who had long experience in the communications services business.

We see in this case, that the knowledge assets of the two proposed companies was critical to their failure or success. The needed knowledge was of how to design and build and provide satellite communication handheld phone services. In the case of Iridium, their knowledge assets were incomplete at the time:

- *Incomplete in technology* by not being able to design workable, small handheld devices with the then state of technology

- *Incomplete in financial strategy* by not being able to deliver the service near competitive wireless phone pricing and for a long enough time to gradually build a large customer base

Incomplete knowledge has often been the reason for well-financed commercial failures. This case illustrates just how very critical are knowledge assets to business strategy.

TYPES OF KNOWLEDGE ASSETS

What kind of knowledge assets are there? These can be found by examining the different kinds of paths in which changes in knowledge can result in improved profits. We can use a technique for mapping strategic perspectives which Robert S. Kaplan and David P. Norton called a "balanced scorecard" approach and focuses upon four strategic perspectives of (Kaplan and Norton, 2000):

- The perspective of knowledge in the business
- The perspective of financial achievements in the business
- The perspective of the customer on the business
- The perspective of the operations of the business

Figure 13.1 summarizes this approach as a chart that specifies the strategic connections of these perspectives in strategy, and the arrows show the directions

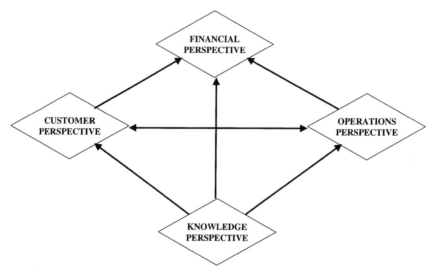

FIGURE 13.1   BALANCED SCORECARD CONCEPT

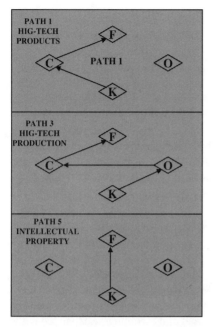
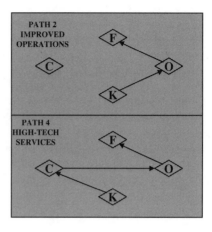

FIGURE 13.1A   PATHS CONNECTING KNOWLEDGE PERSPECTIVE TO FINANCIAL PERSPECTIVE

(or paths) in which any change of knowledge assets in the knowledge perspective might contribute to change of profits in the financial perspective. In the Figure 13.1A, all the logically different kinds of paths that can connect changes of knowledge assets (within the knowledge perspective) to changes of profits (within the financial perspective) are each described.

Path 1: High-Tech Products

This is the application of new knowledge into new products or new services that attract a customer's purchases and contributes to financial revenues. This path has often been called the innovation of high-tech products. An example of this was the introduction of new digital music players, loadable with digital music from the Web.

Path 2: Improved Operations

This is the application of new knowledge to improving business operations that can lower costs by increasing operational efficiency. An example was the innovation of computer-aided design software in the 1980s for the design of hard good products, which greatly increased design efficiency and reduced the cost of designing new products.

Path 3: High-Tech Production

Sometimes the innovation of new operations, such as a new production process, can result in new products for customers, high-tech production. An example was in the integrated circuit (IC) memory chips. In 1970, memory chips had a capacity of 8K; in 1976 16K; in 1981 64K; and in 1985 1M. These kinds of increase in memory chip capacity came as a result not so much of a change in chip design but from changes in production methods that increasingly reduced the size of the transistors on a chip, enabling many more transistors to be engraved in each succeeding generation of chips.

Path 4: High-Tech Services

Sometimes the customer's new knowledge can create the demand for new services, specialized in the latest progress of knowledge, high-tech services. An example of this was the rise of new Web authoring and hosting services that the customers on the Internet needed as the Internet and e-commerce expanded in the late 1990s.

Path 5: Intellectual Property

Sometimes new knowledge can directly result in financial gain when the new knowledge is protected by law in the category called intellectual property. Intellectual property may be sold or licensed. And products protected by intellectual property may have a high financial value as long as the legal protection can be enforced. An example of this are the patents on new drugs developed from molecular biology discoveries.

CASE STUDY: Napster

We next look at intellectual property, wherein a knowledge asset has direct financial value. When the twenty-first century began, one important strategic issue on the Internet was the ability to protect intellectual property of copyrighted material, such as music and other forms of entertainment delivered over the Internet. Then a new dot.com business became a test case for how intellectual property could be defended on the Internet. It was sued by the Recording Industry Association of America (RIAA).

Napster was begun by Shawn Fanning, a college student who was studying computer science at Northeastern University in Boston in the fall of 1998. Earlier he had worked summers for his uncle, John Fanning, whose company,

Netgames, was developing online games. Shawn was bored by college and spent his time at his uncle's office. He had an idea for a search engine to find and share music files over the Internet. He began programming Napster software. He and his uncle saw a commercial potential in the idea. Shawn dropped out of college and completed a test version of the software. His uncle incorporated a new company: "Napster was an instant success. On June 1 of last year (1999), to test the beta version of the software, Shawn gave it to some 30 friends he met through online chat rooms. . . . In just a few days, Napster was downloaded by 3,000 to 4,000 people." (Bruell et al., 2000, p. 114)

With this evidence, John Fanning began to raise capital in August 1999. The demand for Napster soared and Napster hired its first CEO in Eileen Richardson, who was a venture capitalist in Boston:

> Napster raged across the college circuit like a forest fire. College students were discovering Napster, and they couldn't get enough of it. At Oregon State University, Napster was taking up 10% of the school's Internet bandwidth by October, 1999. . . . That fall, it became clear that Napster had a whale by the tail.
> —(Bruell et al., 2000, p. 115)

It became clear to others that the vast trading of music for free on the Internet had serious business implications. One of these was the RIAA, which had earlier sued Diamond Multimedia Systems, Inc., the manufacture of MP3 music players. On December 7, 1999, RIAA sued Napster. Months later on July 26, 2000, a Federal judge of the U.S. District Court in San Francisco ruled against Napster for copyright violation.

Case Analysis

This case illustrates major changes and strategic challenges occurring in the twenty-first century about the traditional intellectual property of copyright.

INTELLECTUAL PROPERTY

In the United States, there are four classes of intellectual property that is recognized by law: (1) patents, (2) copyrights, (3) trademarks, and (4) trade secrets. Patents protect inventions of new and useful ideas. Copyrights protect the expression of ideas but not the ideas expressed. Trademarks are *registered identifications* of products or corporate identities. Trade secrets are commercially important information that is gained of one company by another *without permission and through wrongful means*, such as spying, and so on. In the United States, copyrights are valid for the author's life plus fifty years.

In a patent, the legal right of exclusivity to use the patent is provided for a finite term, in return for the inventor disclosing full details of the invention. The patent concept evolved in England in the eighteenth century in order for details of an invention not to be lost to society due to excessive secrecy by the inventor.

The most common form of patent in the U.S. is called a *utility patent*. It may be granted by the U.S. government to inventors, domestic or foreign, individuals or corporations. Utility patents provide the right to exclude anyone other than the inventor (or those whom the inventor licenses) from making, selling, or using the patented invention for a specific term, normally 20 years.

The steps of acquiring a patent consist of first filing a patent disclosure and patent application with the U.S. Patent Office. The patent office performs a review and grants or denies the patent. In the case of denial, the applicant may appeal. In most countries public disclosure of information about the patent before filing the patent disclosure and application invalidates the patent application. In the application, the claim to invention must include the conception of an idea and the reducing the idea to workable form. "Workable form" may be an actual working device or merely a clear description of a workable form in the patent application (Bell, 1984).

In the U.S. system an invention is patentable only if it satisfies three criteria: utility, novelty, and nonobviousness. To be useful, the courts have generally followed a common notion of someone using it:

> The issue of simple practical usefulness was addressed, and largely settled, in Lowell v. Lewis, heard by Justice Story in 1817. . . . Judge Story held . . . (it) didn't have to be extremely useful or the most useful. . . . It just had to have utility.
> —(Lubar, 1990, p. 11)

Judge Story also added that the patent should not only be useful but it also should not hurt society. Also to be novel, the invention must not simply be a rearrangement of prior art. A revision of the patent law in 1952 added a new standard for patentability of nonobviousness:

> It declared that an invention would be unpatentable 'if the differences between the subject matter as a whole would have been obvious at the time the invention was made to a person having ordinary skill in the art to which said subject matter pertains'.
> —(Lubar, 1990, p. 15)

Since this definition was itself nonobvious, a further change in the patent process occurred:

> In 1982, to help settle this confusion, a whole new court was created, the Court of Appeals for the Federal Circuit, to handle appeals of patent cases. . . . One of the

ways it overcame the problem of defining invention was by putting a greater emphasis on what are called secondary criteria, especially commercial success, in determining the patentability of an invention.

—(Lubar, 1990, p. 16)

Natural phenomena and scientific laws cannot be patented. However, new materials forms can be patented as products, and the processes to produce them may also be patented. New life forms can also be patented as well as the biotechnology process to produce them.

The patent office had earlier distinguished between mathematical algorithms and computer algorithms:

> While patents are not awarded for algorithms, which are considered laws of nature, the Patent Office draws a fine distinction between computer algorithms (which are patentable) and mathematical algorithm's (which are not).
>
> —(Hamilton, 1991, p. 23)

But then computer software became a new area of technology which created changes in patent practices. In 1982 in the United States, the Supreme Court upheld a patent granted for a computerized method of molding tires although it relied upon software algorithms. Thus software patents were permitted when the software algorithms were part of a process. This began a whole new categories of patents. For example:

> Last August (1989), Refac International, Ltd, sued six major spreadsheet publishers, including Lotus, Microsoft and Ashton-Tate, claiming they had infringed on U.S. Patent No. 4, 398, 249. The patent deals with a technique called "natural order recalc," a common feature of spreadsheet calculations that allows a change in one calculation to reverberate throughout a document. . . . Within a few years, software developers have been surprised to learn that hundreds, even thousands, of patents have been awarded for programming processes ranging from sequences of machine instructions to features of the user interface.
>
> —(Kahn, 1990, p. 53)

Another area of patents were added for business processes in 1998 when the U.S. Court of Appeals for the Federal Circuit ruled in favor of a patent on a method for managing mutual funds, filed by the Signature Financial Group. In 1998, over one thousand patents were filed for business methods, and over two thousand filed in 1999. For example:

> Mr. Walker is one of the patent office's best customers. Walker Digital earns money from licensing its inventions—including most notably, the patents for Price-

line.com's name-your-own-price Web site. . . . Walker Digital has its portfolio of 66 patents and 400 pending patents.

—(Angwin, 2000, p. B4)

CASE STUDY: The First Commercial Computers

We next look at the processes of how knowledge assets are created; and to illustrate this we look at the origin of the first commercial computers in the world. Historically, two electronic computer projects directly fostered the innovation of commercial computers:

1. A Mauchly and Eckert project, EDVAC, at the University of Pennsylvania and sponsored by the U.S. Army
2. A Forrester project, Whirlwind, at the Massachusetts Institute of Technology (MIT) and sponsored by the U.S. Air Force

These both occurred in the middle of the twentieth century when the process of knowledge creation and implementation had *institutionally evolved* into an elaborate societal structure including (1) research at universities (e.g., at Penn and MIT), (2) research-intensive industrial firms (e.g., IBM), and (3) large amounts of research funds spent by government agencies (e.g., the U.S. Department of Defense). Together this institutional arrangement was called a "research and development (R&D) infrastructure" of a nation.

The Univac Computer

Central to the invention of the first electronic computers were Von Neumann, Goedel, Turing, Mauchly, and Eckert (Heppenheimer, 1990). John Von Neumann suggested the idea of the world's first stored-program electronic computer but was influenced by the earlier ideas of Kurt Goedel and Allen Turing. Mauchly and Eckert invented one of the world's first all electronic-vacuum-tube-special-purpose-computers. (About the same time, other special-purpose computers were also being independently created in England by Turing and in Germany by Konrad Zuse). Mauchly had also borrowed earlier ideas from John Atanasoff. Mauchly and Eckert were also in turn influenced by Von Neumann's ideas. Also Mauchly and Eckert did create the world's first commercial stored-program, electronic computer, the UNIVAC.

When Mauchly and Eckert built their first vacuum-tube computer called the ENIAC, a major problem for was the need to use of vacuum tubes for active memory in the computer.

Today in personal computers, we use integrated circuit chips called DRAMs (Dynamic Random Access Memory) of a storage capacity of millions of bits (e.g., 64M). But then for each memory bit to the original inventors, they would have needed two electronic vacuum tubes to store just one memory bit. For example, just one electronic sentence in today's personal computers that require 32 bits would then have required 64 tubes. And electron vacuum tubes were the size of a half of a banana, gave off the heat of a light bulb, and burned out frequently. Thus the electron tubes were used for the for the logic circuits of the first general-purpose computers, but they could not be used for the memory circuits.

Eckert and Mauchly invented a memory unit consisting of an electro-acoustic delay line in a longitudinal tube filled with mercury. Their idea was to store a string of bits of ones and zeros morphologically as successive acoustic pulses in a long mercury tube. An electrical transducer at the start end of the tube would pulse acoustic waves into the tube corresponding to a one bit or no pulse corresponding to a zero bit. As the one-bit pulses and the zero-bit spaces between them traveled the length of the tube, the word they represented would be temporarily stored. When the string of pulses began to reach the end of the mercury tube, the acoustic pulse would be re-transduced back into electrical pulses for reading by the computer system, or if not wished to be read, it would then be reinserted electrically back into the front end of the tube as a renewed string of pulses and spaces morphologically expressing the stored word. This storage could repeat in cycle again until the word was read and a new word temporarily stored in the tube.

Eckert and Mauchly then had a conflict with the University of Pennsylvania over the rights to the invention. When Penn would not assign them the rights to the invention, they left the university to build computers on their own. They obtained a $75,000 contract from the Census Bureau to develop a computer and used this contract to start their new company, the Eckert-Mauchly Computer Company. Subsequently they received a $300,000 dollar contract from the Bureau to build the first commercial mainframe computer, which they called the Universal Automatic Computer (UNIVAC). However, even this amount of money was not sufficient for the development costs, and Eckert and Mauchly were forced to solve their financial problems by selling to Remington Rand (a typewriter manufacturer) on February 1, 1950. The mercury storage tube was used in that first UNIVAC built for the Census Bureau and delivered from Remington Rand on March 31, 1951. For permanent storage, the UNIVAC read data and programs from and to magnetic tape and/or punch cards.

Eckert and Mauchly apparently were very good engineers but not as good as managers, for they never really prospered in their business venture. And because a later dispute over patent rights in which Von Neumann testified against them, they did not gain even valuable intellectual property from any computer patents.

Later Remington Rand sold the computer company to Sperry, which even later was put out of the computer business by IBM.

The Sage Computer

The company that really made it big from the early computers was IBM. The story of how IBM got into the computer business goes back to another university-based research project sponsored by a U.S. military agency. While Eckart and Mauchly were building their first commercial computers, there was another very early computer project at MIT, the Whirlwind project directed by Jay Forrester. Historically it was equally important to the history of computers as was the Eckart and Mauchly project. Forrester's computer made the next breakthrough invention required for real progress in active memory part of the computer, a ferrite-core main memory. Also the project Whirlwind accelerated IBM's entry into computers, positioning IBM to become the first mover and dominant player in the mainframe computer industry.

Jay Forrester graduated in engineering at the University of Nebraska in 1939. He went to MIT as a graduate student in electrical engineering, obtaining a doctorate. This occurred during the Second World War, and he worked in military research at the Servomechanisms Laboratory at MIT. During the war in 1944, Forrester participated in studies for an aircraft analyzer and led the Aircraft Stability and Control Analyzer project:

> Jay Forrester was described by people in the project as brilliant as well as cool and distant and personally remote in a way that kept him in control without ever diminishing our loyalty and pride in the project. He insisted on finding and hiring the best people according to his own definition, people with originality and genius who were not bound by the traditional approach.
>
> —(Pugh, 1984, p. 63)

In August 1945, the Servomechanisms Laboratory received a feasibility study contract from the Naval Office of Research and Invention for the aircraft analyzer. Two months later in October, Forrester attended a conference on Advanced Computation Techniques. He wanted to learn about the ENIAC, which Mauchly and Eckert had built. He was interested in using digital electronic computation in his Aircraft Stability and Control Analyzer project. When he saw the digital circuit technologies developed for ENIAC, he thought he could use digital techniques.

With this in mind, Forrester decided to redirect the analyzer project. In January 1946, he went back to the Navy with a new project proposal to design a digital computer and adapt the computer to the aircraft analyzer. The crux of the problem was the main memory subsystem for the computer. The need for response in real-time for control in the aircraft simulator made the use of both the mercury delay line and the rotating magnetic drum technology too slow for this application.

Then international events resulted in a reorientation of the project. After the Soviet Union in Russia exploded an atomic bomb, the United States government decided to build an early warning air defense system.

An Air Defense System Engineering Committee was created in January 1950 under the chairmanship of George E. Vally of MIT to make technical recommendations for such a system. Vally suggested the use of the Whirlwind computer for air defense. Forrester's project was redirected to the new objective of air defense.

Still Forrester had not solved the technical bottleneck in the computer system, active memory, but he did have an idea for a kind of magnetic memory. In April of 1949 he happened to see an advertisement for a new material called Deltamax, and he thought of using it for a novel 3-dimensional magnetic memory array.

Deltamax was made of 50 percent nickel and 50 percent iron rolled very thinly, and had been developed during the war by Germans. After the war, U.S. naval scientists brought samples of the material back to the U.S. and one of the special machines required to make it. They encouraged an American firm, Arnold Engineering (a subsidiary of Allegheny-Ludlum) to make it as a kind of metal tape, called Deltamax. Its important property was its sharp threshold for magnetization reversal when an external magnetic field was applied to it.

Forrester's inventive idea was that he could use the magnetic direction of the metal to store information in binary mathematical form either as a one (in one direction of the magnetization in the material) or as a zero (in the reverse direction of the magnetization in the material). He constructed a rectangular set of magnetic loops as an two dimensional array for storing data. These loops were small magnetic toroids, constructed of Deltamax tape wrapped around each loop to which were connected two sets of electric wires (to carry signals respectively for writing and reading). One wire could be electrically charged with a "write" signal to magnetize the tape in one direction. The other wire was used to sense a "read" signal of the direction of magnetization. If Forrester's computer wanted to store a "one data bit" in a Deltamax tape loop, the computer would send a write signal to the loop to magnetize it in the proper "one" direction. This magnetic direction would stay stored in the loop until the computer wished to read the data bit, which it could at any time by sensing the direction of magnetization in the tape through the read wire. If the computer wished to change the data bit stored on that particular loop, all it had to do at any time was to send a new write signal to reverse the direction of magnetization in the loop (which the computer would interpret as a "zero" rather than a "one" data bit). Each of Forrester's tiny little ferromagnetic tape loops could store a one or zero as a date bit. Taken together a lot of these storage loops could provide an active memory for an electronic computer. And assembled in three dimensional arrays, these tiny toroids would function as a main memory of the first successful mainframe commercial computers. These ferrite-cores memory arrays really made the first computers practical.

When Forrester completed the project, he had built the first computer to use ferrite-core arrays for main memory, with an array of 16 X16. (We no longer use ferrite core memories in computers but transistors in IC memory chips, storing millions of data bits in a single chip, and one can see the vast progress in computer memory knowledge that occurred in the fifty years from 1950 to 2000.)

The next stage of the computer project was then at hand. Although the university research project had designed the computers for the U. S. Department of Defence, an industrial firm would be required to manufacture it in volume. This is how and when IBM got into the computer business.

In June 1952, John McPherson of IBM participated in a committee meeting of a professional society and there talked to Norm Taylor of MIT. Norm Taylor advised that the MIT Digital Computer Laboratory was looking for a commercial concern to manufacture the proposed air defense system. Taylor asked McPherson if IBM interested, and McPherson responded that IBM would indeed be interested. McPherson returned to IBM headquarters and discussed the project with IBM executives. It was the kind of opportunity that Tom Watson had been looking for in order to rebuild IBM's military products division and to improve electronic technology capabilities. IBM told Forrester of their interest. Forrester and his group were reviewing several companies as potential manufacturers of the Air Force computer. They visited Remington Rand, Raytheon and IBM, and Forrester chose IBM to build the computers.

With the deal between MIT and IBM concluded, IBM rented office space at a necktie factory in Poughkeepsie, New York and got to work. The Whirlwind II project was renamed SAGE. By the following summer in 1953, 203 technical and 26 administrative people were working on the IBM part of the project. The system was to have many digital computers at different sites around the country and to be in continual communication with each other. They were to share data and calculate the paths of all aircraft over the country in order to identify any hostile aircraft. These first computers were to use electronic vacuum tube logic circuits and the ferrite core memory. IBM would use the design principles from the Sage project to design and produce their first commercial computer mainframe product line.

Case Analysis

At this time of the innovation of the new computer industry, all the pioneering research had been performed at universities: Aiken at Harvard, Mauchly and Eckert at the University of Pennsylvania, and Forrester at MIT. Moreover, their research was primarily funded by the U.S. Federal government: Aiken by the

Navy (with IBM assistance), Mauchly and Eckert by the Army, and Forrester by the Air Force.

Next the transfer of technology into commercial applications occurred through the formation of new firms and through existing firms entering the new industry. Mauchly and Eckert formed a new company, financed on a government contract from the Bureau of Census; and IBM produced the Sage computer financed by the Air Force.

When IBM built the SAGE computers for the Air Force, it provided IBM with a commercially-important technology advantage. IBM innovated production capabilities to build ferrite core memories, which were the strategic competitive key to the successful early computers.

RESEARCH AND DEVELOPMENT (R&D) INFRASTRUCTURE

As illustrated in this case, the pattern of government support of basic and applied research performed by universities and transferred into industry as applied and development work became a common pattern in the United States after the Second World War. We recall from the earlier case of the innovation of the Internet, that it had followed the same pattern.

> It so happened that in the twentieth century, most new knowledge (pure and applied) for major radical and incremental innovations came from research laboratories—industrial, university, or government research labs.

In a modern institutionalized society, the principle sponsors of R&D are industry and federal governments (with some support by state governments and private philanthropic foundations). The institutional performers of R&D are industrial research laboratories, governmental research laboratories, and universities. Industry has become the principal producer of technological progress, and universities the principal producers of scientific progress. Government laboratories participate in varying degrees by country in performing some technological and scientific progress. Government has become the major sponsor of scientific research and a major sponsor of technological development in some selected areas (such as military technology).

In the second half of the twentieth century, the R&D infrastructure of nations changed dramatically due to the increased direct participation of governments in the support of research. The United States led the way as a kind of superpower in the cold-war political context from 1950 to 1990. The dramatic increase in governmental support of research arose from the experiences of the importance of research to developing military technology that occurred during the Second

World War and also due to a widespread recognition of the importance of R&D to commercial competition.

During the second half of the twentieth century, R&D support in the United States increased to about 2.5 percent of GNP. In 1994, about 62 percent of R&D was sponsored by the federal government, and 38 percent by industry. About 71 percent was performed by industry, 13 percent by universities, and 13 percent by governmental laboratories (NSB, 1996). Up until 1965, half of the funds spent in U.S. industry on R&D was sourced from the federal government. But then industrial expenditure of their own funds on R&D grew, so that by 1995, only 18 percent of U.S industrial R&D expenditures came from the federal government (NSB, 1996).

A second major impact of the Second World War on the U.S. R&D infrastructure was to provide a massive increase in federal funding for university research. Previously, some academic research had always been connected to industry prior to that war (particularly in engineering), but then the federal funds altered the balance. For example, as N. Rosenberg and R. R. Nelson noted:

> One consequence (of World War II) was a shifting of emphasis of university research from the needs of local civilian industry to problems associated with health and defense.
>
> —(Rosenberg and Nelson, 1994, p. 338)

Together the changes in U.S. R&D infrastructure created a kind of division of labor about research between industry and universities:

> R&D to improve existing products and processes became almost exclusively the province of industry, in fields where firms had strong R&D capabilities.... What university research most often does today is to stimulate and enhance the power of R&D done in industry, as contrasted with providing a substitute for it.
>
> —(Rosenberg and Nelson, 1994, p. 340)

Internationally, similar patterns have emerged in all industrialized countries in their R&D infrastructures. What differs from country to country is the level and emphasis of governmental R&D expenditures, modes of governmental support of R&D, roles of governmental laboratories, and research structures of universities.

For example, in the U.S., Germany, Japan, France, and the United Kingdom the ratio of R&D expenditures to GDP ranges from 3 to 2 percent. All other developed countries spend less at a ratio of than 1.5 percent. Also only the U.S., France, and United Kingdom spend government R&D primarily on defense. Other countries distribute government R&D support more evenly over defense, industrial development, health, energy, and space (NSB, 1996).

RESEARCH ORGANIZATION

For creating knowledge assets, a major core competency necessary to keep any firm high-tech is a good corporate research lab and a strategy to innovate new high-tech products, services, production based upon research that transforms knowledge into utility.

The business function whose responsibility is to create and innovate new applied knowledge is the research function of the business. In high-tech and large businesses the research function is organized as research & development laboratories, in three ways:

1. Divisional laboratories reporting to business units
2. Corporate-level laboratory
3. Both divisional laboratories and a corporate-level laboratory

Research in the divisional laboratories is usually focused on next product-model design and on production improvement, whereas research in corporate laboratories is focused on next-generation product-lines and on developing new businesses from new technology.

The kind and number of research units depends on the size and diversity of businesses in a firm. A single-business small firm will likely only have an engineering department. A medium-sized firm will like have an engineering department and divisional laboratories. A large diversified firm will likely have engineering departments and divisional laboratories in business units and also a corporate laboratory for all businesses. Also research organization varies by industry.

Since R&D is an investment in the knowledge assets for the corporation's future, and so should be ultimately evaluated on return on investment. However in practice, this is difficult to do because of the time spans involved. There usually is a long time from funding research to successful research to implementing research as technological innovation to accumulating financial returns from technological innovation. The more basic the research the longer time to pay off and the more developmental the shorter time. For example, the times from basic research to technological innovation have historically varied from seventy years to a minimum of ten years. For applied and developmental research, the time from technological innovation to break even has been from one to five years.

In addition to the varying time spans, the different purposes of research also complicate the problem. Corporate research is aimed at maintaining existing businesses or beginning new businesses or maintaining windows on science. Accordingly, evaluating contributions of R&D to existing businesses requires accounting systems which are activity based and can project expectations of benefits in the future and compare current to projected performance. The evaluation of research needs to be accounted to these purposes of research:

1. R&D projects in support of current business:
   a. The current products are projected as to lifetimes.
   b. This product mix is then projected as a sum of profits.
   c. The current and proposed R&D projects in support of current business are evaluated in terms of their contribution to extending the lifetimes or improving the sales or lowering costs of the projects.
2. R&D projects for new ventures are charted over expected lifetimes.
   a. Projects that result in new ventures are charted over expected return on investment of the new ventures.
3. R&D projects for exploratory research:
   a. These projects are not financially evaluated but treated as an overhead function. They are technically evaluated only on their potential for impact as new technologies.

CASE STUDY: IBM's System 360 Computers

When knowledge is advancing, previous knowledge assets become valueless and even liabilities when a company does not create new generations of products and services to exploit the advancing knowledge. In the 1960s this happened to IBM as its prior knowledge assets of ferrite core memories were no longer viable in the new IC chip memories to replace the early memory technology. Then competitors were catching up with IBM's computer technology and advancing beyond it. IBM had to create a next generation of computers to maintain its competitive lead. How then does a company effectively exploit a national R&D infrastructure for commercial advantage into next generation knowledge and technology products and services? We next look at the case of how IBM came to dominate the world's mainframe computer industry in the latter part of the twentieth century.

We recall that IBM did not invent the computer, but became a first mover in that industry through its manufacturing skill in building ferrite-core memories. Next we will see that IBM faced a major competitive discontinuity in the 1960s when it needed to move its products from using vacuum-tubes in logic circuits to using transistors in its computer logic circuits and also eventually to using IC memory chips.

Initially, IBM had good commercial success with its early computers that used the ferrite-core array memory technology. Also they were getting feedback from customers using the new computers, about the features most needed for advancing applications. The first applications of any new technology are important to refine the focus of the applications and to define the required performance and specifications for the new technology. The result is to define a new model of a high-tech product aimed toward the increasing requirements for improved performance—a next-generation high-tech product. IBM engineers were thinking about how to build a next generation of faster computers.

In January 1955, IBM executives noted that their customers all wanted machines as fast as possible with as much memory as possible. Even in early computer applications, the demand for speed and memory was apparent and still remain dominant requirements. IBM's strategic issue was how to finance further research for a next-generation of high-performance computers. In that meeting, the decision was made for IBM to seek government funds to assist in developing a next generation computer. Next-generation technologies (next-generation knowledges) are expensive and risky. The government funding of the SAGE computers had helped IBM to design its first generation of its computers, and the government contract had helped IBM build its first production capabilities for computers. IBM executives saw direct benefit to them in government-sponsored research and would seek to again use it for its next generation of computers.

IBM learned that Edward Teller (then the associate director of the University of California Lawrence Radiation Laboratory in Livermore, California) was ready to sponsor a high-performance computer. The Livermore Laboratory was a university-administered, government-sponsored research laboratory for hydrogen bomb research. In 1955, both IBM and Sperry Rand submitted proposals to Teller for the project, but Sperry Rand won the contract.

IBM continued to refine its planned next-generation computer, which they called Stretch. The Stretch computer was to be wholly transistorized, replacing the electronic vacuum tubes in the logic circuits. Undaunted, IBM next approached the US. National Security Agency (NSA) with its proposal. In the January 1956, IBM received a contract for memory development and a computer design effort, totalling $1,350,000 dollars. Then IBM won a second contract from the Atomic Energy Commission's Los Alamos Laboratory (to deliver the Stretch computer to the AEC within forty-two months for $4.3 million). As the Stretch project continued, IBM received another contract from the Air Force to prove a transistorized version of its 709 computer.

However, with its radically new technology of transistors, the next generation Stretch project ran into technical difficulties. The project turned out to be costly and take longer than expected. However, IBM under Tom Watson Jr. valued its reputation and delivered the computer at the previously agreed price, taking a loss. Moreover, the project had achieved only half its speed objective. Still there were commercial benefits from the Stretch technology, being the first IBM computer to be fully transistorized.

Government-sponsored research was not always profitable for IBM, but it kept IBM at the cutting edge of knowledge in the early days of the industry. IBM's ability to translate the research into commercial technology provided IBM with economic and competitive advantage.

But IBM had already become large enough that it was beginning to have trouble coordinating innovation within its divisions and research laboratories. On the commercial side, the IBM 7090 computer had been introduced in 1959. Thereafter IBM was selling several computer models from an entry level to advanced models.

Yet none of these models were compatible. When a customer desired to move up from entry level performance, they had to rewrite software applications. By 1960, among the different computers produced by IBM the incompatibility and diversity of the products had become a severe problem. IBM was organized as two divisions—General Products Division for the lower-priced machines (GPD), and Data Systems Division, for the higher-priced machines (DSD). They independently designed their own computer lines. And in addition, two different research laboratories were also designing incompatible computers—the IBM Poughkeepsie Laboratory and the Endicott laboratory. In June 1959, IBM was reorganized into three product divisions:

Data Systems Division (DSD) for high-end computers

General Products Division (GPD) for low-end computers

Data Processing Division for sales and service

The development of the next large computer, the 7070, was thus given to DSD. Responding to this reorganization, the Poughkeepsie Laboratory in GPD, began a frantic effort to regain control of the design of the high-end computers, as Steve Dunwell of the Poughkeepsie lab was certain they could design a computer they would call 70AB, which would be twice as fast and cheaper than the 7070 computer.

We can see that within a large company strategic rivalry can exist between its divisions, over knowledge strategies.

The two research laboratories, Endicott and Poughkeepsie were internal rivals, with different designers and different judgments on best design. At that time of the reorganization, T. Vincent Learson had been made group executive responsible for the three divisions. When Learson learned of Dunwell's decision, Learson ordered that the 7070 be continued, with maximum effort. Yet this still left the Poughkeepsie Laboratory without a product line, and they retargeted their 70AB project toward a lower performance region just above IBM's current 1401 computer. DSD renamed the 70AB project as the 8000 computer series and formally presented their plan to top management.

But one corporate executive did not like what he heard. He was T. Vincent Learson, the new group executive responsible for all three IBM divisions. He was disappointed at the planned performance of the 8000 series. Although some improvements had been planned in circuit technology, yet to get the machine out fast, older technology was still planned on being used in other circuits. Learson was also disturbed at the planned lack of mutual compatibility of computers within divisions and between divisions.

There had been seven incompatible families of IBM computers created in the last decade. As a result, Learson saw that IBM was spending most of its development resources propagating a wide variety of central processors and little development effort was devoted to other areas such as programming or peripheral devices. After the presentation, Learson decided to force IBM to formulate a coherent knowledge strategy. Learson moved Robert Evans (a PhD in electrical engineering) into the Data Systems Division, just below the vice-president of the division but just above the division's chief computer designer, Frederick Brooks (who was responsible for the 8000 series plan). Then Learson instructed Evans to review the 8000 series plans, telling him: "If it's right, build it; and if it's not right, do what's right." (Strohman, 1990, p. 35)

Evans decided it was not right. He wanted a product plan for the whole business, not just the upper-line of the DSD for which the 8000 series had been planned. This brought Evans and Brooks (who was responsible for the 8000 series) into direct conflict. As Brooks (who lost the battle) recalled the conflict:

> Bob (Evans) and I (Brooks) fought bitterly with two separate armies against and for the 8000 series. He (Evans) was arguing that we ought not to do a new product plan for the upper half of the business, but for the total business. I was arguing that was a put-off, and it would mean delaying at least two years. The battle . . . went to the Corporate Management Committee twice. We won the first time, and they won the second time—and Bob was right.
>
> —(Strohman, 1990, p. 35)

Bob Evans then proceeded to get IBM to plan for a wholly new and completely compatible product line from the low-end to the top-end. Evans proceeded toward healing the wounds and getting the teams together. He assembled everyone at the Gideon Putnam Hotel in Saratoga Springs New York to plan out the new strategy for mutually compatible IBM computers. He assigned Brooks senior job of guiding the project, which would be called the System/360.

The disagreement had been an honest one about how important to the company was a better technology versus the time to do it. Technical disagreements and the disagreements about the relative commercial benefits occur in any large, successful organization.

Strategy on next-generation-knowledge products impacts the careers of the different individuals involved, and therefore different knowledge strategies can have different impassioned champions in a large firm.

Evans and Brooks, having bitterly fought the battle, yet respected each other and afterwards worked together. Evans' next job was to get the General Products Division aboard for his total compatibility strategy. For this, Evans needed the help of Donald T. Spaulding, who led Learson's staff group. Spaulding helped Evans by proposing an international top-secret task force to plan every detail of the New Product Line (System /360) and make necessary compromises along the way. Spaulding brought GPD into the new strategy by making John Hanstra of GPD chair of the group, with Evans, vice-chair.

This task force of top technical experts represented all the company's manufacturing and marketing divisions. In November and December 1961, they met daily at the Sheraton New Englander Motel in Greenwich, Connecticut. By the end of December they had a technical plan (knowledge strategy) specifying the requirements of a next-generation of IBM computers, the System/360 product line.

It took a top-level executive to lead a strategic vision on computer knowledge as compatibility across all product lines. It also required a total management commitment (worked out through internal political battles and eventual consensus) to translate that knowledge vision into a concrete technology plan. That knowledge strategy specified seven basic technical points:

- The central processing units were to handle both scientific and business applications with equal ease;
- All products were to accept the same peripheral devices (such as disks and printers).
- In order to reduce technical risk to get the products speedily to market, the processors were to be constructed in micro-miniaturization in transistors (although integrated circuits had been considered).
- Uniform programming would be developed for the whole line.
- A single high-level language would be designed for both business and scientific applications.
- Processors would be able to address up to two billion characters.
- The basic unit of information would be the eight-bit byte.

There were problems in the product development.; but they were all solved and the first 360 computer was shipped in April 1965:

"The financial records speak for the smashing ultimate success of IBM's gamble. In the six years from 1966 through 1971, IBM's gross income more than doubled, from \$3.6 billion to \$8.3 billion."

—(Strohman, 1990, p. 40)

TRANSFORMING KNOWLEDGE INTO UTILITY

As we see illustrated in this case, the transformation of new knowledge strategy into utility of new product strategy (in this case the redesign of IBM's computer lines from tubes to transistors) is not easy but can offer tremendous competitive advantages, if done correctly and timely. Utilization of a national R&D infrastructure can provide a competitive advantage to a business—if the business knows how to effectively transform new knowledge into economically useful products and services. So how is this done? It is done through the core-capabilities in a firm of performing research and development. The heart of knowledge assets in a firm begins in its R&D capability.

Systems in Transforming Knowledge to Value

To understand the complex but important competitive strategy of transforming knowledge into economic assets, we need to look at all the institutional arrangements involved. We can do this by going back to an earlier diagram Figure 8.2. Now to this we can add the sources of industrial knowledge, which goes back into a national R&D infrastructure, as shown in Figure 13.2. There we have sketched the complex sets of interactions that are required to create new knowledge and to transform it into economic utility, including:

- A knowledge infrastructure in a nation
- Industrial and commercial value-adding structures
- Business
- Production
- Product
- Market
- Customer
- Tasks
- Applications

In the modern world, creation of new knowledge begins in the research of a nation's R&D infrastructure, which includes research in universities, industry, and government. Research activities performed or supported in these different institutional contexts interact in a knowledge infrastructure, both nationally and internationally. This research structure creates new knowledge that adds to, elabo-

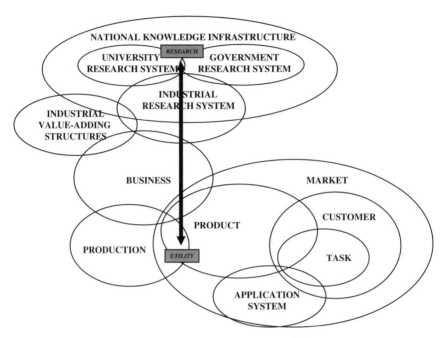

FIGURE 13.2 TRANSFORMING KNOWLEDGE TO VALUE

rates, and deepens knowledge bases. Firms draw upon these knowledge bases in science and in technology to provide the foundations for the business's knowledge assets.

Linear Logic in Basic Knowledge Innovation

The logic by means of which this R&D system operates to transform progress in basic knowledge into economic utility can be understood by focusing on the linear line between research and utility, as symbolized in Figure 13.2. There the bidirectional arrow down the middle emphasizes the knowledge transfer from the research systems of university, government, and industry down into a businesses's high tech products and/or production.

Figure 13.2A next expands that vertical arrow from *Research* into *Utility* as the activities of *Fundamental Research, Applied Research, Technology Development,* and *Commercialization.* These are the necessary steps to transform basic knowledge into useful products, services, and production.

Fundamental research is the pursuit of science to discover and understand nature. For inventions, science provides the scientific feasibility that an invention can actually work. When a basic invention is created, then knowledge moves from *Science* to *Technology* and *Engineering.*

The research line moves also from fundamental research to applied research in developing the invention. These steps can provide a technical feasibility prototype and then a functional application prototype of a new product/process/service that embodies the invention. A first prototype of this new product or process/service can be called an "engineered product prototype."

At this stage, the linear arrow of research moves down from applied research to technology development. In technology development, a design of the new high-tech product (or process or service) is created as an engineering product design. Only then is the knowledge ready to be produced (manufactured for sale). The production of a new product in volume next requires modifying the engineering product design so that the product can be produced in large volume, cheaply and accurately. This requires an additional product design step called a manufacturing product design.

At this stage, the arrow from research to utility has moved into the final stage of commercialization. In this stage a production process to manufacture the new high-tech product may need to be developed and constructed, and then volume production of the product begins and marketing and sales of the product.

By this time, one has moved from science through technology to commercial exploitation. And in this stage of Commercialization, one then needs to continually improve production and periodically improve the product.

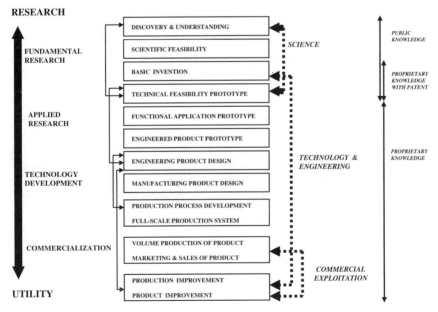

FIGURE 13.2a KNOWLEDGE-UTILITY TRANSFORMATION AS A LINEAR-LOGIC INNOVATION PROCESS

Also knowledge moves from a public domain to a proprietary domain as new knowledge is translated into utility.

One can see that all these stages are complex and innovation of new knowledge into commercially successful utility is a difficult, long and expensive process. The hard facts have been that it usually takes more than ten years to do this and millions of dollars.

For example, in the drug industry, most innovative drugs never make the market, due to lack of effectiveness and toxic side effects. The billions of dollars that the drug industry spends annually on R&D results annually in only a very few innovative new drugs reaching the market.

DEFINING KNOWLEDGE

In industrial fundamental research, research activities are primarily performed by scientists, with technical managers overseeing a firm's corporate research effort. The corporate aim of fundamental research is to understand new knowledge in order to improve the ability to manipulate nature; for it is the manipulation of nature which allows technology. The historical derivation of the term *technical* comes from the Greek word *technikos*, meaning of art, skillful, practical. The portion of the word *ology* indicates a knowledge of or a "systematic treatment of." Thus the derivation of the term *technology* is literally "knowledge of the skillful and practical."

This meaning of technology is one common use of the term. However, it is too indefinite for use in knowledge strategy, since there we need to distinguish between scientific research and technology research—creating "pure" knowledge versus creating "applied knowledge." For this reason we will use a slightly different definition. We can define

1. Technology is the knowledge of the manipulation of nature for human purposes. This definition of technology emphasizes that all practical skills of technique ultimately derive from changes of nature. Technology depends upon a base in the natural world but extends the natural world through manipulation. It is for this reason that the discovery of natural phenomena has often directly resulted in a vision of its technological application (e.g., the discovery of nature's technique of recombining strands of DNA led immediately to a patent for recombinant DNA technology filed by the discovering scientists).

Historically, progress in science has always generated accompanying progress in technology.

Next consider the derivation of the term *science* from the Latin *scientia*, meaning "knowledge." However, the modern concept of scientific research has come to indicate a specific approach toward knowledge, one oriented toward nature, which results in discovery and in explanations of nature. For this reason, we will use the following definition:

2. Science is the discovery and explanation of nature. The link between science and technology is engineering, and we will adopt the following definition as consistent with the preceding definitions:

3. Engineering is the understanding and manipulation of nature for human purposes. In these definitions, one should first note that science is not focused by human purposes, as are engineering and technology.

Science is relevant to human purpose but not focused by it; rather science is focused by the ubiquity and generality of existence of nature.

This is the reason why in R&D basic research is difficult to strategically manage in a corporate environment. Corporations need research focused by purpose, whereas scientists are oriented by their profession to disciplinary classes of ubiquitously existent phenomena.

Although research activities in science focus primarily upon the discovery and understanding, sometimes inventions do occur in scientific research (notably instrument or materials inventions). When inventions occur in basic science, they are usually the basis for radically new technologies (such as the laser or the transistor).

Next, one should note that while technology is the knowledge of how to manipulate nature for human purposes, it does not necessarily imply a deep understanding of nature, which allows the manipulation.

Engineering differs from technology by focusing upon combining the technical knowledge of how to manipulate nature with scientific explanations of nature as to how and why the manipulation works.

Technologically focused research concentrates upon inventions and their development toward product. Research in the engineering disciplines also focuses on advancing the understanding of nature directly relevant to technology.

The better engineers understand nature, the more and clearer ideas they may have for refining technical manipulation.

For this reason engineering research is organized not only on specific technologies but also upon areas of phenomena underlying technological sectors, such as mechanical engineering, electrical engineering, chemical engineering, and civil engineering.

The importance of understanding the relationships between science, technology, and engineering is that research for radical innovation requires *advancing the underlying science base* of the nature manipulated by the technology. (For example, in biotechnology, much engineering basic research has been performed to understand the metabolism of cells, since cells are used in the biotechnology industry as chemical processing units.)

University-based science has often provided the basic invention for a new technology, but industrial research provides the innovation of the new technology. The research spectrum connecting science to technology is the modern basis for university and industrial research cooperation.

Historically, the companies that participated in the early stages of a radically new technology have often been the firms that built strong, large organizations to exploit the new products and new markets of the new technology.

Information technology is a subset of this broader category of applied knowledge called "technology." Information is a subset of the broader category of natural human activity of thinking and communicating.

Information technology is an invention of a way to manipulate information for human purposes.

SUMMARY: USING THE TECHNIQUE OF KNOWLEDGE STRATEGY

Now we summarize the ideas in this chapter as a guide to knowledge strategy.

1. Create a knowledge pathway map for each business
   - Knowledge is implemented into utility at the business level.
   - Strategy for implementing new knowledge needs to identify the specific pathway to contribute to the bottom-line financial perspective of any business.
2. Develop a research and engineering capability for each path
   - New knowledge can not be competitively implemented without a research and engineering capability to customize the knowledge for a businesses activities that match or exceed competitor's similar activities.
3. Develop a research plan
   - Research and engineering capabilities needed to be carefully planned and executed to innovate new products/processes/services effectively and within time and budget targets.

For Reflection

Compare the management of the research & development function in a business to the management of the other business functions, such as marketing, manufacturing, finance, personnel. How does the content of a functional area affect the issues and challenges of management? What are the kinds of knowledge assets are important in each function? How best can these different knowledge assets be kept cutting edge?

BIBLIOGRAPHY

Andrews, Edmund L., 2000. "City of London (and Frankfurt)," *The New York Times*, May 4, pp. C1–C19).

Angwin, Julia, 2000. "Business-Method Patents, Key to Priceline, Draw Growing Protest," *The Wall Street Journal*, October 3, pp. B–4.

Ansoff, H. Igor. 1988. From Chapter 6 in *Corporate Strategy*, McGraw-Hill, 1965, and reprinted in *The Strategy Process*, James B. Quinn, Henry Mintzberg, and Robert M. James (Eds.), Englewood Cliffs, New Jersey: Prentice Hall, pp. 2–8.

Atlas, Riva D. And Jane Tanner, 2001. "Despite Recovery Efforts, Sunbeam Files for Chapter 11," *The New York Times*, February 7, p. C2.

Ayres, Robert U., 1990. "Technological Transformations and Long Waves," Parts I and II, *Technology Forecasting and Social Change*, Vol. 37, pp. 1–37 and Vol. 37, pp. 111–137.

Baird, D.G., 1923. "Eliminating Needless Cost and Confusion," *Industrial Management*, Vol. 65, pp. 334–337.

Baird, D.G.,1935, "Management Coordination Keeps GM Fleet Sales on the Up Grade throughout Depression Years." *Sales Mangement*, Vol. 36, pp. 188–198.

Barboza, David, 2000. "Taking the Starch Out of an American Icon," *The New York Times*, March 19, 2000, Section 3, pp. 1–16)

Barboza, David, 2000a. "Iridium, Bankrupt, Is Planning a Fiery Ending for Its 88 Satellites," *The New York Times*, April 11, pp. C1–2.

Barboza, David, 2000b. "Living and Learning at Dishwasher U." *The New York Times*, September 12, pp. C1–27.

Barrett, Amy, Pamela L. Moe, Diane Brady, Nanette Bynes, and Louis Lavelle, 2000. "Jack's Risky Last Act," *Business Week*, November 6, pp. 41–45.

515

Berman, Dennis, 2000. "Dot.Coms: Can They Climb Back?" *Business Week*, June 19, pp. 101–104.

Betz, Frederick. 1997. *Managing Technological Innovation*. New York: John Wiley & Sons.

Besanko, David and David Dranove and Mark Shanley, 2000. *Economics of Strategy* (2nd ed.), New York: John Wiley & Sons.

Bhattaharya, A. K. and A. D. Walton, 1998. "Strategic Re-engineering at Coats Viyella", *Long Range Planning*, Vol. 31, pp. 711–721.

Brannigan, Martha, 1998. "Sunbeam Audit Finds a Mirage, No Turnround," *The Wall Street Journal*, October 20, 1998, p. A3.

Bradsher, Keith, 2001. " 'What Sales Slump?' Foeign Makers Ask", *The New York Times*, January 13, p B1.

Braudel, Fernand, 1979. *The Wheels of Commerce*, New York: Harper & Row.

Bosomworth, Charles E., and Burton H. Sage, Jr., 1995. "How 26 Companies Manage Their Central Research," *Research-Technology Management*, May–June, pp. 32–40.

Boseman, Glenn, and Arvind Phatak, 1989. *Strategic Management*, New York: John Wiley & Sons.

Bounds, G., et al., 1994. *Beyond Total Quality Management.*, New York: McGraw Hill.

Bower, Joseph, and Thomas M. Hout, 1988. "Fast-Cycle Capability for Competitive Power," *Harvard Business Review*, November–December, pp 110–118.

Brekke, Dan, 2000. "The Future Is Now—or Never", *New York Times Magazine*, January 23, pp. 30–33.

Brown, Dondalson, 1957. *Some Reminisences of an Industrialist*. Reprinted: Ann Arbor, Michigan: University Microfilms, 1981.

Brull, Steven, Dennis K. Berman, and Mike France, 2000. "Inside Napster," *Business Week*, August 14, pp. 112–120.

Brunnel, David, 2000. *Making the Cisco Connection*. New York: John Wiley & Sons.

Burrows, Peter and Peter Elstrom, 1999. "The Boss: CEO Carly Fiorina's Challenge at HP," *Business Week*, August 2, pp. 76–84.

Burrows, Peter and Ronald Grover,1998. "Steve Jobs, Movie Mogul,"

Business Week, 1956. "Selling in an Age of Plenty", *Business Week*, May 5, p. 121.

Business Week, 1997. "SAP AG Corporation", *Business Week*, November 3, pp.

Business Week, 1999. "   " *Business Week*, November 23, pp. 140–154.

Burrows, Peter. 2000. "Apple." *Business Week*, July 31, pp. 102–113.

Byrne, John A., Deidre A. Depke, and John W. Verity, 1991. "IBM—As Markets and Technology Change, Can Big Blue Remake Its Culture?" *Business Week*, June 17, pp. 25–32.

Byrne, John A., 2000. *Chainsaw: The Notorious Career of Al Dunlap in the Era of Profit-At-Any-Price*. New York: Harper Business.

Byrne, John A., 2000. "Management by Web," *Business Week*, August 28, pp. 84–99.

Campbell, Andrew, 1991. "Strategy and Intuition—A Conversation with Henry Mintzberg," *Long Range Planning*, Vol. 24, No. 2. pp. 108–110.

Carey, JOHN, 1991. "The Research is First Class. If Only Development Was Too.," *Business Week*, December 16, p. 118.

Carvajal, Doreen, 2000. "Leading Bertelsman's Race to the Future," *The New York Times*, January 30, Section 3, p. 1.

Chandler, Alfred D., 1990. "The Enduring Logic of Industrial Success," *Harvard Business Review*, March–April, pp. 130–140.

Chandler, Alfred, 1990. "The Enduring Logic of Industrial Success," *Harvard Business Review*, March–April, pp. 130–140.

Chester, Arthur N., 1994. "Aligning Technology with Business Strategy," Research-Technology Management, January–February, pp. 25–32.

Christensen, Clayton M, 1997. *The Innovator's Dilemma*, Cambridge Massachusettes: Harvard University Press.

Christensen, Clayton M., 2000. "Limits of the New Corporation," *Business Week*, August 28, p. 180.

Christensen, Clayton M., and Michael Overdorf, 2000. "Meeting the Challenge of Disruptive Change," *Harvard Business Review*, March–April, pp. 67–76.

Churchill, Winston S. 1950. *The Hinge of Fate*. Boston: Houghton Mifflin.

Churchill, Winston S. 1951. *Closing the Ring*. Boston: Houghton Mifflin.

Churchill, Winston S. 1953. *Triumph and Tragedy*. Boston: Houghton Mifflin).

Clark, Kim B., and T. Fujimoto. 1988, "Overlapping Problem Solving in Product Development," *Harvard Business School*, Working Paper, April.

Cohen, Stephen S. and John Zysman, 1989, "Manufacturing Innovation and American Indusrial Competitiveness," *Science*, Vol. 239, March 4, pp. 1110–1115.

Colvin, Geoffrey, 2001. "Value Driven," *Fortune*, January 8, p. 54.

Cooper, Christopher and Erick Portanger, 2000, "Spooked," *Wall Street Journal*, pp. A1–A8.

Cringley, Robert X.,1997. "High Tech Wealth," *Forbes*, July 7, pp. 296–304.

Crow, C. 1945, *The City of Flint Grows Up*. New York: Harper & Brothers.

Cusumano, Michael A., 1985, *The Japnese Automobile Industry: Technology and Management at Nissan and Toyota*, Cambridge, Massachusetts: Harvard University Press.

Cusumano, Michael A., 1988. "Manufacturing Innovation: Lessons from the Japanese Auto Industry," *Sloan Management Review*, Fall, pp. 29–39.

Davenport, Thomas, 1998. "Putting the Enterprise into the Enterprise System," *Harvard Busienss Review*, July-August, pp. 121–131.

David, Fred R., 1998. *Strategic Management*, Englewood Cliffs, New Jersey: Prentice-Hall.

Deming, W. E., 1986. *Out of Crisis*, Cambridge Massachussetts: MIT Press.

Deutsch, Claudia H., 2000. "Efficiencies Found Online Help Companies Build Profits Offline," *The New York Times*, December 11, p. C4.

Didrichsen, Jon, 1972. "The Development of Diversified and Conglomrate Firms in the United States, 1920–1970," *Business History Review*, Vol. 46, Summer, p. 210.

Drucker, Peter E., 1990, "The Emerging Theory of Manufacturing," *Harvard Business Review*, May–June, p 94–102.

Edvinsson, Leif, 1997. "Developing Inellectual Capital at Skandia," *Long Range Planning*, Vol. 3, pp. 366–373.

Ehrbar, A. F.,1982. "Splitting Up RCA," *Fortune*, March 22, pp. 62–76.

Elliott, Stuart and Jim Rutenberg, 2000. "Weak Retail and Dot-coms Take a Toll," *The New York Times*, Decmber 11, p. C1.

Engardo, Pete, 2000. "The Barons of Outsourcing," *Business Week*, August 28, pp. 177–178.

Epstein, R. C., 1928. *The Automobile Industry*. Chicago: A. W. Shaw. Reprinted: New York: Arno Press, 1972.

Feder, Barnaby J., 2000. "Can Craig McCaw Keep His Satellites From Crashing?" *The New York Times*, June 4, Section 3, pp. 1–7.

Feder, Barnaby J., 2000a. "I.B.M. Introduces Top of Line In New Family of Computers," *The New York Times*, October 4, p. C4.

Forbes, B. C., 1924. "We Face the Future without Fear, with Faith—Sloan," *Forbes*, March 29, 13, p. 759.

Friedland, Jonathan, 2000. "How a Need for Speed Turned Guadalajara Into a High-Tech Hub," *Wall Street Journal*, March 2, pp. A1–A8.

Ghosh, Shikhar. 1998. "Making Business Sense of the Internet," *Harvard Business Review*, March–April, pp. 125–135.

Gimein, Mark, 2001. "Who Turned The Lights Out?" *Fortune*, February 5, pp. 111–116.

Goldhar, J. D., M. Jelinck, and T.W. Schlie, 1991. "Flexiblity and Competitive Advantage—Manufacturing becomes a Service Business," *International Journal of Technology Management*, May, pp. 243–259.

Goldman, Jacob E., 2000. "Innovation Isn't the Mircosoft Way," *New York Times*, June 10, p. A27.

Gottwald, Floyd D., 1987. "Diversifying at Ethyl," *Research Management*, May–June, pp. 27–29.

Graham, Alan K., and Peter M. Senge, 1980. "A Long-Wave Hypothesis of Innovation," *Technological Forecasting and Social Change*, Vol. 17, August, pp. 283–312.

Green, Heather, Catherine Yang, Pau Judge, 2000. "The Dot.Coms Falling to Earth," *Business Week*, January 17, pp. 38–39.

Gunther, Marc, 1998. "The Internet is Mr. Case's Neighborhood," *Fortune*, March 30, pp. 69–80.

Gunther, Marc, 2000. "These Guys Want It All," Fortune, February 7, pp. 71–78.

Gustke, Constance, 2000. "Back to the Future," *Worth*, February, pp. 41–44.

Haddad, Charles, 2001. "Sunbeam's Sole Ray of Hope," *Business Week*, February 19, p. 62.

Hamm, Steve, Peter Burrows, Andy Reinhardt, 2000. "Is Windows Ready to Run E-Business?" *Business Week*, January 24, pp. 154–160.

Hammer, M. And J. Champy, 1993. *Re-engineering the Corporation*. New York: Harper Business.

Hansell,Saul, 2000a. "President of Amazon Quits for VerticalNet," *The New York Times*, July 26, p. C4.

Hansell, Saul, 2000. "Amazon Posts Big Losses in Quarter," *The New York Times*, July 27, pp. C1–10.

Hauben, Michael, 1993. "History of ARPANET" ⟨http://www.dei.ise.ipp.pt/docs/arpa.html⟩ (haruden@columbia.edu) December 2.

Heppenheimer, T. A., 1990. "How Von Neumann Showed the Way," *Invention and Technology*, Vol. 6, No. 2, Fall, pp. 8–17.

Hoffman, Donna L., and Thomas P. Novak, 2000. "How to Acquire Customers on the Web," *Harvard Business Review*, May–June, pp. 179–188.

Hutcheson, P., A. W. Pearson, and D. F. Ball, 1996. "Sources of Technological Innovation n the Network of Companies Providing Chemical Process Plant and Equipment," *Research Policy*, Vol. 24, pp. 25–41.

Iacoca, L., with W. Novak, 1984. *Iacocca: An Autobioraphy*. New York: Bantam Books.

Jaffe, Greg, 2000. "Pentagon to Pay Iridium for Use Of Its Network," *The Wall Street Journal*, December 7, p. A18.

Kahn, Brian, 1990. "The Software Patent Crisis," *Technology Review*, April, pp. 53–58.

Kaplan, Robert S., and David P. Norton, 2000. "Having Trouble with Your Strategy? Then Map It," *Harvard Business Review*, September–October, pp. 167–176.

Kapner, Suzanne, 2000. "Boo.com, Online Fashion Flop, Is ready to rise rom Ashes," *The New York Times*, October 17, p. C8.

Kirkpatrick, David D., 2000. "Book Advance for G.E. Chief Is $7.1 Million." *The New York Times*, July 14, pp. C1–C2.

Kenny, David and John F. Marshall, 2000. "Contextual Marketing," *Harvard Busienss Review*, November–December, pp. 119–125.

Klevorick, Alvin K., Richard C. Levin, Richard R. Nelson, and Sidney G. Winter, 1995. "On the Sources and Significance of Interindustry Differences in Technological Opportunities," *Research Policy*, Vol. 24, pp. 185–205.

Kocaoglu, Dundar F., M. Guven Iyigun, and Chuck Valcesschini, 1990. "New Product Development Cycle," Presented at the TIMS/ORSA Conference, Nashville, Kentucky, August, 1990.

Krantz, Michael, 1999. "Steve's Two Jobs," *Time*, October 18, pp. 62–68.

Kuhn, Arthur J., 1986. *GM Passes Ford, 1918–1938*: University Park, Pennsylvania: the Pennsylvania State University Press.

Laing, Johnathan R., 1998. "Dangerous Games," *Barron's*, June 8, pp. 17–18.

Lawton, Thomas C., 1999. "The Limits of Price Leadership: Needs-Base Positioning Strategy and the Long-term Competitiveness of Europe's Low Fare Airlines," *Long Range Planning*, Vol. 32, No. 6, December, pp. 573–586.

Leslie, S. W., 1983. *Boss Kettering: Wizard of General Motors*. New York: Columbia University Press.

Lev, Baruch, 1996. SEC Workshop on "The Reporting of Intanglible Assets," SEC: Washington, DC, April 11–12.

Lipin, Steven, Jathon Sapsford, Paul Becket and Charles Gasparino, 2000. "Chase Agrees to Buy J.P. Morgan & Co. In a Historic Linkup," *The Wall Street Journal*, September 13, pp. A1–A18.

Lohr, Steve, 2000. "Broad Reorganizaion at IBM Hints at Successor to Gerstner," *The New York Times*, July 25, pp. C1–C8.

Loomis, Carol J., 1998. "Long-Term Capital: a House Built on Sand," *Fortune*, October 26, pp. 111–118

Loomis, Carol J., 2000. "AOL+TWX=???," *Fortune*, February 7, pp. 81–84.

Lubar, Steven, 1990. "New, Useful and Nonobvious," *Invention and Technology*, Spring-Summer, pp. 8–16.

Mack, Tim, 2000. "Electronic Marketing: What you Can Expect," *The Futurist*, March–April. pp. 40–44.

Markoff, John, 1999. "An Internet Pioneer Ponders the Next Revolution," *New York Times*, December 20, p C38.

McNamara, Cathy Rae, 2000. *Integration of the Deltek Financial Enteprise System at AlliedSignal Technical Services Inc.* College Park, Maryland: University of Maryland University College.

Markoff, John, 2000. "Pondering the Impact of a Breakup," *The New York Times*, May 1, pp. C1–C4.

Marquis, Donald G., 1982. "The Anatomy of Successful Innovations," *Innovation*, November, Reprinted in *Readings in the Management of Innovation*, M. L. Tushman and W. L. Morre (Eds.) Marshfield, Massachusettes: Pitman.

Marshall, Eliot, 1991. "The Patent Game: Raising the Ante," *Science*, Vol. 253, July 5, pp. 20–24.

McGeehan, Patrick and Andrew Ross Sorkin, 2000. "Chase is on Verge of Buying Morgan for $30.9 Billion," *The New York Times*, September 13, pp. A1–C14.

Mensch, Gerhard, 1979. *Stalemate in Technology*. Cambridge, Massachusettes: Ballinger.

Merrills, Roy, 1989. "How Northern Telecom Competes on Time," *Harvard Business Review*, July–August, pp. 198–114.

Miller, William, 1999. *Fourth Generation R&D*, New York: John Wiley.

Mintzberg, Henry, 1988. "Opening Up the Definition of Strategy", in *The Strategy Process* James B. Quinn, Henry Mintizberg and Robert M. James (Eds.), Englewood Cliffs, New Jersey: Prentice Hall, pp. 13–20.

Mintzberg, Henry, 1990. "The Design School: Reconsidering the Basic Premises of Strategic Management," *Strategic Management Journal*, Vol. 11, March–April, pp. 171–195.

Mintzberg, Henry and Joseph Lampel, 1999. "Reflecting on the Strategy Process," Sloan Management Review, Spring, Vol. 40 No. 3, pp. 21–30

Morgenson, Gretchen, 2000. "First Jolts of the Internet Shakeout," *The New York Times*, January 16, Section 3, p. 1.

Morgentson, Gretchen, 2000. "Anaysts Talk and Amazon.com Shares Reel", *The New York Times*, June 24, pp. B1–B3).

Munk, Nina, 1999. "Title Fight," *Business Week*, November 13, pp. 50–53.

Musashi, Miyamoto, 1982. *Book of Five Rings*. New York: Bantam.

National Science Board, 1984. *Industry/University Research Cooperation*. Washigton, DC: National Science Foundation.

Nocera, Joseph, 1995. "Cooking with Cisco: What Does It Take to Keep a Hot Stock Sizzling?" *Fortune*, December 25, pp. 114–122.

Nocera, Joseph, 2000. "The Men who Would Be King," *Fortune*, Feburary 7, pp. 66–69.

Nocera, Joseph and Tim Carvell, 2000. "50 Lessons," *Fortune*, October 30, pp. 136–137.

Norris, Floyd, 2001. "Amazing Amazon: Losses Grow as They Seem to Shrink," *The New York Times*, February 2, p. C1.

Nulty, Peter, 1981. "A Peacemaker Comes to RCA," *Fortune*, May 4, pp. 140–153.

O'Harrow Jr., Robert, and Kathleen Day, 2000. "Cool Bankers Strike a Hot Deal," *The Washington Post*, September 14, p. E1–3.

Oppel, Richard A., 2000. "The Higher Stakes of Business-to-Business Trade," *The New York Times*, March 5, Section 3 p. 3.

O'Reilly, Brian, 2000. "The Power Merchant," *Fortune*, April 17, pp. 148–160.

Pittau, Joseph, 1967. *Political Thought in Early Meiji Japan*, Cambridge, Massachusetts: Harvard University Press.

Porter, Michael, 1985. *Competitive Strategy*, New York: The Free Press.

Prahalad, C. K. And Gary Hamel, 1990. "The Core Competence of the Corporation," *Harvard Business Review*, May–June, pp. 79–91.

Pugh, Emerson, 1984. *Memories That Shaped an Industry: Decisions Leading to IBM System/360*. Cambridge, Massachusetts: The MIT Press)

Quick, Rebecca, 2000. "A Makeover That Began at the Top," *Wall Street Journal*, May 25, pp. B1–B4)

Richtel, Matt, 2000. "Agent's Role in Music Site May Be Shift In Rights War," *The New York Times*, May 22, pp. C1–C2.

Roberts, Edward B., 1980. "New Ventures for Corporate Growth," *Harvard Business Review*, July–August, pp. 134–a42.

Rosenberg, Nathan, and Richard R. Neson, 1994. "American Universities and Technical Advance in Industry," *Research Policy*, Vol. 23, pp. 323–348.

Quinn, James Brian, 1988. "Strategies for Change," in *The Strategy Process*, James B. Quinn, Henry Mintizberg, and Robert M. James (Eds.), Englewood Cliffs, New Jersey: Prentice Hall, pp. 2–8.

Quinn, James Brian, Jordan J. Baruch, and Penny C. Paquette, 1988. "Exploiting the Manufacturing-Services Inerface," *Sloan Management Review*, Summer, pp. 45–55.

Quinn, James Brian, and Penny C. Paquette, 1988. "Ford: Team Taurus," Amos Tuck School, Dartmouth College, Case Study.

Quinn, James Brian, Thomas L. Doorley, and Penny C. Paquettte, 1990. "Technology in Services: Rethinking Strategic Focus," *Sloan Management Review*, Winter pp. 79–87.

Ray, George F., 1980. "Innovation and the Long Cycle," in *Current Innovation*, Bengt-Arne Vedin (Ed.) Stockholm: Almqvist and Wiksell.

Rebello, Kathy, Peter Burrows, and Ira Sager, 1996. "The Fall of an American Icon," *Business Week*, February 5, pp. 34–42.

Reingold, Jennifer, Marcia Stepanek, and Diane Brady, 2000. "Why the productivity Revolution Will Spread," *Business Week*, February 14, pp. 48–51.

Roberts, John G., 1989. *Mitusi: Three Centuries of Japanese Business*. New York: Weatherhill.

Rosenberg, Geanne, 2000. "Insiders Get A Sturdy Tool To Rake In Stock Gains," *The New York Times*, September 27, pp. C1–10.

Roth, Daniel, 1999. "Dell's Big New Act," *Fortune*, December 6, pp. 152–156.

Rowe, Alan J., Richard O. Mason, Karl E. Dickel, and Niel H. Snyder,1989. *Strategic Management*, New York: Addison-Wesley Publishing Co.

Russell, Orland D., 1939. *The House of Mitsui*. Boston: Little, Brown & Co.

Ruykeser, M. S., 1927. "General Motors and Ford: A Race for Leadership." *American Review of Reviews*, Vol. 76, pp. 372–379.

Sanderson, Susan and Mustafa Uzumeri, 1995. "Managing Product Families: The Case of the Sony Walkman," *Research Policy*, 24, pp. 761–782.

Sanderson, Susan, and Mustafa Uzumeri, 1995a. "A Framework for Model and Product Family Competition," *Research Policy*, Vol. 24, pp. 583–607.

Sanderson, Susan, and Mustafa Uzumeri, 1995b. "Managing Product Families: The Case of the Sony Walkman," *Research Policy*, Vol. 24, pp. 761–782.

Schiesel, Seth, 2000. "AT&T Board Seeks Ways to Lift Stock," *The New York Times*, September 25, pp. C1–12.

Schifrin, Matthew, 1998. "The Unkindest Cuts," *Forbes*, May 4, pp. 44–45.

Schlender, Brent, 1999. "The Real Road Ahead," *Fortune*, October 25, pp. 138–152.

Schmenner, Roger W., 1988. "The Merit of Making Things Fast," *Sloan Management Review*, Fall, pp. 11–17.

Schneider, Greg, 2000. "4 Defense Contractors to Join Forces," *Washington Post*, March 23, pp. E1–E6.

Schwartz, Evan I., 1990. "The Dinosaur That Cost Merrill Lynch a Million a Year," *Business Week*, November 26, p. 122.

Segal, David, 2000. "MP3.com Is Loser In Copyright Case," *Washington Post*, April 29, p. E1.

Sewer, Andy, 2000. "Chase Banks on TECH," *Fortune*, February 7, pp. 123–126.

Shaw, Gordon, Robet Brown, and Philip Bromley, 1998. "Strategic Stories: How 3M Is Rewriting Business Planning," *Harvard Business Review*, May–June, pp. 41–50.

Shannon, Henry, 2000. "Kinko's Makes an Original," *Washington Post*, March 2, pp. E1–E6.

Sherman, Stratford P., 1989. "The Mind of Jack Welch," *Fortune*, October 30, pp. 84–94.

Siklos, Richard, and Catherine Yang, 2000. "Welcome to the 21st Century," *Business Week*, January 24, pp. 37–44.

Sloan, Alfred P., Jr., 1926. "Make Smaller Profits Pay," *Factory*, Vol. 37, pp. 993–997.

Sloan, Alfred P., Jr., 1927. " 'Getting the Facts' Is the Keystone of General Motors Success," *Automotive Industries*, Vol. 57, pp. 550–551.

Sloan, Alfred P., Jr., 1929. "Sloan of General Motors Predicts a Revolution in Distribution," *Printers' Ink*, February 7, pp. 88–98.

Sloan, Alfred P., Jr., 1964. *My Years With General Motors*. Garden City, New York: Doubleday & Co.

Steele, Lowell W., 1989. Managing Technology, New York: McGraw-Hill.

Strom, Stephanie, 2000. "Taking a Warlord's Advice to Shake Up the Marketplace," *The New York Times*, June 7, E-commerce Part 2, p. 32.

Symonds, William C., 1986. "Big Trouble at Allegheny: Lavish Perks, Poor Invesments— And a Board That Let It Happen," *Businss Week*, August 11, pp. 56–61.

Taguchi, Genichi, and Don Clausing, 1990. "Robust Quality," *Harvard Business Review*, January–February, pp. 65–75.

Takanaka, Hideo, 1991. "Critical Success Facors in Factory Automation," *Long Range Planning*, Vol. 24, No. 4, pp. 29–35.

Tanner, Jane, 2000. "Bringing New Order to Sunbeam's Chaos," *The New York Times*, August 27, p. B6.

Taylor, Bernard, 1997. "The Return of Strategic Planning—Once More with Feeling," *Long Range Planning*, Vol. 30, No. 3, pp. 334–344.

Tedeshi, Bob, 2000. "How to Lure Prestigious Beauty Goods to Cyberspace," *The New York Times*, January 17, pp. C1–8.

Teitelman, Robert, 2001. "The Imperiled High-Tech CEO", *IEEE Spectrum*, January, pp. 30–31.

Thompson, Arthur A., and A. J. Strickland, 1998. *Strategic Management*, New York, Richard D. Irwin, Inc.

Thurm, Scott, 2000. "Under Cisco's System, Mergers Usually Work; That Defies the Odds," *Wall Street Journal*, March 1, pp. A1–A12.

Tichy, Noel, and Ram Charan, 1989. "Speed, Simplicity, Self-Confidence: An Interview with Jack Welch," *Harvard Business Review*, September–October, pp. 112–120.

Tully, Shawn, 2000. "The B2B Tool That Really is Changing the World," *Fortune*, March 20, pp. 132–145

Turban, Efraim, and Ephriam McLean, and James Wetherbe, 1999. *Information Technology for Management*. New York: John Wiley & Sons.

Tushman, Michael and Anderson, Philip, 1986. "Technological Discontiuities and Organizational Environments," *Administrative Science Quarterley,* 31, pp. 439–465.

Tushman, M., and Romanelli, E, 1985. "Organizational Evolution: A Metamorphosis Model of Convergence and Reorientation" in L. Cummins and B Saw (Eds.) *Research in Organizational Behavior* Greenwich, Connecticut: AI Press 1985.

Uchitelle, Louis, 1999. "The 1.2 Trillion Spigot," *The New York Times*, December 30, p. C1.

Uchitelle, Lous, 2000. "Inflation Spreading Beyond Fuel and Energy," *The New York Times*, April 15, pp B1–B3)

Veraldi, Lew, 1988. "Team Taurus—A Simultaneous Approach to Product Development", A speech given at the Thayer School of Engineering, Dartmouth College,

Vesey, Joseph T.,1991. "Speed-Market Distinguishes the New Cometitors," *Research-Technology Management*, November–December, pp. 33–38.

Vizard, Michael and Katherine Bull, 1999. "Fiorina talks e-services," *Infoworld*, November 15, p. 8.

Welch, David, 2000. "Consumers to GM: You Talking to Me?" *Business Week*, June 19, pp. 213–215.

Wheelwright, Steven C., and W. Earl Sasser, Jr., 1989. "The New Product Development Map," *Harvard Business Review*, May–June, pp. 112–125.

Wheelwright, Steven, and Kim Clark, 1992. *Revolutionizing Product Development*, New York: The Free Press.

White, George, and Margaret W. Graham, 1978. "How to Spot a Technological Winner," *Harvard Business Review*, March–April, pp. 146–152.

Wise, Richard and David Morison, 2000. "Beyond the Exchange: The Future of B2B," *Harvard Business Review*, November–December, pp. 86–96.

Worth, 2000. "Inside the Mind of a CEO," *Worth*, May, pp. 180–186.

Wysocki, Bernard, 2000. "Power Grid," *Wall Street Journal*, July 7, pp. A1–6.

WuDunn, Sheryl, 1999. "Capitalizing on Asian Doldrums," *The New York Times*, Tues, September 14, pp. C1–C22.

INDEX